T0309778

SCIENCE AND TECHNOLOGY OF POLYMER NANOFIBERS

SCIENCE AND TECHNOLOGY OF POLYMER NANOFIBERS

Anthony L. Andrady

Research Triangle Institute

A JOHN WILEY & SONS, INC., PUBLICATION

Published by John Wiley & Sons, Inc., Hoboken, New Jersey
Published simultaneously in Canada

Library of Congress Cataloging-in-Publication Data:

Andrady, A. L. (Anthony L.)
Science and technology of polymer nanofibers / Anthony L. Andrady.
p. cm.
Includes bibliographical references.
ISBN 978-0-471-79059-4 (cloth)
1. Nanofibers. 2. Electrospinning. 3. Synthetic products. 4. Polymers.
I. Title.
TA418.9.F5A53 2008
677′.02832–dc22 2007027362

Printed in the United States of America

10 9 8 7 6 5 4 3 2 1

To my wife Lalitha

CONTENTS

PREFACE

Since its inception six years ago, federal R&D investment in the national nanotechnology initiative focused on exploring the "inner space" has been well over $6.5 billion, matched by about the same amount from private industry. It is an impressive budget, but one that still pales in comparison to the hundreds of billions of dollars spent on outer-space exploration, our most visible science project. In terms of rewards, however, the few billions spent have opened the portals of nanoscale phenomena to hundreds if not thousands of researchers worldwide (as opposed to the few fortunate astronauts who trod the lunar soil or enjoyed the breathtaking view of Earth from outer space.) In terms of pure science and educational dividends, the investment is already a resounding success. In Feynman's words, there certainly "is plenty of room at the bottom" to accommodate all those curious minds and those yet to follow. Rewards in terms of products reaching the marketplace, however, have been slower to come. The relatively young nanotechnogy effort appears to be paying off in terms of the emerging nano-enabled products already entering the marketplace. If the projections are correct, the estimated market value of "nano-goods" resulting from the R&D effort in the near future is indeed staggering. The National Science Foundation estimates a market of US$240 billion per year for nanomaterials. If there were unambiguous definitions of what constitutes "nano" or "nano-enabled," then one might even be able to count and trend these technologies. Therein lies a fundamental and very practical question: what constitutes a nanomaterial and specifically a nanofiber?

One can conveniently invoke the familiar and accepted technical criterion: "nano" being 10^{-9}th of a meter, an object that is 100 nm or smaller in at least one of its dimensions is a nanomaterial. It is an arbitrary size range in any event, and reliable techniques to even assess if a particle is slightly over or under this limit do not exist. Real-world materials with particle sizes that are several hundred nanometers, a micron, or even several microns are loosely referred to as "nanomaterials." Textile fibers that are as large as 500 nm in diameter are by convention referred to as nanofibers in the industry. The marketplace boasts of hundreds of nanomaterials and nano-products ranging from the familiar inorganic reinforcing fillers, composites, and coatings to the exotic quantum dots. A less-rigorous working definition of what constitutes a nanomaterial can be particularly useful given the wide range of products in the marketplace claiming to be nano-enabled. Also, there is the issue of macro-scale objects carrying nanoscale features that provide them with functionality (nanoporous polymer foams); certainly they are nano-enabled materials, but are they distinct from nanomaterials?

Going by the restrictive scientific definition, one can envision classes of nanomaterials based on their dimensionality, counting the non-nanoscale dimensions associated with an object. A nanoparticle such as a quantum dot where all dimensions fall within a defined nano regime (say <100 nm for the sake of discussion) is clearly a zero-dimension (0-D) nanomaterial. A material where two of the dimensions are not nanoscale (only a single nano-dimension) will then be a two-dimensional (2-D) nanomaterial, and would include ultrathin coatings or plate-like fillers. Nanofibers or nanowires where a single dimension falls outside the nano regime will be classified as one-dimensional (1-D) material according to this scheme. Electrospun nanofibers are 1-D nanomaterials based on this taxonomy. However, in the electrospinning literature, nanofibers (along with nanorods, nanowires, and nanobelts) are sometimes referred to as 2-D structures. This is based on the alternative convention of counting only the nanoscale dimensions of a material. The length scale of 1-D nano-object can take any value outside the nano regime and therefore includes fibers, nanotubes, most nanoribbons, and high-aspect-ratio particles.

Reducing the size of a particle will eventually force its characteristics to change. The classical paradigms that apply in macro-world will cease to describe its behavior and will need to be replaced by quantum mechanical descriptions. The size scale where the gradual change from the classical to quantum behavior occurs encompasses classical nanomaterials, with their unexpected, unusual characteristics. Even at dimensions where classical rules continue to apply, particle size reduction and the ensuing increased fraction of atoms at the interface (based on dimensionality) will bring about dramatic changes in material properties. It is the exploitation of these two sets of tunable materials characteristics that the nanotechnologists typically work with. The so-called "molecular Lego set" of nano-engineering is nothing more than an exceptionally economical, bottom-up approach to engineering design that replaces the convention of turning out devices (and waste) from large chunks of materials.

Nature was the first nanomaterials foundry, producing nanoparticles in natural geological phenomena, mainly in volcanic eruptions and in forest fires. As the human population density increased along with their increasingly energy-intensive lifestyles, nanoparticles from the burning of fossil fuel, dust from industrial processes, and fines exhausted into the environment from transportation also increased. Ultrafines and their negative impact were identified as far back as the mid-1970s with an appreciation of the particularly damaging effects of the smallest of these ultrafine particles. The PM-10, PM 2.5, and PM-1 program focus by the United States Environmental Protection Agency (USEPA) in the 1980s and 1990s did not quite encompass the nano regime, but that was mainly because of limitations in the available monitoring equipment at the time.

THE PRESENT

Interest in producing smaller-diameter textile fibers came about long before interest in engineered nanomaterials surfaced in recent years. The first microdenier fibers (denier <1) in the United States were spun in 1989 by the DuPont Company. Several ingenious textile techniques such as the spinning of bicomponent polymer fibers through islands-in-the-sea dies followed by extraction of the soluble component, melt-spinning of splittable bicomponents, and melt-blowing have since been used to obtain fibers with average diameters in the range of hundreds of nanometers (even sub-100 nm fibers have been claimed) and commercial fibers that are considerably finer than silk. Electrospinning, however, introduces a new level of versatility and a wider range of materials into the micro/nanofiber range. An old technology rediscovered, refined, and expanded into nontextile applications in recent years, electrospinning is unique among nanofiber fabrication techniques in terms of process control, materials combinations, and the potential for scale-up. This has led to it being recognized as a key platform technology that will yield products for a broad range of uses including electronics, drug delivery, chemical sensors, tissue scaffolding, filtration, and solid-state lighting applications.

This renaissance is partly a result of the availability of key tools such as scanning probe microscopy and high-resolution electron microscopy to enable facile exploration of the size-scale involved. However, it is mainly the rediscovery of the nanoscale nature of electrospun fiber and an appreciation of the unique behavior typical of nanomaterials that has spearheaded the resurgence of electrospinning. It is this same expectation that encourages research on nanofibers in nontextile uses (as the process is hardly cost competitive with conventional spinning in comfort-fiber applications) such as in sensors, scaffolds, and electronic devices. High-value applications, mainly biomedical applications, account for the majority of patents associated with the technology. A consideration in scaling up the process comprises the environmental and safety attributes of electrospinning.

With the solvent-electrospun mats (as opposed to the melt-electrospun fibers) having more controllable and finer morphologies, the environmental issues of scaled-up electrospinning in a textile setting can be as prohibitive as with conventional dry spinning. In nontextile applications, however, the volumes of material processed can be small by comparison and the same concerns can be better addressed. In filter applications, for instance, commercial electrospinning operations processing moderate volumes of nanofiber are already in commercial operation; for example, multiple-needle spin heads for pilot plant and scale-up operations are beginning to be advertised. The sole high-volume application for nanofibers at the present time is in the area of air filtration. With the present emphasis on homeland security, effective filtration is indeed a critical application.

Ultimately, however, the value of the technology lies in the smallest fiber diameters that can be fabricated and manipulated under practical conditions. Research literature claims 1–2 nm nanofibers electrospun from solution.[1] These, however, are very small samples, which can be imaged microscopically, but these cannot as yet be consistently electrospun as large homogeneous mats of fiber to be used in practical applications. The high degree of process and material control needed to fabricate these is not compatible with high-speed manufacturing environments. Yet, mats comprising nanofibers that are a few hundred nanometers in diameter and of consistent variability appear to be achievable even in large-scale electrospinning. With improvements in rapid characterization technologies for mats, more robust stable power supplies and tighter process control, innovative scale-up possibilities for the technology should definitely increase.

THE FUTURE

Future advancements in nanofiber technology will be fueled primarily by (1) improvements in electrospinning technology and process control to allow consistent production of nanofiber mats with single-digit fiber diameters, and (2) the potential to combine several physical, chemical, and biological functionalities into a single fiber to make multipurpose fiber mats and smart materials a reality. The functionalities considered need to move well beyond the simple passive effects of biocidal effectiveness (for instance by incorporating nanosilver), superhydrophobicity by surface

[1]Nanofibers that are only 1.6 nm diameter, electrospun from nylon-4,6 in 99% formic acid (2% nylon with 0.44% pyridine) have been reported (Huang, C. B., et al. 2006a). A 1.2-nm diameter cylinder theoretically accommodates only 6–7 nylon molecules!

texturing, or simple breathable biodegradable wound dressings. Future nanofibers are likely to deliver far more advanced multiple functionalities, and will likely be active devices that perhaps enable impressive disruptive technologies. These will include fabric-based computing/communications capabilities (integration of circuitry and transponders into nanofibers), disposable physiology/environment monitoring in apparel (disposable sensors, alarms, and on-demand countermeasures integrated into fabric), rapid physiological testing arrays (automated or on-demand bedside clinical testing), fibrous photovoltaic technologies (solar sails for space exploration and batteries in nanofiber geometry); they will also provide tunable highly efficient photo- and electroluminescent solid-state illumination. The enabling base technologies for all these are already on the horizon as far as material choices go, but design, integration, and productization has yet to be carried out. Refinements in electrospinning technology that will support these innovations broadly fall into two categories: innovations in process/materials and the recognition of new cross-disciplinary applications for electrospun materials.

Recent electrospinning of phospholipids, genetic materials, and biomimetic proteins into electrospun fibers, as well as the potential of nanofibers as controlled delivery vehicles for plasmid DNA or large protein drugs, and, autologous stem-cell scaffolding studies also suggest exciting directions for future advances. Recent fiber-level innovations include core–shell bicomponent fibers that can be used in drug delivery, nanoparticle-reinforced nanoscale fibers for composite applications, nanofibrous scaffolding for complex tissue replacement, and the development of inorganic oxide nanofibers for efficient sensors or catalysis. Also, the adoption of nanofibers as composites containing quantum dots, the fabrication of semiconductor "quantum fibers," the use of conducting polymer nanofibers with quantum confinement properties (see FET studies[2]), and nanowire circuitry show great future promise. A significant breakthrough that will overcome the limitations in temperature sensitivity and aging issues in organic polymers is the recent advancement in sol–gel spinning to yield inorganic nanofibers. Catalysis and some mechanical application often require the nanofibers to be exposed to high temperatures and solvents, which affect the organic polymer nanofibers.

As an integral and key component of the nanomaterials revolution, organic and inorganic nanofibers remain an increasingly versatile class of nanomaterials that promises to touch upon and improve different aspects of the

[2]The fabrication of an electrospun regio-regular poly(3-hexylthiophene-2,5-diyl) nanofiber-based field effect transistor (FET) was reported (González and Pinto 2005).

human condition, from the improvement human health to playing a key role in the drive for energy production.

ANTHONY L. ANDRADY PHD

Research Triangle Institute
March 2008

ACKNOWLEDGMENTS

This volume would not have been possible if not for the support of a host of individuals. Partial funding for this work provided via an internal professional development grant and permission to use illustrative research data, by RTI International (RTP, NC 27709) is greatly appreciated. I would like to specially thank Drs David Myers (VP, Engineering Technology Division, RTI), David Ensor (Senior Fellow and Director, Center for Aerosol Technology, RTI) and Dr. Lynn Davis (Program Manager, Nanotechnology) for their encouragement, critical reviews, and valuable discussions. Dr. Ensor's valuable suggestions for the section on filtration are particularly appreciated. It is a pleasure to acknowledge the help given by numerous colleagues. Dr. Teri Walker (RTI), who compiled and managed the comprehensive database of research literature, a key resource for this work, and also provided critical comments on several of the chapters, also deserves special thanks. The critique and suggestions on various chapters by Drs Lynn Davis, Teri Walker, Li Han, Howard Walls, Michael Franklin, Reshan Fernando, and Ms. Jenia Tufts (all RTI) are gratefully acknowledged.

I appreciate the many helpful suggestions by Carlos Nunez (USEPA, RTP), Dr. Li Zhang (USEPA, Washington DC), Drs S. Ramakrishna and W. E. Teo (The National University of Singapore), Professor John van Aalst (Department of Surgery, UNC Chapel Hill), Professor Saad Khan (Department of Chemical Engineering, NCSU), and Dr. Peter Ingram (Duke University). It is a pleasure to acknowledge the encouragement and helpful suggestions by Rosalyn Farkas (John Wiley and Sons) in making this book a success.

A. L. A.

1

INTRODUCTION

Fueled by the promise of lucrative returns, nanotechnology has enjoyed unprecedented global research and development support over the last few years. Among the many facets of this unique technology, nanomaterials appear to be the first, albeit relatively low-technology, product to have reached commercialization. Nanomaterials enjoy the advantage of an existing sophisticated microscale technology for producing bulk micropowders, fibers, and thin film in enhancing their utility as high-performance smart materials in a myriad of applications. Their unusual physicochemical characteristics are primarily governed by their very high surface area to volume ratio (or the ratio of surface atoms to the interior atoms in the cluster). Material characteristics that determine catalysis, optical properties, certain mechanical properties, and even biological phenomena generally have a length scale in the 100 nm range. Nanomaterials can have very different geometries — they might be nanoparticles or clusters, nanolayers or nanofilms, nanowires, and nanodots. Building on existing robust fine-powder technology, nanoparticle materials have been among the first nanoscale products to be commercialized and are already creating a significant impact in diverse industries. These include their use in catalytic converters, oxides in sunscreens, nanoclay reinforcing fillers, abrasion-resistant oxides (e.g., alumina or zirconia-based oxides) coatings, ferrofluids, and conductive inks.

Furthermore, those materials that fall into the strict nano-regime (where one of their dimensions is <100 nm) may display unique and controllable

---

Science and Technology of Polymer Nanofibers. By Anthony L. Andrady

properties governed by quantum constraint effects (He, J. H., et al. 2007b). For instance, nanoparticles of semiconductor CdSe behave as pseudo-atoms with molecular orbitals delocalized over the entire cluster. The associated quantized energy levels[1] allow these (quantum dots) to display, on excitation, well-defined size-dependent fluorescence emissions at visible wavelengths. The bandgaps of the semiconductor nanoparticles vary with particle size. As the particle sizes of the quantum dots vary from 2 nm to 6 nm, the emission wavelength changes from blue to red when excited at $\lambda = 290$ nm. Other properties such as the ionization potential, melting temperature, catalytic activity, glass transition temperature, magnetic susceptibility are all size-dependent properties of nanomaterials.

Nanofibers, especially organic nanofibers, constitute a particularly interesting and versatile class of one-dimensional (1-D) nanomaterial. The more exotic of the conventional textile fiber technologies include "microdenier fibers" (0.2–1.5 denier per filament), produced using multistep fabrication techniques such as melt spinning using "islands at sea" type extrusion dies. Further refinement of these textile industry techniques to obtain nanoscale fibers (that are several orders of magnitude smaller in diameter) is not practical, cost-effective, or scalable. Several techniques unrelated to electrospinning were reported in early literature for the laboratory preparation of nanofibers. Self-assembly of polymers under certain conditions and drawing of polymer melts can produce small samples of polymer nanofibers.

Electrostatic spinning or electrospinning, however, remains the most convenient and scalable technique for nanofiber production. The process has been successfully scaled up and is already used in the production of industrial products such as air filter media. Fibers with a diameter in the range $d = 50$–900 nm can readily be electrospun into mats; at $d \sim 50$ nm about 10,000 polymer chains, each up to a length of 100 μm, pass through the cross-section of the nanofiber (Reneker and Chun 1996). Electrospun nanofibers are orders of magnitude smaller in diameter compared to synthetic textile fibers and common natural fibers (Table 1.1). Electrospun nanofibers with diameters as small as 3–5 nm have been reported (Zhou et al. 2003); however, these cannot be generated consistently in quantity, even at the laboratory scale. The smallest of the nanofibers, with diameters of only several nanometers, can be selected for imaging from an ensemble of nanofibers electrospun usually from dilute solutions of a high-molecular-weight polymer under carefully controlled conditions.

[1]Small nanoparticles with quantized energy levels are sometimes referred to as "artificial atoms." Although there is no central nucleus holding the electrons, a parabolic potential well holds the electrons, which can move in a two-dimensional plane in the well.

TABLE 1.1 Comparison of natural and textile fibers

Fiber	Diameter (μm)	Coefficient of Variation (%)
Spider silk	3.57	14.8
Bombyx mori silk	12.9	24.8
Merino wool	25.5	25.6
Human hair	89.3	17.0
Cotton	10–27	2.5
Polyester	12–25	4–5
Nylon	16–24	3–6

1.1 HISTORICAL BACKGROUND

The first documented accounts of electrostatic spinning of a polymer solution into nanofibers were described in 1902 by J. F. Cooley and by W. J. Morton (see Table 1.2). Figure 1.1 shows Cooley's diagram of the electrospinning equipment as it appears in his 1902 U.S. patent # 692,631 (note that the

TABLE 1.2 Chronological development of electrospinning patents

Year	Persons	Description
1902	Cooley, J. F.	U.S. pat. # 692,631
1902	Morton, W. J.	U.S. pat. # 705,691
1903	Cooley, J. F.	U.S. pat. # 745,276
1934–1944	Formhals, A.	U.S. pat. #s 1,975,504; 2,077,373; 2,109,333; 2,116,942; 2,123,992; 2,158,415; 2,158,416; 2,160,962; 2,187,306; 2,323,025; 2,349,950
1929	Hagiwara, K.	U.S. pat. # 1,699,615
1936	Norton, C. L.	U.S. pat. # 2,048,651
1939	Gladding, E. K.	U.S. pat. # 2,168,027
1943	Manning, F. W.	U.S. pat. # 2,336,745
1966	Simons, H. L.	U.S. pat. # 3,280,229
1976	Simm, W., et al.	U.S. pat. # 3,944,258
1977/1978	Martin, G. E., et al.	U.S. pat. # 4,043,331; 4,044,404; 4,127,706
1978	Simm, W., et al.	U.S. pat. # 4,069,026
1980	Fine, J., et al.	U.S. pat. # 4,223,101
1980/1981	Guignard, C.	U.S. pat. # 4,230,650; 4,287,139
1982	Bornat, A.	U.S. pat. # 4,323,525
1985	How, T. V.	U.S. pat. # 4,552,707
1987	Bornat, A.	U.S. pat. # 4,689,186
1989	Martin, G. E., et al.	U.S. pat. # 4,878,908
1991	Berry, J. P.	U.S. pat. # 5,024,789
2000	Scardino, F. L. and Balonis, R. J.	U.S. pat. # 6,106,913
2004	Chu, B., et al.	U.S. pat. # 6,713,011

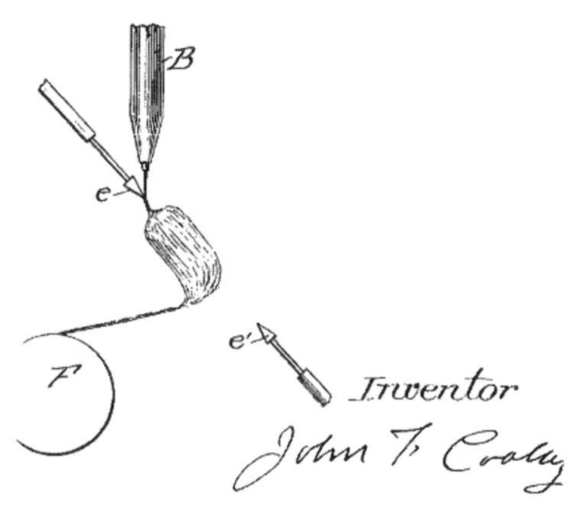

Figure 1.1 A solution of polymer (e.g., collodion or cellulose nitrate in ether or acetone) delivered into the high-voltage direct current (DC) electric field via tube *B* to form electrospun nanofibers collected on a drum *F*. (*Source*: Cooley 1902, Fig. 5 of U.S. patent 692, 631.)

static electricity generator connected to the electrodes is not shown).[2] These patents teach the deposition of a viscous polymer solution on a positively charged electrode (a roughened brass sphere) held close to an electrode of opposite charge to obtain electrostatic spinning. The spun fibers were collected as "a cob-web like mass" on the negatively charged electrode. The process was described as being the result of "electrical disruption of the fluid." A closely related patent issued a year later in 1903 to Cooley also addressed electrospinning. The claims in the latter patent included the introduction of the viscous polymer solution near the terminus of a charged electrode, but not necessarily in contact with it, to yield electrospun fibers. These early patents emphasize the need for the polymer solution to be of adequate viscosity and used, as a specific example, the electrospinning of nitrocellulose. Interestingly, the fundamental features of the process, as described in these century-old patents, have changed little with time.

Anton Formhals, a quarter century later in 1934, patented an improved version of the electrospinning process and apparatus. His first patents on electrospinning of cellulose acetate from acetone used a fiber collection system that could be moved, allowing some degree of fiber orientation during spinning. He recognized the importance of adequate drying of the fibers prior to the nanofibers being collected on a grounded surface. By 1944, he had filed four more patents on improved processes and claimed methods to electrospin even multi-component webs that contained more than one type of nanofiber.

[2]The first reported electrostatic spraying of a liquid was described by Jean-Antoine Nollet in 1750, long before the term electrospraying was even coined.

In 1936, C. L. Norton (see Table 1.2) used a plate collector electrode in conjunction with a static electricity generator in his design to provide a "transverse intermittent electromotive force" to improve fiber quality and collection.

Sir Geoffrey Taylor's contribution in the 1960s towards the fundamental understanding of the behavior of droplets placed in an electric field helped further develop the technique (Taylor 1964, 1969). In 1966, H. L. Simons (see Table 1.2) described the production of nonwoven nanofiber mats of a variety of thermoplastics including polycarbonate and polyurethane using metal grids to obtain a variety of patterned mats with uneven fiber density. His patent identifies viscosity, dielectric constant, conductivity, and volatility of the solvent as the key process parameters. His work explicitly identified the role of viscosity of the polymer solution in obtaining finer continuous fibers. Peter Baumgarten, working with an acrylic copolymer/ dimethylformamide (DMF) system, described the dependence of fiber diameter on viscosity (and hence on concentration) of the solution as well as on the magnitude of the electric field (Baumgarten 1971). His experiment included a high-voltage power supply as well as a positive displacement pump.

Similar data for electrospinning polyolefins in the melt were reported by Larrondo and St. John Manley (1981a, 1981b, 1981c), with obtained fiber diameters being somewhat larger than those of solvent-spun nanofibers. Increasing the melt temperature and therefore decreasing melt viscosity resulted in smaller fiber diameters. Melt electrospinning can be an important approach, especially with common thermoplastics such as polyethylene (PE), polypropylene (PP), poly(ethylene terephthalate) (PET), and nylon (PA), which do not dissolve in common solvents (Dalton et al. 2006; Larrondo and St. John Manley 1981a, 1981b, 1981c; Lyons et al. 2004). Melt spinning, however, has to be carried out at high temperatures (usually $>200^\circ C$), requires larger electric fields (compared to electrospinning solutions), and is usually carried out in a vacuum.

Although this early work laid down the basic technique of electrostatic spinning, the present understanding of the process is mainly due to more recent work, especially that carried out within the last 10–15 years. Recent contributions towards understanding fluid dynamics (Hohman et al. 2001a, 2001b) and electrostatics (Shin et al. 2001a, 2001b; Spivak and Dzenis 1999) associated with electrospinning were fundamental to the resurgence of interest in the technique. Doshi and Reneker (1995), Jaeger et al. (1998) and Reneker et al. (2000) in the 1990s quantified the reduction in electrospun jet diameter as a function of distance away from the Taylor's cone for poly(ethylene oxide) (PEO) in water. In a systematic study, Doshi and Reneker (1995) established a viscosity window for successful electrospinning of PEO solutions (applicable of course to the particular average molecular weight of polymer used). Hayati et al. (1987) recognized the relationship between the solution conductivity and the whipping instability (as well as the likelihood of electrospray behavior). Early attempts at electrospinning polymers were beset with experimental difficulties, the most important among them being "bead" formation. Deitzel et al. (2001a), as well as Doshi and Reneker (1995), studied bead formation in nanofibers, relating their frequency of occurrence to the applied voltage and recognizing the influence of the changes in shape of droplet with electric field in yielding beaded fibers.

Present-day laboratory electrospinning equipment is quite similar to that used in the approaches described above. The basic hardware components remain the same, especially in research electrospinning apparatus. However, the availability of more stable power supply units and pulse-free pumps to regulate the delivery of polymer solution to the charged electrodes now allows for better nanofiber quality. Minor modifications to the basic experimental setup have been described. Controlling the nonlinear whipping instability during electrospinning by modifying the geometry of the applied electric fields has been attempted. Warner et al. (1999) and others (Shin et al. 2001a, 2001b), for instance, claimed to improve the uniformity of the electric field by using a disc electrode of about the same diameter as the collector at the capillary tip resulting in a parallel-plate electrode design. Others have used a second ring electrode (Jaeger et al. 1998) or auxiliary plate electrodes to control and focus the electrospun fiber on the collector plate. Using a ring electrode at the same potential as the main electrode improved stability in the initial part of the jet (close to the droplet); however, the whipping instability, which occurs closer to the fiber collection region, was not substantially improved. Most of these innovations, however, can be traced back to aspects of the very early disclosures on the technique; auxiliary electrodes and rotating collectors, and solid tips were all featured in the very earliest patents on electrospinning. For example, several early patents such as U.S. patents # 4,043,331 (1977, Martin, G. E., et al.), # 4,127,706 (1978, Martin, G. E., et al.), # 4,878,908 (1989, Martin, G. E., et al.) and # 3,994,258 (1976, Simm, W., et al.) described rotating or moving-belt type collectors for the electrospun fiber mats.

The bulk of the reported early research on electrospinning focused on a limited number of polymer/solvent combinations. Naturally, these were the polymers that were easy to electrospin under laboratory conditions. These likely included those polymers that dissolved in common solvents that are "good solvents" for the polymer, where the chain-like polymer molecules adopt open, extended macromolecular conformations (as opposed to compact globular geometries) that allow adequate entanglement of polymer chains. With potential for future scale-up in mind, solvents that are both economical and also environmentally acceptable were preferred.

These considerations encouraged water-soluble polymers such as PEO to be popularly studied in early research on electrospinning. Only limited work on electrospinning of polymers such as polyamides was reported in the early literature because of the requirement for expensive and/or hazardous solvents (e.g., formic acid for nylon-6,6).

1.2 BASIC EXPERIMENTAL APPROACH

The minimum equipment requirements for demonstration of simple electrospinning in the laboratory are as follows:

1. A viscous polymer solution or a melt.
2. An electrode (hollow tubular or solid) that is maintained in contact with the polymer solution.
3. A high-voltage DC generator connected to the electrode.[3]
4. A grounded or oppositely charged surface to collect the nanofibers.

Figure 1.2 is a schematic representation of the equipment generally used in laboratory electrospinning of polymer solutions.

A simple experimental setup may consist of a glass pipette drawn into a capillary at one end, carrying a few milliliters of a viscous solution of a high polymer (for example a 20% w/w solution of polystyrene (PS) dissolved in methylene chloride). The viscosity of the solution is high enough to prevent it dripping from the vertical pipette under gravity. The tube is mounted vertically a few inches (6–10 inches) above a grounded metal (e.g., aluminum) plate or drum. A metal wire electrode that dips into the solution in the tube is connected to the positive terminal of a high-voltage DC power supply unit.[4]

The power is switched on and the voltage increased to 10–20 kV using the controls on the power supply. At a certain threshold voltage (depending on a number of factors to be discussed later), a droplet of the liquid is drawn out of the tube into a cone-shaped terminus and sprays downwards

[3]Alternating current (AC) potentials can also be used in electrospinning. He and colleagues developed a mathematical model for electrospinning using an AC potential (He and Gong 2003; He, J.-H., et al. 2005a). A comparison of PEO mats spun from DC and AC potentials showed the latter to suppress whipping of the jet and result in better alignment of the nanofibers (Kessick et al. 2004). The charge build-up on the collector is likely to be less of a problem with AC voltage compared to DC voltage.

[4]All that is needed is a strong enough electric field, not necessarily an electrode in contact with the polymer solution. Electrospinning an 8 wt% solution of poly(acrylonitrile) (PAN) in dimethylformamide (DMF) using ionized field charging with a noncontacting ring electrode was recently reported (Kalayci et al. 2005).

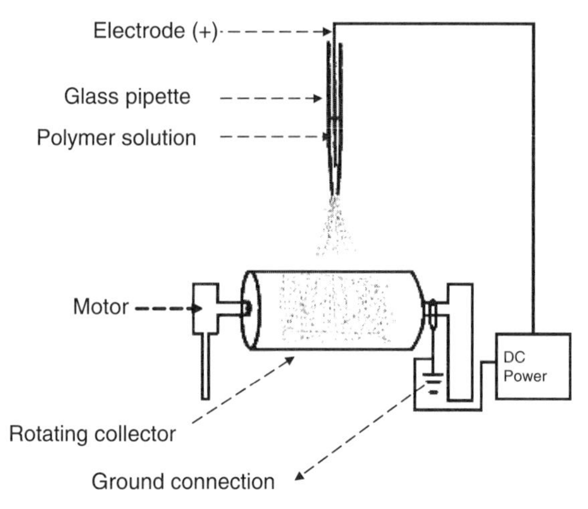

Figure 1.2 A schematic of a simple electrospinning experiment. Reprinted with permission from J.-S. Kim and Reneker (1999b). Copyright 1999. John Wiley & Sons.

as a jet towards the grounded plate as a barely visible nanoscale fiber. The high charge density on the surface of the fine jet leads to electrical instability of the electrospinning fiber, making it whip about rapidly. This splaying of the nanofiber often gives the appearance of a multiplicity of nanofibers being sprayed from the single droplet suspended from the capillary tip of the glass tube. High-speed photography, however, has demonstrated that, in general, a single nanofiber is spun out of the droplet, and its rapid movement generates the appearance of a multiplicity of fibers (Reneker et al. 2000). Consistent with this observation, one rarely observes fiber ends in high-resolution microscopic images of the nanofiber mats collected on the grounded surface. The mat is generally composed of a single long fiber arranged randomly on the collector surface. The solvent, which often accounts for more than 80% of the solution, evaporates rapidly from the surface of the spinning jet. It is desirable to select a solvent, gap distance, and temperature that would ensure that the electrospun fiber is completely dry by the time it reaches the grounded plate. Any residual surface charge on the nanofiber is rapidly dissipated on contact with the grounded metal plate, and the nanofiber mat can be peeled off it. Samples of nanofiber for microscopic examination are conveniently obtained by placing a sample collection stub over the grounded surface.

Shenoy et al. (2005a) pointed out the similarities between conventional pressure-driven dry spinning and electrospinning. Although both fiber-forming processes use polymer solutions and rely on rapid removal of the solvent to generate the fiber, the mechanisms responsible for the initial formation of the cylindrical fiber geometry and the subsequent "drawing" or thinning of the fiber are too different in the two processes to consider

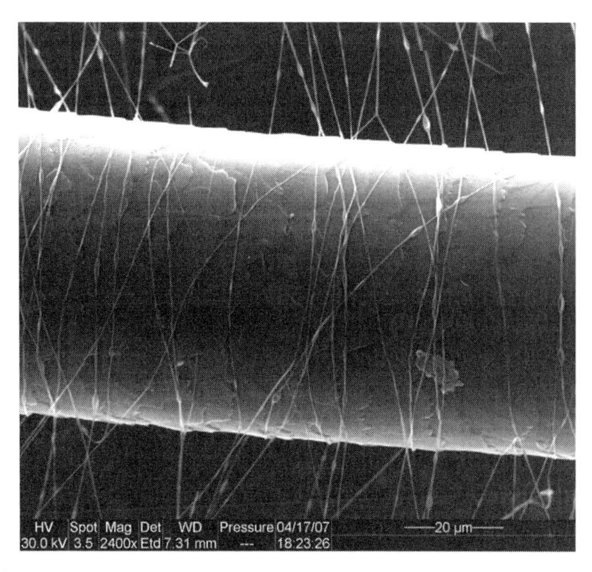

Figure 1.3 SEM image comparing the diameter of a human hair with that of PMMA nanofibers electrospun from DMF solution. (Courtesy of RTI International.)

electrospinning a special case of dry spinning. The quality of nanofibers produced in any electrospinning experiment is affected by a variety of material and process variables. From a practical standpoint, two such sets of variables might be identified — equipment-related and material-related variables. Each set includes a number of different and interrelated variables. These include the solution (or melt) temperature, concentration of solution, feed rate, electric field applied, volatility of solvent, gaseous environment about the spinning fiber, solvent vapor pressure, dielectric properties of the system, conductivity of the solution, surface tension of the solution, and the molecular weight of the polymer. These will be discussed in some detail in Chapter 4. Changing any of these can not only change fiber morphology and mat structure of the nanofiber formed, but in some instances can even determine if electrospinning occurs at all. Comprehensive predictive models that encompass all pertinent variables have not been developed as yet. Only qualitative general guidelines are available on the effect of these on fiber or mat quality, making electrospinning as much an art as it is a science.

Figure 1.3 shows a scanning electron microscopy (SEM) image of electrospun nanofibers of poly(methyl methacrylate) (PMMA) on a human hair.

1.3 DESCRIPTION OF ELECTROSTATIC SPINNING

The main features of the electrospraying process are common to electrospinning, and the former has been studied in some detail over several decades.

The differences between the two processes center on chain entanglement and the resulting elongational viscous forces that operate in polymer solutions undergoing electrospinning. This results in the extraction of a fiber, as opposed to the production of droplets, from the coulomb explosion of a supercharged drop of solution at the end of the tip or a capillary.[5] When either a dilute solution of a polymer or solutions of a low-molecular-weight polymer are electrospun, it is common to obtain a mix of electrosprayed particles along with malformed uneven short nanofibers. Reneker and Fong (2006) separated the electrospinning process into several key stages for convenience of description: launching of the jet; elongation of the straight segment; development of whipping instability; and solidification into a fiber. The same is used here for ease of description but with the first stage subdivided into a droplet generation stage and a Taylor's cone formation stage.

1.3.1 Droplet Generation

Variants of electrospinning that do not rely on droplets being produced at the capillary or the end of a needle are known. In most laboratory studies, however, the charging of a droplet of polymer solution is the initial step in electrospinning. Typically, a polymer solution is pumped at a low flow rate into a capillary tip. In the absence of an applied electric field, the droplets form at the end of the capillary[6] and fall off under the influence of gravity. Assuming the surface tension of the liquid, γ, and the gravitational force, F_G, to be the only two forces acting on the meniscus of the droplet, the radius of the droplet, r_0, produced by the capillary of internal radius R is

$$r_0 = (3R\gamma/2\rho g)^{1/3}, \tag{1.1}$$

where ρ is the density of the liquid and g is the gravitational constant.

This "dripping" regime may continue even in the presence of low electric fields. When a high enough voltage is applied and the liquid has finite conductivity, the electric force F_E, as well as the gravitational force, will work against the capillary surface forces (i.e., $F_\gamma = F_E + F_G$) and the sustainable droplet size at the capillary tip will be reduced to r ($r < r_0$).

In laboratory electrostatic spraying or spinning, where a capillary[6] carrying a positive voltage V is held at a distance L from a grounded metal surface,

[5] The term "capillary" or "tip" is preferred over the term "spinneret," as the latter can be confused with spinnerets encountered in the spinning of textile fibers in conventional fiber manufacture.

[6] A hollow needle-like capillary is not essential for electrospinning — a droplet on a solid electrode behaves similarly. Ultrafine droplets picked up by an atomic force microscopy (AFM) tip, nanofabricated microfluidic channels (Kameoka and Craighead 2003), or by a dip-pen type tip (Sun et al. 2006) have been electrospun successfully.

F_E for the system can be expressed as in equation (1.2) (Bugarski et al. 1994; DeShon and Carson 1968; Lee 2003). The expression is based on that for an electric field at the tip of a metal point and a grounded plate as proposed by Loeb et al. (1941):

$$F_E = (4\pi\varepsilon V^2)/[\ln(4L/R)^2], \tag{1.2}$$

where ε is the permittivity of the medium (air in most experiments) and V is the applied voltage. Bugarski et al. (1994) obtained the droplet radius r for such a system as

$$r = \{(3/2\rho g)[R\gamma - (2\varepsilon V^2)/(\ln(4L/R))]\}^{1/3} \tag{1.3}$$

As V increases, r becomes progressively smaller until droplet instability sets in at a value of the electric field $V = V_C$, and electrostatic spraying occurs.

Due to the electric field, charge separation will take place in a droplet that is electrically conductive. Where the capillary is positively charged, for instance, the positively charged species migrate to the surface of the droplet and the negatively charged species accumulate in its interior until the electric field within the liquid droplet is zero. Charge separation will generate a force that is countered by the surface tension within the droplet. The velocity at which these ionic species move through the liquid is determined by the magnitude of the electric field and the ionic mobility of the species. For an electric field of $\sim 10^5$ V/m typical of electrospinning, the drift velocity has been estimated to be ~ 0.15 m/s (Reneker and Chun 1996). However, the velocity achieved by the jet itself in electrospinning tends to be much higher, reaching values of 10 m/s in typical runs. Ionic species therefore must move at comparable velocities and in the direction of the jet.

The stability of an electrically charged droplet at the end of a capillary requires the inward surface tension forces to exceed the outward repulsion forces of like charges accumulating on the droplet surface:

$$F_E \leq g\rho[(r^2/\beta) - V], \tag{1.4}$$

where g is the gravitational constant, V is the volume of the droplet, ρ is the density of the liquid, and β is the shape factor for the droplet.

However, the maximum surface charge Q_R that the surface of a droplet can accommodate in vacuum is limited by the Rayleigh condition (Rayleigh 1882):

$$Q_R = 8\pi(\varepsilon\gamma r^3)^{1/2} \tag{1.5}$$

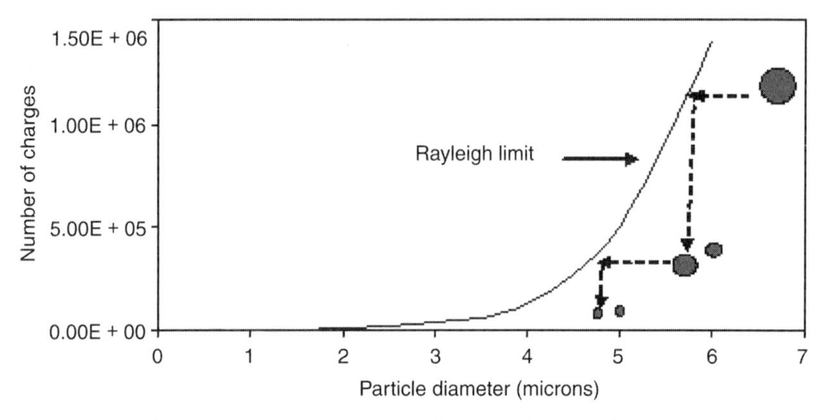

Figure 1.4 Schematic representation of the explosion of charged droplets.

At $|Q| > Q_R$ the droplet first deforms and then explodes into a number of smaller droplets due to coulombic repulsion of positive surface charges crowded on its surface. In practice, the limit can be reached by either gradually increasing the electric field or by allowing the liquid droplet under a constant electric field to reduce its diameter via evaporation. Evaporation of a charged droplet does not discernibly reduce its surface charge (Abbas and Latham 1967). However, charge transfer to ambient moisture in the spinning environment cannot be ruled out (Kalayci et al. 2005).

With low-molecular weight liquids, this build-up of electrical pressure results in primary asymmetric fission of the droplet, giving rise to smaller highly charged sibling droplets (note, spherical shape minimizes the surface) that in turn subdivide again with continued evaporation (Fig. 1.4). In the absence of long-chain polymer molecules that are long enough to undergo entanglement,[7] break-up of the jet into individual droplets is inevitable. Under proper conditions the process may even continue until single ions result (soft electrospray ionization used in mass spectrometry and ion mobility spectrometry relies on this process to generate individual analyte ions). Negatively charged droplets will continue to disintegrate to a size where spontaneous electron emission occurs (10–100 nm depending on the value of γ). Except under very high electric fields, positive ion emission is unlikely and positively charged droplets might be expected to disintegrate down only to molecular dimensions. In addition to this fission mechanism (Dole et al. 1968), direct ion evaporation from a supercharged

[7]There are situations where other nonbonded interactions between molecules or aggregates of molecules such as micelles are strong enough to obtain electrospinning into nanofibers of solutions where no polymer is present. For example, lecithin electrospun from 35 wt% solutions in $CHCl_3$/DMF (70/30) was shown to yield nanofibers (McKee et al. 2006b).

droplet (Iribarne and Thompson 1976) has also been proposed. Although both mechanisms are feasible, their relative importance under various experimental conditions is not clear.

1.3.2 Taylor's Cone Formation

Deformation of relatively small charged droplets under an electric field, from a sphere to an ellipsoid, is well known (Macky 1931). The effect diminishes as r increases, because the electric field just outside the droplet varies inversely with r^2. For droplets of water, such deformation has been observed at fields exceeding 5000 V/cm. The elongated droplet assumes a cone-like shape and a narrow jet of liquid is ejected from its point (Taylor 1964, 1969; Melcher 1972). This "Taylors cone" is formed at the critical voltage V_C applied to a droplet at the end of a capillary of length h and radius R (Taylor 1969):

$$V_C^2 = (2L/h)^2(\ln(2h/R) - 1.5)(0.117\pi RT). \qquad (1.6)$$

Observing the process in a range of different liquids, Taylor determined the equilibrium between surface tension and electrostatic forces to be achieved when the half angle of the cone was 49.3°, a value verified later by others (Larrondo and St. John Manley 1981a, 1981b). This value can, however, be different for different polymer solutions and melts. For instance, with molten PP, a half angle of 37.5° has been observed (Rangkupan and Reneker 2003). It is the change in shape of the droplet into this conical shape that defines the onset of extensional force initiating droplet/fibril formation that eventually leads to electrostatic spraying and spinning. At the minimum spraying voltage some liquids display a pulsation of the droplet, with spray being associated with these pulses. With high-viscosity liquids such as polymer solutions, a smooth transition to a Taylor's cone geometry is generally obtained. This cone-like shape is not necessarily maintained throughout the electrospinning process — it can change depending on the ratio of the feed rate to the mass transfer rate away from the droplet (Wang, Z.-G., et al. 2006). Zeleny (1935), following the work of Rayleigh, studied electrospraying from a glass capillary where the liquid was electrified (using a set of 15 Leyden jars as the generator) via an electrode. He reported multiple jets emanating from a single droplet under certain conditions. Recent high-speed imaging observations of levitated ethylene glycol droplets in an electric field showed deformation and disintegration into fine jets (often called Rayleigh jets), as predicted by Rayleigh (Duft et al. 2003). The jets disintegrated into fine droplets amounting to about 0.3% of the mass, and carrying about one-third of the total charge of the mother droplet.

Equation (1.6) suggests high-surface-tension liquids to require high electric fields V_C for electrostatic processing that may possibly lead to corona discharge. However, neither the conductivity nor the viscosity of the liquid that forms the droplet is taken into account in the above equation or other similar expressions for V_C (Hendricks et al. 1964). In practice, however, both parameters heavily influence cone formation in electrostatic spraying and spinning and can be readily varied using additives. Although this description is based on droplet geometry, electrospinning can also occur from an essentially flat surface of liquid subjected to a strong enough electric field (Yarin and Zussman 2004).

Kalayci et al. (2005) recently described the charging process in the electrospinning solution. The mobility of an ion in the viscous solution depends upon the electrostatic force $F_E = qE$ and the viscous drag force $F_d = 6\pi\eta r'\mu E$, which work against each other (r' is the hydrodynamic radius of the ion, q is the ionic charge, η is the solution viscosity, μ is the ionic mobility, and E is the electric field strength). Their expression for the sum of electrostatic forces can be reduced to the following (Kalayci et al. 2005):

$$\sum F_{(\text{electrostatic})} = (n_1 qE) - (n_1 6\pi\eta r'_+ \mu_+ E) - [n_1(1-y)qE] + [n_1(1-y)6\pi\eta r_- \mu_- E], \quad (1.7)$$

where n_1 is the number of ions in a solution of mass m, $(1-y)$ is the fraction of negative ions in the droplet, and the subscripts + and – refer to values for the positive and negative ions. The geometric representation of the jet in the Taylor's cone area from Kalayci et al. (2005) is reproduced in Fig. 1.5. V_1 and V_2 refer to the volume of the conical frustum and the volume of the space in which the jet is contained, respectively. They assumed $V_1 = 2/3\,V$ in their analysis.

1.3.3 Launching of the Jet

Due to copious entanglement of polymer chains in concentrated solution, the (outward) force available to a droplet via coulombic repulsion will generally be insufficient to explode it. However, the surface area has somehow to be increased to accommodate the charge build-up on the jet surface, and this occurs through the formation of fibers. A slender fibril emanates from the cone to create additional surface area needed to accommodate surface charges, and it initially travels directly towards the grounded collector. The effect of charge repulsion is not unlike the mechanical stretching experienced by a jet in conventional fiber spinning (Burger et al. 2006; Shenoy et al. 2005a). Studying the mobility of particles in electrospinning jets, Deitzel et al. (2006) suggested this jet initiation to occur from the surface layers of

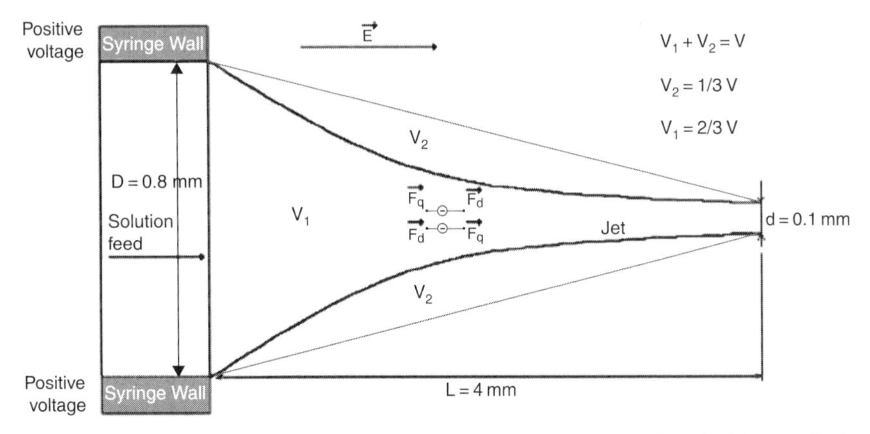

Figure 1.5 Geometric model of the Taylor's cone region. Reprinted with permission from Kalayci et al. (2005). Copyright 2005. Elsevier.

the cone. This is partly due to surface shear forces generated by the potential difference between the base and the tip of the Taylor's cone. Quasi-stable multiple jets emanating from the same droplet have been observed with some systems (Figs. 1.6 and 1.7). The tendency is for one of these to become stable while the others die off, without affecting the total current flow in the system (Koombhongse et al. 2001). Electrospinning a segmented polyurethane urea from DMF solutions (2.5–17% w/w) using an electric field of 6 kV/cm, Demir et al. (2002) reported as many as six jets emanating from a single droplet at low concentrations of polymer, the average number

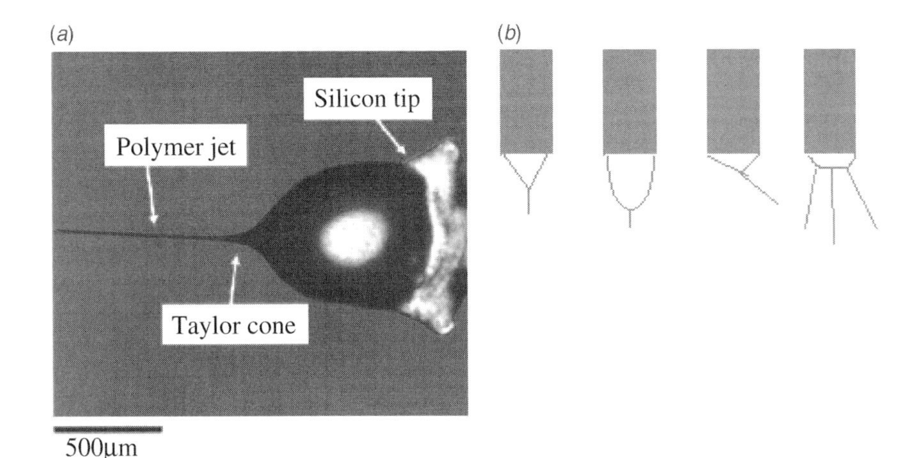

Figure 1.6 (*a*) Optical image of the Taylor's cone and tapering linear segment of a jet emanating from a microfabricated silicon tip. Reprinted with permission from Kameoka et al. (2003). Copyright 2005. American Institute of Physics. (*b*) Diagram of different geometries of Taylor's cone obtained in practice.

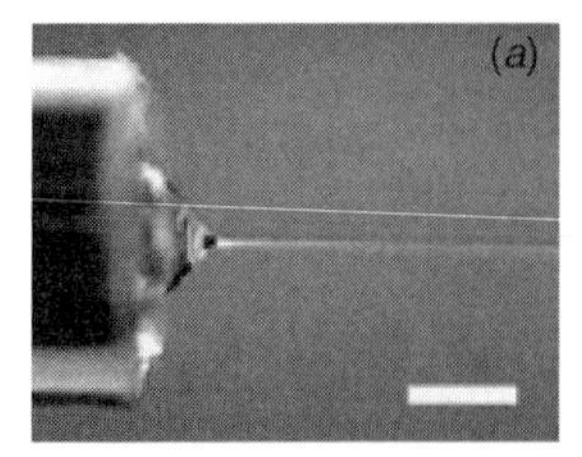

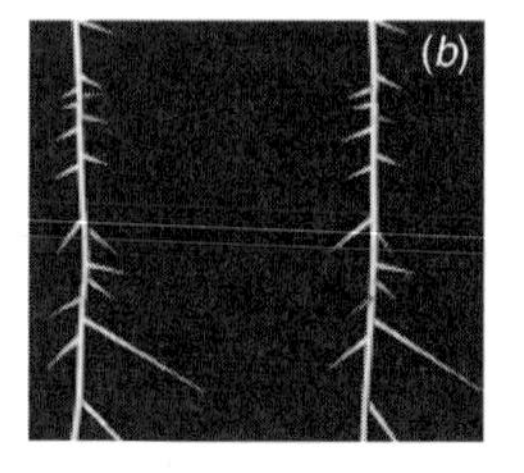

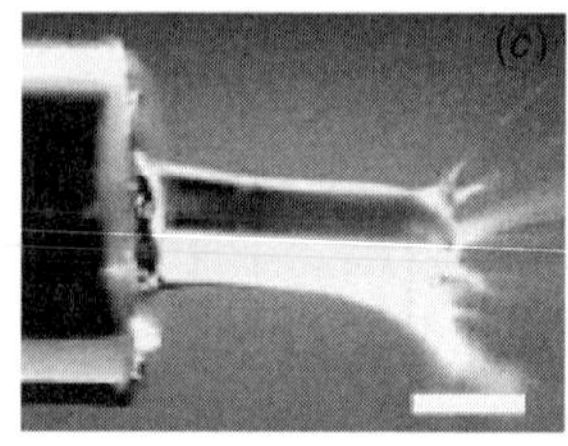

Figure 1.7 (*a*) Taylor's cone and straight jet formed during electrospinning of a 10 wt% solution of PLA in methylene chloride at an applied field of 1.2 kV/cm. Reprinted with permission from Larsen et al. (2004b). Copyright 2004. John Wiley & Sons. (*b*) Branching of jet during electrospinning captured in a high-speed photograph. Reprinted with permission from Yarin et al. (2005). Copyright 2005. American Institute of Physics. (*c*) Multiple jets emanating from a single elongated droplet at 1.2 kV/cm, but without N_2 flow coaxial to the jet. Reprinted with permission from Larsen et al. (2004b). Copyright 2004. John Wiley & Sons.

of jets increasing linearly with the electric field (kV/cm). Figure 1.6 illustrates the Taylor's cone.

Once launched, the jet can be described by considering the conservation of mass in the system:

$$\text{Feed rate} = (\pi d^2 \rho u)/4, \tag{1.8}$$

where $d = 2r$ is the diameter of the jet, ρ is the density, and u its velocity. Similarly, the conservation of charge for the jet yields the following (He, J.-H., et al. 2005a, 2005b):

$$\pi d Q u + (k \pi d^2 E)/4 = I, \tag{1.9}$$

where E is the applied electric field, I the current flowing through the jet, k the dimensionless conductivity of the solution, and Q is the surface charge.

1.3.4 Elongation of Straight Segment

Jet initiation occurs almost instantaneously on application of a voltage exceeding V_C to the polymer solution. The coulombic repulsion of surface charges on the jet has an axial component that elongates the jet in its passage towards the collector. Laser doppler velocimetry experiments (Buer et al. 2001) reveal the velocity of the jet as well as the variance in jet velocity to increase with the distance away from the Taylor's cone. As a result, the jet diameter decreases rapidly due to both extension and evaporation of the solvent. The initially straight jet tapers down as it accelerates towards the collector, and the tapering is pronounced in the region below the Taylor's cone. As the jet thins, the surface area per unit mass of jet material increases while the surface charge per unit area decreases. Loss of charge by adventitious discharges due to air-borne charged species and ions, which neutralize the surface charge on the fibers, become increasingly likely as the surface area increases. However, solvent evaporation continually increases the surface charge per unit area, driving the increase in surface area through extension.

It is the extensional modulus of the rapidly drying jet (due to chain entanglement) that prevents the onset of capillary instability and yields a stable jet. Recent work on electrospinning of PEO solutions containing low-molecular-weight poly(ethylene glycol) (PEG) emphasizes the role of elasticity of the jet in obtaining electrospinning (Yu et al. 2006). For these systems the ratio of the fluid relaxation time to the time for growth of instabilities (the Deborah number) was shown to correlate with arrest of Raleigh instability and with electrospinning. Bunyan et al. (2006) reported the length of the linear jet to increase when the electric field was changed by increasing the diameter of the collector disc electrode used at the spinneret.

Xu and Reneker (2006) measured the diameter of spinning jet at different points below the Taylor's cone using interference colors generated when a beam of light impinges on the jet. The technique allows diameters in the range of 500 nm to 15 μm to be conveniently measured in real time using a single camera and a light source. The diameters can also be measured by laser velocimetry (Warner et al. 1999) or by optical imaging near the Taylor's cone (Deitzel et al. 2006), but the procedure becomes increasingly difficult when nanofiber diameters taper down to dimensions close to the wavelengths of light. As the jet diameters dip below 100 nm, very significant chain orientation (made evident by changes in birefringence) occurs. The relative modulus at the surface of electrospun nanofibers was recently measured using scanning probe microscopy (SPM) techniques (Ji et al. 2006b), and was found to increase with the decrease in fiber diameter. This is likely due to the shear-induced orientation of chains (Jaeger et al. 1996). The length of the linear portion of the jet, as well as the rate at which its diameter is reduced due to drawing, is determined by the solution feed rate and the strength of the electric field. It is useful to learn the composition of the jet at different distances from the Taylor's cone to be able to quantify the drying rate of the jet. Raman spectroscopy has been successfully used (although only on relatively thick jets) to obtain a polymer : solvent ratio in the electrospinning jet (Stephens et al. 2001).

Modeling the behavior of a jet in this linear regime, before the onset of whipping, appears to be relatively straightforward and has been attempted (Feng 2003). Experimental results appear to be in reasonable agreement with the models. The simpler one-dimensional models that assume the solution to be a leaky dielectric provide good numerical dimensions of the jet in this region (Hohman et al. 2001a, 2001b). Recently, J. H. He et al. (2005a, 2005b) used a simple approach based on Cauchy's inequality to model the straight region of the jet and to predict the length L of the jet segment.

1.3.5 Whipping Instability Region

The initially straight jet segment invariably becomes unstable and displays bending, undulating movements during its passage towards the collector. Early-stage varicose instability that promotes jet extension to accommodate surface charges can be modeled reasonably well. Bending of the jet invariably increases surface area and therefore tends to reduce the density of charges. Theoretical studies on electrically forced jets by Hohman et al. (2001b), Reneker et al. (2000), Yarin et al. (2001a, 2001b), and others (Spivak and Dzenis 1998; Spivak et al. 2000) modeled the whipping jet as being the result of competition between several different modes of instability. These modes of instability (that incidentally also occur in nonviscoelastic solutions) are Raleigh instability, axissymmetric instability, and bending mode instability (Shin et al. 2001a, 2001b). The mode of instability obtained is dependent on the electric field, with stronger fields favoring whipping instability. The jet in this region exhibits components of electrostatic repulsive forces that are not predominantly axial (Hohman et al. 2001b). As a result, it whips about within a conical envelope, still symmetrically arranged about the axis of its straight segment. Figure 1.8 shows an image of the whipping region of a jet and also illustrates two modes of instability in the jet. It is the whipping instability that dramatically increases the surface area of the jet and rapidly lowers the surface charge density.

High-speed imaging studies of the jet by Reneker et al. (2000) concluded the jet to be invariably thrown into a series of loops of increasing diameter, spiraling down towards the collector. The cone-shaped envelope of the unstable jet typical of electrospinning is created by the rapid symmetric movement of a single jet. The axis of the straight jet is maintained and the additional envelope volume that contributes to the loops of larger diameter is generated via extension of the jet along the perimeter of the loops. Reneker's images of larger loops closer to the collector show higher order bending instability where the jet being looped forms right- and left-handed coils (Reneker et al. 2000). Both the rate of increase in surface area during whipping instability and the solvent evaporation rate are high in this regime, further reducing the jet diameter. Interplay between the increasing charge density on the one hand, and the viscous and surface tension forces that resist elongation on the other determines the intricacies of the instability obtained. Even more complicated modes of whipping instability resulting in particularly complex curved

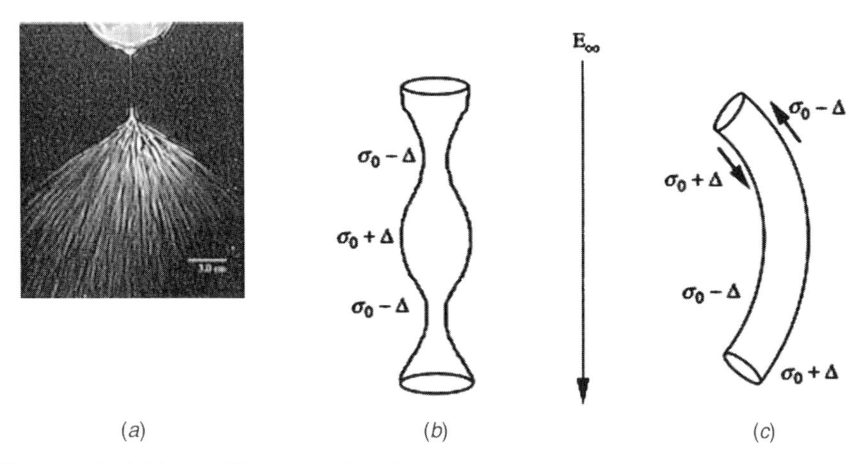

Figure 1.8 (*a*) Image illustrating the whipping region of a typical electrospinning jet (Reneker et al. 2000). (*b*) Axisymmetric instability and (*c*) bending instability in a fluid jet carrying a surface charge, placed in an electric field. Reprinted with permission from Shin et al. (2001b). Copyright 2001. Elsevier. $\Delta\sigma$ denotes the perturbation of the surface charge density and arrows indicate the direction of local torque responsible for bending.

trajectories of electrospinning jets have been observed in practice (Reneker et al. 2002).

From the above description it is clear that whipping instability is the primary mechanism responsible for reducing nanofiber dimensions during electrospinning. However, as Dzenis (2004) pointed out recently, suppressing this instability using either a secondary electric field or a short gap distance (between the tip of the needle and the collector) did not result in substantially thick nanofibers being generated. Understanding of the process is incomplete, and all the factors that govern fiber formation are not well understood.

Whipping instability is a rapid process and it is possible for surface charges to be nonuniformly distributed and to result in sections with high local charge density. This may give rise to secondary jet initiation, resulting in the formation of somewhat less frequent branched nanofibers (see Fig. 1.7b) (Yarin et al. 2005). Branching allows a means of rapidly increasing surface area to accommodate local concentrations of charges. Initiation of such splaying (Deitzel et al. 2001a; Fang and Reneker 1997; Hsu and Shivkumar 2004a, 2004b) and the presence of branched fibers in mats have been observed in practice (Koombhongse et al. 2001; Krishnappa et al. 2003). Inducing splaying by using co-solvents of higher conductivity and/or dielectric constant can therefore result in smaller average fiber diameters as observed for poly(ε-caprolactone) (PCL) in $CHCl_3$/DMF system (Hsu and Shivkumar 2004b; Lee et al. 2003b).

Deitzel et al. (2001a) attributed the presence of smaller-diameter nanofibers in the bimodal distribution of fibers from electrospinning PEO in water solutions to splaying of the jet.

It is useful to review the different forces acting on the whipping charged jet during electrospinning (Wannatong et al. 2004):

1. Gravitational force F_G (towards the collector plate in a vertically arranged apparatus). The force is dependent on density of solution (usually ignored in models). $F_G = \rho \pi r^2 g$, where ρ is the density of the liquid and g is the acceleration due to gravity.
2. The electrostatic force F_E, which extends the jet and propels it towards the grounded collector. The force is determined by the applied electric field and material characteristics. $F_E \propto E$.
3. Coulombic repulsion forces F_C on the surface of the jet, which introduce instability and whipping motions. The magnitude of F_C is dependent on the characteristics of the polymer and solvent.
4. Viscoelastic forces, which work against elongation of the jet in the electric field. This depends on the polymer molecular weight, the solvent, and the type of polymer.
5. Surface tension forces, which work against the stretching of the jet. This depends on solvent type, polymer, and additives.
6. Frictional forces between the surface of the jet and the surrounding air or gas.

The interplay of these different forces (a simple expression is the sum of these forces) determines the diameter of the jet. Some of these change very rapidly in time due to solvent evaporation and charge dissipation, making any quantitative description of the process particularly difficult. Consequently, no entirely satisfactory mathematical models describing jets undergoing whipping instability are available.

Models that are applicable to the onset of instability in the linear instability region are available and primarily interpret instability in terms of surface charge density, viscosity, and inertia. Particularly interesting would be models that accurately predict the envelope of the undulating or whipping nanofiber as well as its change in diameter with time. Fridrikh et al. (2003, 2006) developed a model of the nonlinear behavior of jets in electrospinning of non-Newtonian fluids. Their model takes strain-hardening also into account and suggests that during the later stages a terminal amplitude for the instability and a corresponding terminal jet radius, h_f, are reached by the jet (provided the gap length is sufficiently large). Despite approximations used in their model to account for drying of the jet, agreement between experimental data (on electrospinning PCL solutions) and model predictions on the

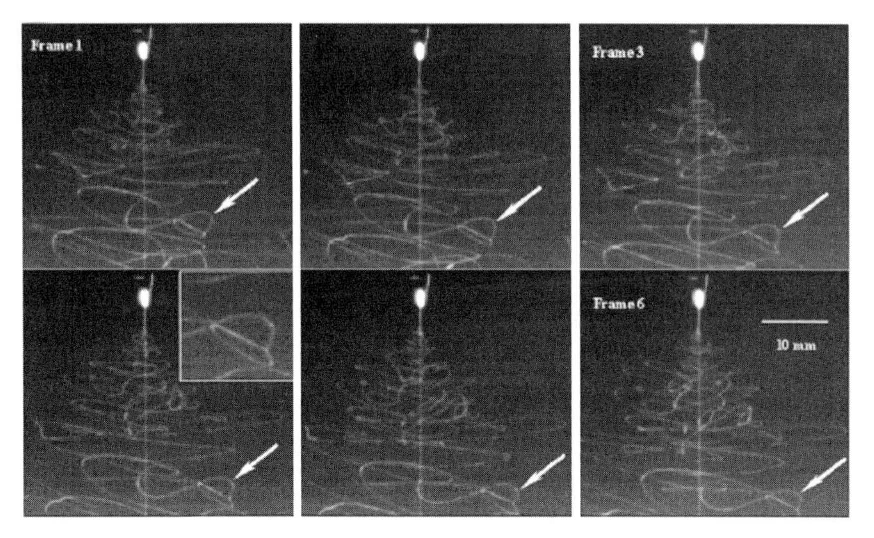

Figure 1.9 Complicated trajectories of the jet in the whipping instability region during the electrospinning of PCL in 15 wt% acetone solution (applied voltage is 5 kV and gap distance is 14 cm). Reprinted with permission from Reneker et al. (2002). Copyright 2002. Elsevier.

dependence of h_f on the inverse charge density of the jet, Σ^{-1}, are impressive. The scaling $h_f \approx \Sigma^{-2/3}$ appears to hold, at least for the narrow range of fiber diameters for which data are available. Figure 1.9 shows the complex jet trajectories obtained in whipping instability.

The high strain rate experienced by the jet results in a degree of polymer chain orientation in the nanofibers. High axial strain rates of about 10^5/s expected (Reneker et al. 2000) in electrospinning should be sufficient to extend the conformations of polymers with even the shortest relaxation times. Although this elongation of the jet is sufficient to induce a considerable degree of chain orientation in the polymer nanofiber, it is generally not expected to result in any chemical degradation by chain scission. Gel permeation chromatographic (GPC) studies on PS before and after electrospinning from tetrahydrofuran (THF) solutions did not show a significant difference in molecular weight (Casper et al. 2004).

Often, the jet dries too rapidly to allow extensive crystallization, but some orientation can still result. With PEO electrospun from a 10 wt% water solution, X-ray studies (WAXD pattern) shows broad diffused peaks as opposed to characteristic powder patterns for the polymer (Deitzel et al. 2001b, 2001c). The effect of macromolecular strain on the secondary structure of the nanofiber is particularly important in processing biological polymers. Changes in secondary and tertiary structure in biopolymers can result in corresponding loss of activity. Nylon-6 and nylon-12 electrospun from 15 wt%

HFP solutions when investigated using Raman spectroscopy showed evidence of changes in macromolecular conformation due to electrospinning (see also Chapter 6). In the case of nylon-6, the crystalline structure changed from the α-form to the γ-form, implying high strain on the fibers during their formation (Stephens et al. 2004). The conformation of macromolecules and the type of crystallinity obtained in electrospun fibers are therefore different from those in cast films of identical material (Stephens et al. 2005).

1.3.6 Solidification into Nanofiber

The duration available to the jet to undergo whipping instability is also governed by the rate of evaporation of the solvent, which occurs at increasing rates on a mass basis as the jet area dramatically increases during whipping. With a solvent of high vapor pressure, the elongational viscosity of the jet may reach levels too high to achieve any further deformation quite early in the whipping instability stage, yielding thick nanofibers. Solvent volatility is therefore a key consideration in controlling fiber diameter. With appropriate selection of solvents and process parameters, extremely fine nanofibers[8] can be electrospun. For instance, increasing the volume fraction of the less volatile DMF in a THF/DMF solvent mixture yielded decreasing nanofiber diameters for electrospinning of PVC solutions (Lee, K. H., et al. 2002). As reported for the case of PS (Megelski et al. 2002) and PC/polybutadiene blends electrospun from THF/DMF (Wei et al. 2006a), the microstructure of the nanofibers and hence their mechanical integrity is also governed by the volatility characteristics of the solvent mixture. A quasi-one-dimensional model that describes

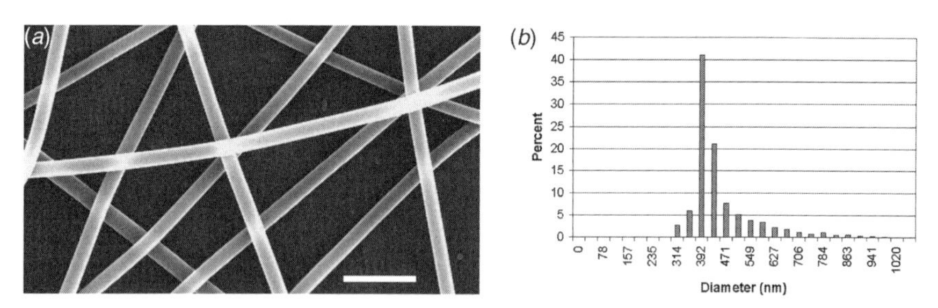

Figure 1.10 (*a*) Scanning electron micrograph of PS nanofibers electrospun from a 22.5 wt% solution in DMF using a 0.15-mm diameter Teflon tube as the capillary tip. Scale bar is 2 μm. (*b*) Nanofiber diameter distribution derived from the image. (Courtesy of RTI International.)

[8]Nanofibers with diameters in the 1–2 nm range have been electrospun from solutions of nylon (Huang, C. B., et al. 2006a). Burger et al. (2006) estimated that a nanofiber 100 nm in diameter stretched from the Earth to the Moon (a distance of 380,000 km) would have a mass of only ~3 g.

the volume change in the jet and incorporates evaporation has been proposed (Yarin et al. 2001a). However, the models presently available do not fully take into account the kinetics of drying of the nanofiber and the consequent changes in rheology that affect the finer dimensions and deposit patterns.

The fibers obtained under the best electrospinning conditions are generally of circular cross-section, continuous, and bead free. However, the literature on electrospinning reports other geometries of nanofibers (Koombhongse et al. 2001; Larsen et al. 2004a; Reneker et al. 2002). Figure 1.10 show defect-free nanofibers of PS electrospun from methylene chloride solution.

1.4 NANOFIBER APPLICATION AREAS

Nanofiber-related publications and patents appear to have grown in number rapidly over recent years. An analysis of patent activity in particular allows an overall summary of the commercial potential of electrospun nanofibers and affords the identification of application areas where the technology might play a key role. A large majority of the patents issued on the technology are U.S. patents, with about two-thirds being related to biological or medical application of nanofibers. The second largest group deals with application of nanofibers in filtration, followed by other applications such as sensors, composites, and catalysis. Figure 1.11 by Huang et al. (2003) illustrates the diversity of applications where nanofibers might be used.

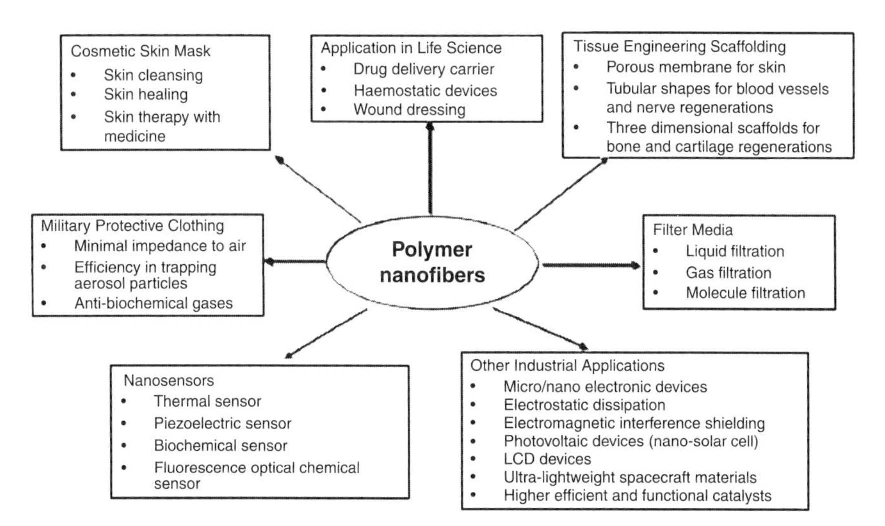

Figure 1.11 The diversity of applications proposed for polymer nanofibers. Redrawn with permission from Huang et al. (2003). Copyright 2003. Elsevier.

The following discussion identifies the major application areas for nanofibers reported in the literature. However, numerous examples of other possible applications such as magnetoresponsive fiber materials (Li, D., et al. 2003a; Tan, S. T., et al. 2005; Wang, M., et al. 2004b; Zhu et al. 2006b), electrical applications such as carbon nanofiber-based supercapacitors (Kim and Yang 2003; Kim, C. et al. 2004a, 2004c, 2004e; Kim, C., 2005), nanofiber photovoltaic devices (Drew et al. 2002; Onozuka et al. 2006; Tomer et al. 2005), catalysis applications (Demir et al. 2004; He and Gong 2003; Li, D., et al. 2004b; Wang, Z.-G., et al. 2006), and superhydrophobic surfaces (Acatay et al. 2004; Jiang, L., et al. 2004; Ma, M. L., et al. 2005a, 2005b; Singh et al. 2005; Ying et al. 2006; Zhu et al. 2006c) have been reported in the literature.

1.4.1 Filtration and Protective Apparel

As the efficiency of particle capture in an air filter increases with decreasing fiber diameter in a mat, using nanofiber filters for air or gas filtration (Liu and Rubow 1986; Park and Park 2005; Qin and Wang 2006) as well as in liquid filtration (Shin et al. 2005; Wang, X. F., et al. 2005; Yoon et al. 2006) are promising applications. The very low resistance to air flow afforded by nanofiber mats makes them especially good candidates as filter media. Commercial air-filter manufacturers such as Donaldson Inc. (Minneapolis, MN) have developed the technology for well over two decades. In recent years, several key patents claiming constructs where nanofibers are integrated with conventional filter media have been issued to Donaldson Inc. Using nanofibers in conjunction with (e.g., formed on the surface of) conventional filter media as described in these patents offers a practical advantage because of the relative fragility and difficulty of handling unsupported polymer nanofibers. Donaldson's Ultra-Web® nanofiber filters, commercialized in 1981, are used in industrial air cleaning. With the U.S. air-filter market alone estimated at $7.5 billion (estimate by the McIlvaine Company, November 2005), there is continued corporate interest in filtration applications. The need for low-cost, high-efficiency particulate air filters (HEPA-grade) for the homeland security and military markets also contributes to the growth of nanofibers filter development now and in the coming years.

Demand for light-weight protective apparel for military personnel has helped the development of nanofiber materials for future textile applications (Gibson and Schreuder-Gibson 2000, 2006; Schreuder-Gibson et al. 2002; Tsai et al. 2002). These require high permeability to moisture and gases to ensure the breathability of the fabric, and should be able to effectively filter out biological particles and ultrafines in air. Nanofibers may in principle be used alone or in combination with other nonwoven materials for protective garments. High-strength and high-temperature nanofibers will be particularly appealing in this particular application (Huang, C. B., et al. 2006b). Also of interest will be reactive textile fibers that carry specific additives that interact with chemical threat agents in air [e.g., the nanosized MgO filler to remove organophosphorous agents (Ramaseshan et al. 2006)]. These may offer a substitute for the currently used garments based on charcoal absorption technology. Nanocrystalline (magnesium oxide–PEO) composite nanofibers have been reported by Ramkumar and colleagues (Subbiah et al. 2005) to be effective against nerve agents (Sarin, Soman, and VX agents) as well as organophosphorous agents. Nanocrystalline magnesium oxide is particularly effective as a destructive adsorbent, breaking P–O and P–F bonds and immobilizing the resultant fragments (Gibson et al. 1999; Hussain et al. 2005). Poly(vinyl pyrrolidone) (PVP) nanofibers filled with oxides of tungsten and molybdenum have been investigated as gas sensor elements (Sawicka et al. 2005). Highly porous nanofiber mats and their potential for chemical modification via their high surface area therefore make them particularly good candidates for the application (see also Chapter 8).

1.4.2 Tissue Scaffolding and Drug Delivery

There is a growing need for bioresorbable three-dimensional tissue scaffolding matrices (Murugan and Ramakrishna 2006; Yoshimoto et al. 2003; Zhang, Y. Z., et al. 2005a; Zong et al. 2005), for artificial organ design (Venugopal and Ramakrishna 2005; Zhong et al. 2006), and as drug delivery platforms for therapeutic agents (Luu et al. 2003; Zeng et al. 2003b; Zhang, C. X., et al. 2005a) such as peptides and nucleic acids. Both application areas find the very high specific surface area of nanofibers to be an advantage in designing the next generation of devices. The finding that biodegradable polymers can be electrospun into nanofibers and that different cell types have been shown to adhere and proliferate on the fibrous scaffolding encourages applications research in this area. Particularly exciting is the finding that mammalian stem cells survive and proliferate on the nanofiber surfaces (see also Chapter 7).

1.4.3 Nanocomposites

The use of reinforcing fillers and fibers in polymers to improve their mechanical properties is commonly encountered in polymer technology. Conventional fibers such as carbon fibers, glass fibers, gel-spun polyethylene fibers, and aramids are routinely used in composites of a range of different polymers (Chronakis 2005). The improvement in modulus and strength achieved by using even low levels of a reinforcing fiber in a composite is impressive. Some of this improvement is due to the properties at the fiber/matrix interface and therefore dependent on the surface area of the interface. Nanofibers, with their very high specific surface area, should therefore deliver particularly good composite characteristics. For instance (poly(2,2′(m-phenylene)-5,5′-dibenzimidazole)) (PBI) electrospun nanofiber filler in epoxy EPON 828 (Shell Chemical Company) and rubber matrices has been studied by Kim and Reneker (1999a). Even at the 10 phr level, the nanofibers increased the modulus of styrene-butadiene rubber (SBR) tenfold! Nanofibers of nylon-4,6 used in an epoxy matrix that yielded a transparent composite have also been reported (Bergshoef and Vancso 1999). This book does not include a detailed discussion of nanofiber-filled polymer composites; most of the published work on the topic appears to deal with the use of carbon nanofibers and nanotubes as fillers. However, Chapter 6 discusses composite nanofibers (electrospun or post-treated to yield nanofibers made up of polymer/filler materials).

1.4.4 Sensor Applications

Nanofibers are attractive sensor materials because their high specific area allows them to sorb and/or react rapidly with low levels of analytes in the air (Aussawasathien et al. 2005; Dersch et al. 2005; Ding et al. 2005b; Virji et al. 2004). It is reasonable to therefore expect better performance from nanofiber sensors. Examples of chemical sensors based on a change in electrical resistance have recently been reported. For instance, using nanofibers ($\sim$100 nm) of polyaniline (PANI), several workers were able to detect NH_3 levels down to 0.5 ppm (Liu et al. 2004). Generally, the nanofiber geometry appears to improve sensitivity as well as the response time of chemical sensors compared to similar chemistries used in thin-film geometries (see also Chapter 8).

2

INTRODUCTION TO POLYMER SOLUTIONS

A solid polymer generally consists of a collection of randomly oriented macromolecular chains held together by strong intermolecular attractive forces. It is this long chain-like molecular geometry and the van der Waals attractive forces between them that are primarily responsible for the superior and often unique mechanical properties of polymer materials. Where their chemical structure and methods of synthesis permit, the morphology of polymers may include chain branching, partial crystallinity, and even crosslinking, which further contribute to their superior mechanical attributes.

When placed in a good solvent for the polymer, the solvent molecules interact with the repeat units of the polymer chain, first swelling the material, and ultimately dissolving it into a homogeneous solution. Qualitatively, the polymer–solvent interactions in solution needs to be stronger relative to the polymer–polymer interactions in the solid phase, to facilitate solvation. The ease with which this is achieved and the time needed for dissolution depend on a number of factors including the choice of solvent, the average molecular weight of the polymer, and the temperature. Where the polymer is crosslinked, however, the dissolution is precluded by the presence of covalent linkages between the chains (ideally, the crosslinked polymer is a single giant reticular molecule.) Crosslinked polymers will merely swell even in the best of solvents without dissolving and therefore cannot be electrospun. Crosslinked (or thermoset) polymer nanofibers may be fabricated via crosslinking post-treatment of the electrospun nanofibers (Zeng et al. 2005b;

Zhang, Y. Z., et al. 2006c). It is also possible to carry out photocrosslinking of nanofibers concurrently with electrospinning (in reactive-mode electrospinning) (Kim, S. H., et al. 2005). Polymers with a high degree of partial crystallinity also similarly resist dissolution, as most solvents cannot access the crystalline regions to solvate the chain segments within them. Dissolution is generally promoted at higher solvent temperatures as the entropy of the system increases allowing the polymer chains to assume a wider range of configurations. Polyethylene, with a crystalline melting point T_m of 135°C, for instance, will only dissolve in hot solvents (such as trichloroethane at 87°C). The presence of a sorbed solvent lowers the melting point of the polymer crystallites. A polymer, particularly a semicrystalline polymer in poor or moderate solvents, may display partial solubility, yielding a soft swollen gel, but such a gel cannot be electrospun.

As intermolecular interactions invariably depend on the chain length, higher molecular weight linear polymers generally have unusual and interesting solution properties. Often, these are obtained only when the average molecular weight exceeds a threshold value. This is true of electrospinning as well. As elaborated in Chapter 3, at a given concentration a critical molecular weight needs to be exceeded to obtain a polymer solution that is amenable to electrospinning as opposed to electrospraying.

2.1 AVERAGE MOLECULAR WEIGHT

The key solution properties of polymers are for the most part determined by the length of the macromolecular chain or by the number of repeat units comprising the chain-like molecule. Depending on the chemical route used in its synthesis, the polymer consists of a mix of chains of different chain lengths. Although the chemical structure of each repeat unit in all such chain molecules (in a pure sample of the polymer) is the same, the number of such units per chain and therefore the chain length will be variable. Two consequences of this polydisperse chain-like composition of polymers (that do not apply to most other organic compounds) are therefore (1) the lack of a unique molecular weight for a given type of polymer and (2) the need to describe polymer molecular weights in statistical terms. For instance, samples of linear polyethylene may consist of polymer chains made up of repeat units having the chemical structure —(—CH_2—CH_2—)— with a repeat unit molecular weight of $M_o = 28$. Molecular weight M_n of a single polyethylene chain, for example, is $M_n = M_o x_n$ where x_n is the "degree of polymerization" (DP) (or the number of repeat units in the chain). However, a single sample of synthetic polymer will have a statistical distribution of chain lengths (all polymer chains in the sample will not have the

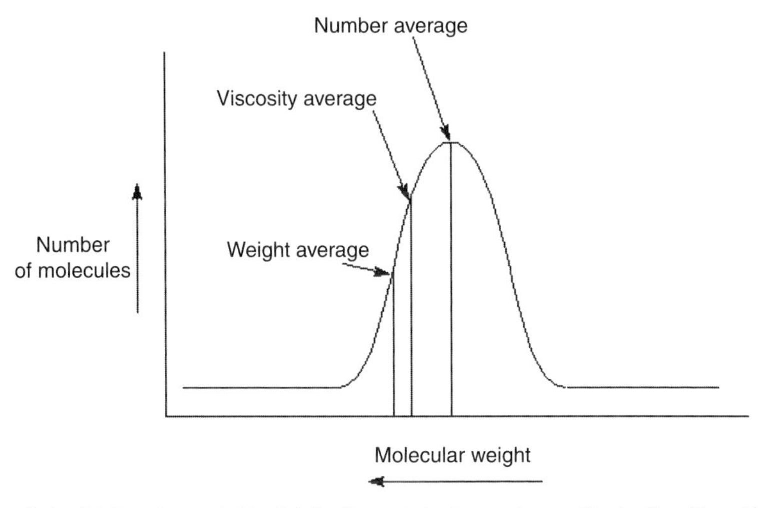

Figure 2.1 Molecular weight distribution plot of a polymer illustrating the different averages.

same length). Therefore, the molecular weight of a sample of the polymer can only be described in terms of an average value $\overline{M}$ (g/mol). The numerical value of the average molecular weight $\overline{M}$ will then depend on the type of average used to compute it. Assuming a statistical distribution function for the DP of chains, average molecular weights can be calculated.[1]

The number average molecular weight designated $\overline{M}_n$ and the weight average molecular weight designated $\overline{M}_w$ are the most commonly used averages (Fig. 2.1). A viscosity average molecular weight $\overline{M}_v$ is also sometimes quoted in the literature and can be useful in understanding solution properties of the polymer. These averages for a single polymer sample can have very different numerical values depending on the breadth and symmetry of the length distribution of polymer chains. Values of $\overline{M}_n$ correlate well with those solution properties that vary with the number of macromolecules in solution, such as the concentration of end-groups or the osmotic pressure of solutions. The $\overline{M}_w$ values correlate well with those characteristics such as light scattering in solutions that depend on the size of the polymer molecules in the system. Viscosity average molecular weight $\overline{M}_v$ is mostly used in

[1]Adopting a binomial distribution model where p is the probability that a randomly selected repeat unit is linked to the next one (or has reacted with it), average $\bar{x}_n$ is given by $\bar{x}_n = \sum_{x=1}^{\infty} p^{x-1}(1-p) = 1/(1-p)$ and $\overline{M}_n = M_o/(1-p)$. Similarly, $M_w = M_o x_w$ can be also expressed as $\bar{x}_w = \sum_{x=1}^{\infty} x w_x = \sum_{x=1}^{\infty} x^2 p^{x-1}(1-p)^2 = (1+p)/(1-p)$ and $\overline{M}_w = M_o\,[(1+p)/(1-p)]$.

discussions of the solution rheology of polymers. For an ensemble of chains where M_i is the molecular weight and N_i is the number of chains with a molecular weight of M_i, the molecular weight averages are defined as follows.

Number average molecular weight:

$$\overline{M}_n = \frac{\sum N_i M_i}{\sum N_i} \tag{2.1}$$

Weight average molecular weight

$$\overline{M}_{\mathrm{w}} = \frac{\sum N_i M_i^2}{\sum N_i M_i} = \frac{\sum w_i M_i}{\sum w_i}, \tag{2.2}$$

where w_i is the weight concentration of molecules with a molecular weight M_i.

Viscosity average molecular weight

$$\overline{M}_{\mathrm{v}} = \left[\frac{\sum N_i M_i^{1+a}}{\sum N_i M_i}\right]^{1/a}, \tag{2.3}$$

where a is a constant.

If a polymer sample did comprise of macromolecular chains of equal length, it can be shown that $\overline{M}_n = \overline{M}_{\mathrm{w}} = \overline{M}_{\mathrm{v}}$. In real polymer samples, however, $\overline{M}_n < \overline{M}_{\mathrm{v}} < \overline{M}_{\mathrm{w}}$.

It is the average molecular weight that determines the physical characteristics of polymers both in solid phase and in solutions. For example, polyethylene, with an average molecular weight $\overline{M}_{\mathrm{w}}$ of 3000 (g/mol), is a soft waxy material used as a lubricant, but the ultra-high molecular weight grade of polyethylene (UHMW-PE) with identical repeat unit chemistry but with an average molecular weight of 3–6 million (g/mol) is an exceptionally strong tough polymer.[2] Bulk properties as well as solution properties of polymers are also influenced by the breadth of the molecular weight distribution. A convenient measure of the latter is the polydispersity index $P = \overline{M}_{\mathrm{w}}/\overline{M}_n$. Polydispersity is determined by the synthetic route used in manufacturing. Condensation polymerization, for instance, generally yields a narrow distribution, but free-radical polymerization yields a broad distribution of chain lengths. The kinetic features of the polymerization process are generally

[2]UHMW-PE has the highest abrasion resistance of any thermoplastic polymer and excellent impact strength under a wide range of temperatures.

controlled in manufacture to obtain polymers of the desired average molecular weight and polydispersity.

2.2 SELECTING SOLVENTS: SOLUBILITY PARAMETER

A convenient practical measure of the intermolecular forces between polymer chains that might be used in selecting solvents for specific polymers is the solubility parameter. Solubility parameter δ is related to the heat of vaporization, ΔH_{vap} (cal/cm^3), also a measure of the same intermolecular attractive force.[3] The cohesive energy density c of a liquid is the energy needed to overcome intermolecular forces to separate the molecules from each other and is given by the expression

$$c = (\Delta H_{\text{vap}} - RT)/V_{\text{m}} \tag{2.4}$$

where V_{m} is the molar volume, R is the gas constant, and T is the temperature. The Hilderbrand solubility parameter δ, which quantifies the tendency towards solubility, is defined in terms of the cohesive energy density c:

$$\delta = \sqrt{c}\,(\text{cal/cm}^3)^{1/2}. \tag{2.5}$$

Conversion to SI units yields $\delta\ (\text{MPa})^{1/2} = 2.0455\delta\ (\text{cal/cm}^3)^{1/2}$.

Common solvents can be ranked using the Hilderbrand solubility parameter δ to reflect their solvent effectiveness. It correlates reasonably well with other solubility scales (such as the Kauri–Butanol value) used in practice. In general, a polymer with a given value of δ will readily dissolve in a solvent that also has about the same value of δ. In the case of solvent mixtures, the components contribute to δ (mixture) according to their volume fraction in the mixture. A limitation of this simple single-parameter measure is that it is based solely on the dispersive forces due to induced dipoles (London or van der Waals forces) and ignores specific interactions such as hydrogen bonding and the polarity of solvent. Although in simple polymers such as polyethylene the dispersive forces in fact solely determine solubility characteristics, the same is not true for polymers of greater structural complexity. Polymers such as nylons, polyalcohols, and poly(carboxylic acids), for instance, have strong hydrogen-bonded interactions. Hansen (1967a, 1967b) proposed the more comprehensive three-parameter expression

[3] As polymers cannot be vaporized the ΔH values are measured indirectly using equilibrium swelling or cloud point measurements.

for the solubility parameter to also take into account these intermolecular interactions in polar and hydrogen-bonding systems:

$$(\delta_T)^2 = (\delta_D)^2 + (\delta_P)^2 + (\delta_H)^2, \tag{2.6}$$

where the subscripts D, P, and H refer to dispersive, polar, and hydrogen-bonded contributions to the total Hansen solubility parameter, δ_T. This formalism is particularly useful with mixed solvents. The values are determined empirically based on experimental observations. Often in electrospinning experiments a mixture of solvents are used to dissolve the polymer as the nature of solvent determines ease of electrospinnability.

It is convenient to express components of δ_T as fractional contributions of,

$$f_D = \delta_D/(\delta_D + \delta_P + \delta_H) \tag{2.7a}$$

$$f_P = \delta_P/(\delta_D + \delta_P + \delta_H) \tag{2.7b}$$

$$f_H = \delta_H/(\delta_D + \delta_P + \delta_H) \tag{2.7c}$$

and

$$(f_D + f_P + f_H) = 1. \tag{2.7d}$$

These contributions are empirically determined, and the corresponding values of δ_T obtained should in theory be the same as the Hilderbrand parameter δ. Table 2.1 lists solubility parameters for common solvents and for selected polymers. Although it is an improvement on the single-parameter values of δ, the Hansen δ_T also fails to accurately and completely describe the solution thermodynamics of a significant number of the polymer-solvent systems.

A given solvent might then be uniquely defined in terms of the three fractional values (equations 2.7a–d) on a triangular plot (referred to as a Teas plot; Teas 1968) as shown in Fig. 2.2. The data points on the plot represent the values for different solvents (Burke 1984). If a series of solubility tests are carried out in each of these solvents with a given polymer at a constant temperature and concentration, a region within the Teas plot where the polymer is soluble might be demarcated by an area as shown in the figure. Solvents that fall within the circular region in the plot will therefore dissolve the polymer, but those at or near the envelope only swell the polymer. Nonsolvents for the polymer will lie well outside the circle. The plot is essentially a convenient means of presenting empirical solubility data. Despite the lack of an adequate theoretical justification for their use, Teas plots offer a

TABLE 2.1 Hansen solubility parameters of representative polymers and solvents

Polymer or Solvent	δ_D (MPa)$^{1/2}$	δ_P (MPa)$^{1/2}$	δ_H (MPa)$^{1/2}$	δ_T (MPa)$^{1/2}$	Reference
		Polymers			
Polyamide 66	18.62	5.11	12.28	22.87	Rigbi (1978)
Poly(acrylonitrile)	18.21	16.16	6.75	25.27	
Poly(ethylene terephthalate)	19.44	3.48	8.59	21.54	
Poly(vinyl chloride)	18.82	10.03	3.07	21.54	
Poly(methacrylic acid)	17.39	12.48	15.69	26.8	Ho et al. (1991)
Poly(methacrylonitrile)	18.00	15.96	7.98	25.37	
Poly(methacrylonitrile-*co*-methacrylic acid)	17.39	14.32	12.28	25.78	
Poly(4-acetoxy styrene)	17.8	9.00	8.39	21.69	Arichi and Himuro (1989)
Poly(4-hydroxy styrene)	17.6	10.03	13.71	24.55	
		Common Solvents			
Acetone	15.5	10.4	7.0	20.1	Zeng (2007)
Chloroform	17.8	3.1	5.7	19.0	
Cyclohexanol	17.4	4.1	13.5	22.5	
Ethyl acetate	15.8	5.3	7.2	18.2	
Methanol	15.1	12.3	22.3	29.7	
Dimethylformamide	17.4	13.7	11.3	24.8	
Methylene chloride	18.2	6.3	6.1	20.3	
Carbon tetrachloride	17.8	0	0.6	17.8	
Tetrahydrofuran	16.8	5.7	8.0	19.4	

convenient and useful means of describing solubility relationships in polymers. The representation does not of course work well with all polymer or solvent systems. With solvents such as alkanes, their positions on the graph are not in line with the empirical data.

The plots (or their simpler two-dimensional form, which is a plot of f_H vs f_P, in effect assuming f_D to be invariant) have considerable practical utility in solvent selection (Burke 1984). The solubility of a polymer is often achieved using a mixture of solvents. As a general rule, a mixture of solvents will dissolve a polymer if the solubility parameter of that mixture lies close to that of a known good solvent for the polymer. In designing mixed solvent systems the Teas plot allows the solubility characteristics of solvent mixtures to be predicted to some extent. In Fig. 2.3, the solvents carbon tetrachloride (CCl_4) and methanol (CH_3OH) are clearly nonsolvents for poly(methyl methacrylate)

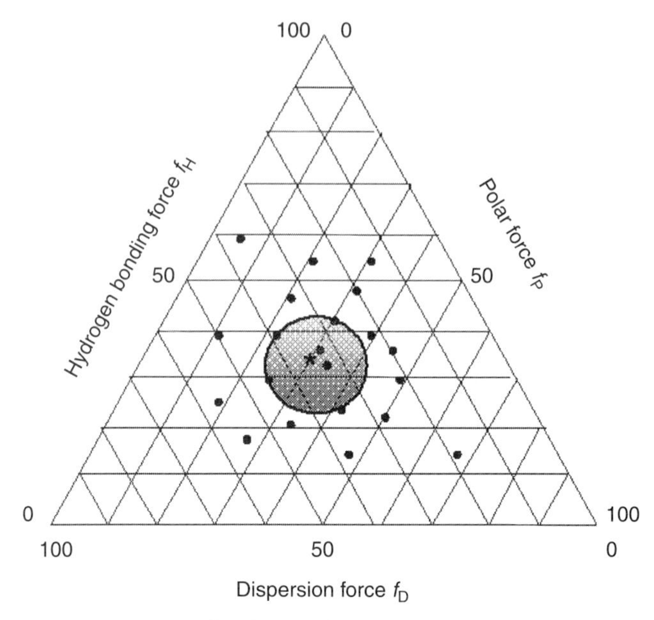

Figure 2.2 Representation of solvents in terms of the fractional contribution to Hansen solubility parameter on a Teas plot. Dots represent different solvents and the asterisk the polymer. The circle is an idealized area on the plot where solvents within it dissolve the polymer.

(PMMA), as they lie outside the region of solubility for the polymer. However, specific mixtures of the two nonsolvents (that lie on the line connecting the data points on the plot) do fall within the circular region and should therefore dissolve PMMA; this agrees with experimental observations for this particular polymer/solvents system (Deb and Palit 1973). Similarly, mixtures of diethyl ether and acetone dissolve polystyrene (average molecular weight 110,000, tested at 0°C), although each separately is a nonsolvent for the polymer (Wolf and Molinari 1973).[4] Using a second solvent affects the conformation of polymer chains dissolved in the mixture and alters the solution properties such as the surface tension or the dielectric properties that influence ease of electrospinning. These effects also need to be considered in developing mixed solvents for electrospinning.

High-surface-area nanofibers are sometimes made via selective dissolution of one polymer component from a bicomponent polymer nanofiber mat (Li and Hsieh 2006; You et al. 2006b; Zhang, Y. Z., et al. 2006a) (see Chapter 9). Selecting a solvent mixture that dissolves away just the one polymer component

[4]Interestingly, it is also possible to have the opposite phenomenon. Polystyrene dissolves well in dimethylformamide (DMF) and in cyclohexane (CH) solvents, but does not dissolve in some mixtures of the pair of solvents (Wolf and Willms 1978).

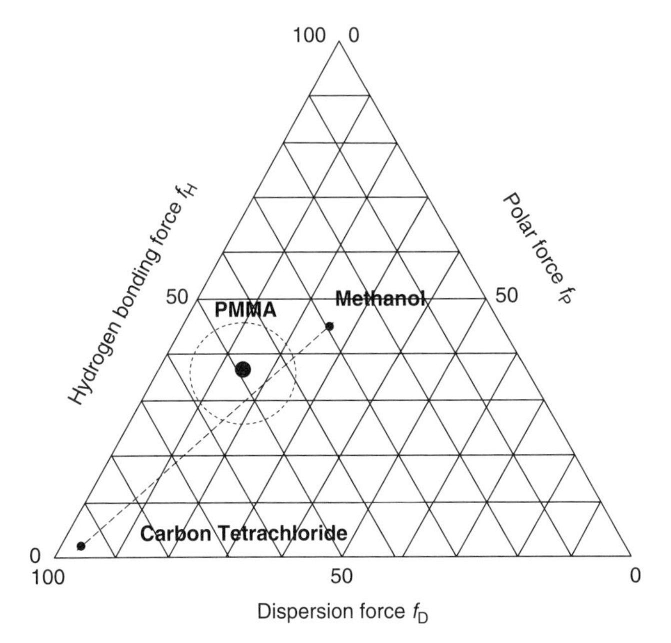

Figure 2.3 Representation on a Teas plot of a mixture of nonsolvents for PMMA that dissolves the polymer.

is relatively easy using a Teas plot. However, it is important to appreciate that the solubility of polymers also depends on the average molecular weight and the concentration range, as the entropy term will depend on the number of molecules dispersed in solution. Therefore, solubility (and Teas plots) at one concentration (say 10 wt%) may not guarantee solubility at a lower concentration. This is because at the lower concentrations the entropy of the dissolved polymer in solvent may decrease to an extent where the polymer may phase separate. The fractionation of a polymer into samples of different average molecular weight can be carried out using its solubility in solvent mixtures.

2.3 THERMODYNAMIC CRITERION FOR SOLUBILITY

The change in Gibbs free energy of mixing, ΔG, of a polymer dissolved in a solvent is given by the simple relationship that includes the change in enthalpy of mixing (ΔH_{mix}) and the change in entropy (ΔS_{mix}) on mixing, at a temperature T:

$$\Delta G_{\text{mix}} = \Delta H_{\text{mix}} - T\Delta S_{\text{mix}}. \tag{2.8}$$

Solubility is achieved only if $\Delta G_{\text{mix}} < 0$; the starting components (polymer and solvent) should have a higher free energy relative to that of the solution. Both T and ΔS_{mix} are positive quantities (as the molecules of any solute will be far more disordered after dissolution). In general, increasing T, and therefore $T\Delta S_{mix}$, will help dissolution. The condition $\Delta G_{mix} < 0$ generally implies a homogeneous (single-phase) solution of the polymer at temperature T to be more stable than the polymer and solvent taken alone. However, phase separation cannot be necessarily ruled out at $\Delta G_{mix} < 0$, as a two-phase system can sometimes be more stable than the single-phase homogeneous solution. An additional criteria for miscibility is that $(\partial^2 \Delta G_{mix})/\partial x_1^2 > 0$ (i.e., the plot of G_{mix} vs x_1 be concave upwards if the components are miscible).

If a nonsolvent such as methanol is gradually added to a homogeneous solution of a polymer in a good solvent (such as polystyrene in chloroform) the negative value of ΔG will decrease correspondingly, until the special situation of $\Delta G = 0$ (i.e., $\Delta H_{mix} = T\Delta S_{mix}$), called the (theta) θ-condition, is reached. Under such conditions ΔS_{mix} is minimized and the magnitude of polymer–polymer and polymer–solvent interactions become equal. The θ-condition for a polymer can be achieved by carefully adjusting the composition of a mixed solvent system, the molecular weight of the polymer, or the temperature. The temperature at which the θ-state is reached (at constant concentration) is called the θ-temperature or the Flory temperature for a given polymer solution. Alternatively, at a given temperature, a solvent or a mix of solvents (called the θ-solvent) can be identified where the system will be in the θ-state.

At the macromolecular level, θ-temperature can be understood in terms of polymer chain configurations. Statistically averaged configurations of polymer chains in a solvent are invariably dictated by both short-range and long-range interactions. Short-range interactions are those between neighboring repeat units on the same polymer chain; long-range interactions occur between segments that are spatially close to each other but well separated on the polymer chain. Under θ-conditions, the average configuration of the chain will be dictated solely by the short-range interactions.

To determine ΔG_{mix} of a polymer/solvent mixture, expressions for the change in entropy, ΔS_{mix}, and change in enthalpy, ΔH_{mix}, need to be derived.

2.3.1 Change in Entropy

The Boltzmann equation is generally used to obtain an expression for ΔS of simple mixtures (mixtures of solvent–solvent or solvent–simple solute molecules) from the number of different arrangements Ω (or the thermodynamic probabilities) of the solute and solvent molecules in the system. For simple systems, the volume elements of solution are modeled by a three-dimensional lattice, where solute or solvent molecules can occupy any cell within the lattice (see Fig. 2.4). From the Boltzmann equation,

$$\Delta S_{mix} = k[\ln \Omega_{12} - (\ln \Omega_1 + \ln \Omega_2)] = k \ln(\Omega_{12}/\Omega_1\Omega_2), \tag{2.9}$$

where Ω_1, Ω_2, and Ω_{12} are respectively the total numbers of distinguishable spatial arrangements of the molecules in pure solvent 1, pure solvent 2, and in an ideal mixture of the two solvents. As for a pure solvent $\Omega_1 = \Omega_2 = 1$, it can be easily shown that[5]

$$\Delta S_{mix} = k \ln \Omega_{12} = -R[n_1 \ln x_1 + n_2 \ln x_2], \tag{2.10}$$

where k is the Boltzmann constant, $x_1 = n_1/(n_1 + n_2)$ and $x_2 = n_2/(n_1 + n_2)$ and x_1 and x_2 are the mole fraction of one solvent (i.e., $x_1 + x_2 = 1$), and $n = (n_1 + n_2)$ is the total moles in the system. The above treatment assumes comparable molecular sizes for solvents 1 and 2.

A polymer solution might also be similarly treated. The chain-like polymer molecules, however, are very much larger in size than solute molecules and the assumption of comparable molecular size is not a realistic one. Assigning a single repeat unit of the polymer to each lattice site or cell occupied by polymer species, however, yields a convenient approximation.[6] A mixture that has N_1 solvent molecules and N_2 polymer molecules will therefore occupy $(N_1 + N_2X)$ lattice sites, where X is the number of segments in the polymer chain. The Flory–Huggins model allows the estimation of Ω_{12}, the number of arrangements of the components in such a mixture, assuming $\Delta H_{mix} = 0$ (i.e., the intermolecular interactions between solvent–solvent and solvent–polymer segment are the same).

An equation corresponding to equation (2.10) can be written for the polymer solution:

$$\Delta S_{mix} = k \ln(\Omega_{12}/\Omega_2). \tag{2.11}$$

As the polymer can take many different conformations, $\Omega_2 > 1$; both Ω_{12} and Ω_2 have to be evaluated in order to determine ΔS_{mix}.

Consider the two-dimensional lattice shown in Fig. 2.4, where the cells within it can be occupied either by a single repeat unit of the polymer molecule or by a solvent molecule. The first segment of a polymer chain can be placed at any of the $(N_1 + N_2X)$ lattice positions. The second segment

[5]Note that $\Omega_{12} = (N_1 + N_2)!/(N_1!\,N_2!)$, and Stirling's approximation (i.e., $\ln N! = N \ln N - N$ for large values of N) is used in deriving equation (2.10).

[6]The "segment" of the polymer chain used in the lattice theory is strictly the section of a polymer chain that occupies the same volume as a solute molecule, and need not be the same as a monomeric unit.

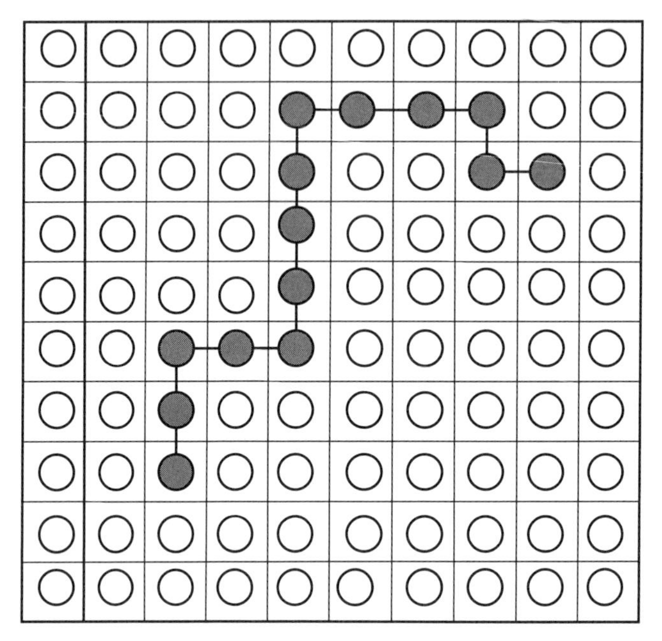

Figure 2.4 Arrangement of a polymer chain on a two-dimensional lattice.

of the chain, however, being covalently linked to the first, will be restricted to the Z adjacent lattice points (Z being the lattice coordination number). The third and subsequent segments will have a choice of $(Z - 1)$ lattice points and so on. The total number of arrangements, υ_1, available to a single polymer chain within the lattice, can be shown to be equal to

$$\upsilon_1 = (N_1 + N_2X)\,Z(Z-1)^{X-2}\varepsilon_j^{X-4}, \tag{2.12}$$

where $\varepsilon_j \approx [N_1 + X(N_2 - j)]/(N_1 + N_2X)$ computed for the chain. Similar expressions can be written for a second and subsequent polymer chains introduced into the lattice.

In general, for the $(j + 1)$th chain in the system,

$$\upsilon_{j+1} = [N_1 + (N_2 - j)X]Z(Z-1)^{X-2}\varepsilon_j^{X-1}. \tag{2.13}$$

Approximating $Z \approx Z - 1$, the above can be rearranged to obtain a simple expression for Ω_{12} for the polymer/solvent mixture:

$$\Omega_{12} = \frac{(N_1 + N_2X)!}{N_1!N_2!}\;[(Z-1)/(N_1 + N_2X)]^{N_2(X-1)} \tag{2.14}$$

For the pure polymer,

$$\Omega_2 = [(XN_2)!/N_2!][(Z-1)/XN_2]^{N_2(X-1)}. \tag{2.15}$$

For the pure solvent,

$$\Omega_1 = (N_1)!/N_1!0!) = 1 \text{ and } \Delta S_{\text{solvent}} = 0$$

$$\Delta S_{\text{mix}} = S_{\text{mix}} - (S_{\text{solvent}} + S_{\text{polymer}}) = k[\ln \Omega_{12} - (\ln \Omega_1 + \ln \Omega_2)]. \tag{2.16}$$

Substituting for quantities in equation (2.16), it can be shown that

$$\Delta S_{\text{mix}} = -k(N_1 \ln \phi_1 + N_2 \ln \phi_2), \tag{2.17}$$

where ϕ_1 and ϕ_2 ($\phi_1 + \phi_2 = 1$) are the volume fractions of the polymer and of solvent in the mixture, $\phi_1 = (XN_2)/(N_1 + XN_2)$ and $\phi_2 = N_1/(N_1 + XN_2)$.

Despite its rather tedious derivation, equation (2.17) for ΔS_{mix} of the polymer–solute mixture is of the same form as that for two solvents (in equation 2.10), except that the value is decreased by a factor of $1/X$ (to account for the decreased entropy due to connectivity of X segments into a single molecule). As X, the number of segments per chain, is large, this amounts to a significant reduction. It implies that the higher the molecular weight of the polymer, the smaller will be the value of ΔS (and therefore the smaller the likelihood of dissolution).

2.3.2 Change in Enthalpy (ΔH_{mix})

Change in enthalpy allows the derivation of an expression for the intermolecular interactions in solution. During the dissolution process, cohesive interactions between the repeat segments of polymer P and between the solvent molecules S are replaced by cross-interaction between the two species (P–S).

$$[\text{P–P}] + [\text{S–S}] \ldots\ldots\ldots 2\,[\text{P–S}]$$

Using the same lattice model, the change in enthalpy ΔH_{mix}, for replacing one species by the other in adjacent cells might be derived. The van der Waals type interactions involved may arise from permanent or induced dipole–dipole and dispersion mechanisms.

Three types of short-range interactions are possible between a solvent molecule and a segment on the polymer chain. The corresponding change in energy of interaction, ΔW, is given by

$$\Delta W_{12} = W_{12} - (W_{11} + W_{22})/2. \tag{2.18}$$

where the subscript 1 and 2 refer to the polymer and solvent respectively.

Each polymer segment in the lattice is surrounded by $Z\phi_2$ polymer segments and $Z\phi_1$ solvent molecules. The total number of solvent–segment (or 1–2) contacts possible in solution, p_{1-2}, depends on Z, the coordination number of the lattice, and is given as

$$p_{1-2} = N_2 XZ\phi_1 = N_1\phi_2 Z. \tag{2.19}$$

The value of ΔH for the solution is then given by

$$\Delta H_{\text{mix}} = N_1\phi_2 Z\Delta W_{12}. \tag{2.20}$$

Setting $(Z\Delta W_{12}) = \chi kT$, where χ is the Flory–Huggins interaction parameter,

$$\Delta H_{\text{mix}} = N_1\phi_2\chi kT \tag{2.21}$$

where χ is a dimensionless quantity that characterizes the interaction energy per solvent molecule divided by kT. Combining equations (2.17) and (2.21) and simplifying, the Flory–Huggins equation for the change in Gibbs free energy on mixing, ΔG, for the polymer–solvent system is obtained as follows (the numbers of segments and molecules have been converted back to mole fractions n_1 and n_2):

$$\Delta G_{\text{mix}} = \Delta H_{\text{mix}} - T\Delta S_{\text{mix}} = RT(n_1\phi_2\chi + n_1 \ln \phi_1 + n_2 \ln \phi_2). \tag{2.22}$$

The first term on the right-hand side of the equation is the enthalpic contribution to the free energy. Because of approximations used in its derivation the equation is strictly applicable to moderately concentrated solutions of monodisperse, flexible polymer chain where the mixing of components is random. Also, any concentration-dependence of χ was ignored.

As might be expected, the Flory–Huggins parameter χ is related to the Hilderbrand solubility parameter δ. From solution theory,

$$\Delta H_{\text{mix}} = V\phi_1\phi_2(\delta_1 - \delta_2)^2, \tag{2.23}$$

where V is the volume of the segment. Therefore

$$\chi = (V_m/RT)(\delta_1 - \delta_2)^2, \tag{2.24}$$

where V_m is the molar volume of the solvent.

Thus for complete solubility at all compositions (i.e., $\chi \leq 0.5$)

$$(\delta_1 - \delta_2)^2 = \chi RT/V_m = [(0.5)(8.3\,\text{J mol/K})\,298\,\text{K}]/[V_m\,\text{m}^3/\text{mol}]. \tag{2.25}$$

This works out to $(\delta_1 - \delta_2) \sim 3$ MPa and suggests that the smaller the magnitude of difference $(\delta_1 - \delta_2)$, the better will be the solvent; this is indeed found to be reasonable, as seen from Table 2.1. A difference of less than 2 or 3 generally yields solubility.

The solubility parameter χ indicates the quality of the solvent; $\chi < 0.5$ indicates a good solvent, $\chi > 0.5$ indicates a poor solvent, and $\chi = 0.5$ indicates a Θ-solvent and the smaller the value of χ, the better will be the solvent in general, (negative values of χ generally indicate a strong polar interaction between the polymer and solvent). However, χ is not an inherent property of a solvent, but decreases with temperature and increases with the concentration of the polymer in solution. It is not significantly affected by changing the molecular weight of the polymer. Typical values of the polymer–solvent interaction parameter are given in Table 2.2. A fairly comprehensive collection of values was tabulated by Orwall and Arnold (2007).

2.4 MACROMOLECULAR MODELS

The simplest representation of the dimension of an unperturbed polymer chain molecule in solution or in the melt is given by its mean square end-to-end distance $\langle r^2 \rangle$. This value, the square of the vector sum of the end-to-end distance averaged over all possible configurations of the chain, is always smaller than the contour length of the chain. Provided the chain is long enough, the value of $\langle r^2 \rangle$ for an ideal chain[7] is obtained by adopting a model for the polymer chain. The simplest such model is the "freely jointed" chain (Fig. 2.5) of backbone bonds consisting on n bonds linked by fully flexible joints. The angle between two adjacent bonds, θ, can take any value (i.e., $-1 > \cos\theta > 1$). The statistics of the freely jointed chain can be described by the random walk expression, which yields an unrealistic but simple

[7]There is no unique value of the end-to-end distance that can be assigned to a chain, as its conformations change due to bond rotation over time. The $\langle\,\rangle$ enclosing r^2 indicates that the value is averaged over time. The quantity $\langle r^2 \rangle^{1/2}$ (the root-mean-square end-to-end distance) is the quantity that best describes the dimension in statistical mechanics.

TABLE 2.2 Typical values of polymer–solvent interaction parameter

Polymer	Acetone	T (°C)	Chloroform	T (°C)	Toluene	T (°C)
Poly (ε-caprolactone)	0.46–0.54	100–120	−0.40 to −0.22	100–120	0.08	100
Polyethylene	—	—	0.41	135	0.34	120–145
Poly(ethylene oxide)	0.47	100	−0.55	100	0.26	100
Poly(isobutylene)	1.90	100	0.78	100	0.60	100
Poly(methyl acrylate)	0.40	100	−0.10	100	0.53	100
Polystyrene	1.08	40	0.13	40	0.19	40
Poly(tetramethylene oxide)	0.73	100	−0.38	100	0.04	100
Poly(vinyl acetate)	0.31–0.39	30–50	−0.09 to −0.17	80–135	0.40–0.56	80–140
Poly(vinyl methyl ether)	0.75	40	−0.92	40	0.14	40
Poly(vinyl chloride)	0.53–0.77	125–140	0.91	120	0.41–0.45	125–140

Source: Orwall and Arnold 2007.

expression for $\langle r^2 \rangle_f$ (the subscript denotes freely-jointed chain). However, it turns out to be a particularly good approximation for a polymer solvated under Θ-conditions or one that is in the melt state. Assuming a chain of n links, each of length l and therefore of contour length of nl, the distribution of $\langle r \rangle$ in an ensemble of polymer chains is given by

$$\langle r \rangle_f = \int_0^{\infty} r^2 w(r) 4\pi r^2 \cdot dr = nl^2, \tag{2.26}$$

where $w(r)$ is the probability density function for a Markov chain,

$$w(r) = \left[\frac{3}{2\pi nl^2}\right]^{3/2} \exp\left[\frac{-3r^2}{2nl^2}\right].$$

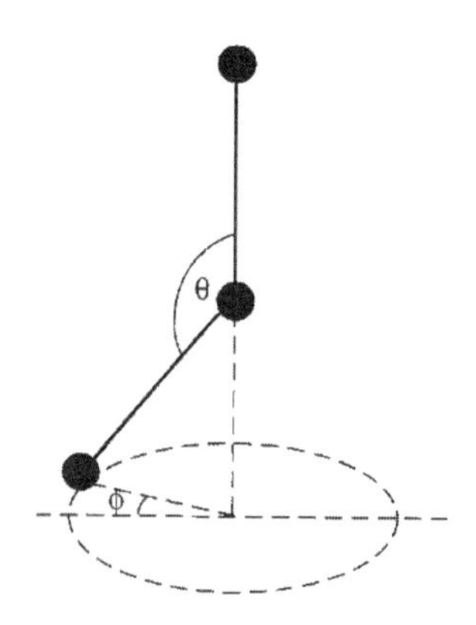

Figure 2.5 Representation of the freely-rotating chain.

This probability monotonically decreases with increasing values of the end to end distance r for a chain fixed at the origin and extending out into space. The random walk model, however, does not recognize the physical nature of polymer segments. Real polymer molecules also have geometric restrictions due to the fixed bond angle between the backbone bonds in the chain. This excludes a large fraction of the conformations counted in the freely jointed chain model. Restricting the freely jointed chain by introducing a fixed bond angle θ (e.g., 109.5° for C—C—C and $\cos\theta \approx -0.33$ for carbon–carbon bonds in vinyl polymers) in the chain improves the model and changes the equation (2.26) into the form shown in equation (2.27).[8] A given bond (or chain segment), however, can still freely rotate about the atom (usually a carbon atom) where it is attached to its neighbor despite this angular constraint on the bonds (Fig. 2.4). This improvement (i.e., resulting in the values of $[\langle r^2\rangle]_{\text{restricted bond angle}} > [\langle r^2\rangle]_{\text{freely jointed}}$, gives

$$\langle r^2\rangle_{\mathrm{f}\theta} = nl^2\frac{1-\cos\theta}{1+\cos\theta}, \tag{2.27}$$

where θ is the skeletal bond angle. Depending on the structure of polymer, the right-hand side of the equation may be multiplied by a parameter to take steric effects into consideration. Clearly, $\langle r^2\rangle_{\mathrm{f}\theta} > \langle r^2\rangle_{\mathrm{f}}$, as the angle θ generally lies between 90° and 180°. Introducing the concept of the torsional or dihedral angle ϕ is also therefore critical as it allows the polymer chain to be three-dimensional. As shown in Fig. 2.5, the bonds are still able to rotate about the carbon atom maintaining the skeletal bond angle. The expression can be further refined as follows by restricting the value of this torsional angle ϕ, by recognizing that all values of ϕ are not equally probable because of different functional groups attached to the chain (or because of steric effects). This yields the following

$$\langle r^2\rangle_{\mathrm{o}} = nl^2\frac{1-\cos\theta}{1+\cos\theta}\cdot\frac{1+\cos\phi}{1-\cos\phi} = C_\infty nl^2. \tag{2.28}$$

If all torsional angles were assumed to be equally probable, $\cos\phi = 0$ and the equation (2.28) reduces to equation (2.27). The characteristic ratio C_∞ therefore takes into account all local or short-range steric interactions and is a measure of the flexibility of the polymer chain. Very flexible chains will have values of C_∞ close to unity. Typical values of C_∞ are 6.7 for polyethylene, 10.2 for polystyrene, and about 600 for DNA — much higher than the

[8]Some texts define θ as the complement of the skeletal or bond angle (instead of the bond angle as used here). Using that convention the equation comes out to be $\langle r^2\rangle_{\mathrm{f}\theta} = nl^2[(1+\cos\theta)/(1-\cos\theta)]$.

value of unity expected of an ideal chain. The rotational isomeric state (RIS) theory accounting for rotation of adjacent bonds yields calculated values of C_∞ that are close to the experimentally determined values.[9]

A second useful measure of the size of polymer molecule in solution (irrespective of the shape) is the radius of gyration $\langle s^2 \rangle$. This is the average distance from the center of gravity of the chain to the chain segment[10] and, as with $\langle r^2 \rangle$, its value also depends on the solvent in which the polymer is dissolved. Unlike the value of $\langle r^2 \rangle$ the value of $\langle s^2 \rangle$ can be easily measured experimentally by techniques such as light scattering or gel permeation chromatography.[11] For a long enough Gaussian chain the two quantities are related as follows (equation (2.29); where both values are experimentally available, equation (2.29) might also be used to test if Gaussian statistics might be used to describe the system:

$$\langle s^2 \rangle_o / \langle r^2 \rangle_o = 1/6. \tag{2.29}$$

The foregoing introductory discussion applies to any polymer chain regardless of its chemical structure or how the polymer interacts with the solvent. With real polymer chains, however, the effects of physical volume of the chain segments (referred to as the excluded volume), must be taken into account; those physically impossible conformations where a segment occupying a volume element already occupied by another segment has to be excluded (this restriction did not apply to freely jointed models). A convenient means of including these long-range interactions into the equations is to adopt an empirical factor α with a numerical value that depends on the average molecular weight of the polymer, to take the excluded-volume effects into account:

$$\langle r^2 \rangle = \alpha^2 \langle r^2 \rangle_o. \tag{2.30}$$

An important determinant of $\langle r^2 \rangle$ in real polymer solutions is the quality of the solvent (Fig. 2.6). With a very good solvent the solvent–repeat unit interactions are maximized, resulting in a relatively expanded free-draining polymer chain. Conversely, in a very poor solvent the polymer chains are close to their most compact average conformation, behaving similarly to rigid spheres suspended in solution. Both the expanded and contracted

[9]Even all the sterically allowed bond torsional angles may not be available to a polymer chain because of long-range interference due to neighboring chain segments.

[10]$s^2 = (1/n)\sum_{i=1}^{n} (\vec{r}_i - \vec{r}_o)^2$ where $\vec{r}_o$ is the center of mass of the polymer chain and $\vec{r}_i$ is the coordinate of the ith monomer unit.

[11]Values of $\langle r^2 \rangle$ can be obtained from the hydrodynamic radius values obtained from intrinsic viscosity measurements.

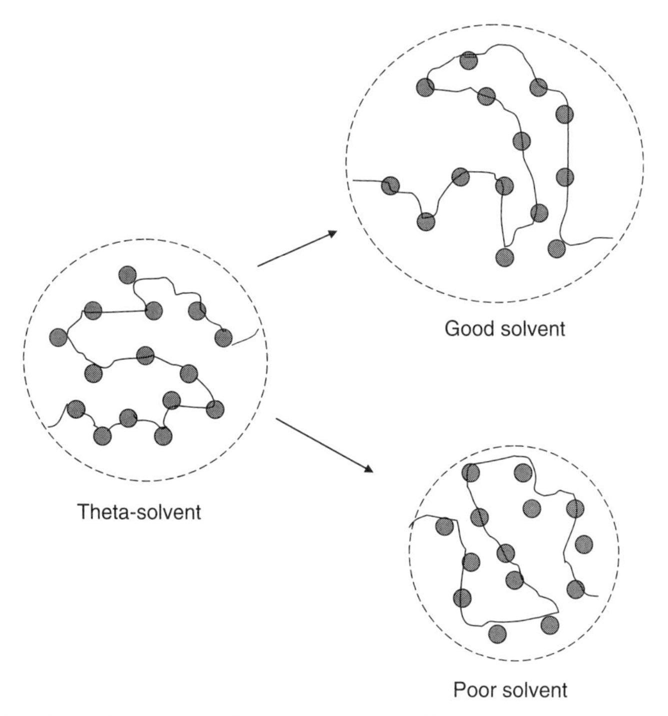

Figure 2.6 Illustration of the effect of solvent quality parameter on chain geometry.

chains, however, behave very differently from ideal chains and the excluded-volume effects change with the solvent. A convenient way to characterize the solvent effects is to simply define α as the effective excluded volume of the system. The effect of solvent quality (accounting for the excluded volume effects) is quantified by introducing a solvent quality parameter α^2 into equation (2.28):

$$\langle r^2 \rangle = C_\infty \alpha^2 n l^2 \quad \text{and} \quad \alpha^2 = \{\langle r^2 \rangle / \langle r^2 \rangle_\Theta\}^2. \tag{2.31}$$

The expression in equation (2.31) reduces to that given by the random-walk model when $\alpha^2 = 1$ in equation (2.28).

2.5 VISCOSITY OF DILUTE POLYMER SOLUTIONS

Dissolving even a small quantity of a high molecular weight polymer in a good solvent results in a marked increase in solution viscosity. Solution viscosity depends on the nature of polymer, its molecular weight, concentration of the solution, and the temperature. Viscometry is therefore a convenient practical experimental method to determine an average molecular weight ($\overline{M}_v$) of polymers. Despite the experimental simplicity of viscometry, the viscosity-average molecular weight $\overline{M}_v$ is invariably solvent dependent. It is therefore less precise than the averages, $\overline{M}_n$ or $\overline{M}_w$, determined by other methods. Most polymers are reported to be electrospun from solution — a solution of high enough viscosity is essential to obtain continuous electrostatic spinning (as opposed to electrospraying). An introductory discussion of solution viscosity and nomenclature is therefore pertinent here. A detailed quantitative discussion, however, is outside the scope of this chapter and the reader is referred to other excellent reviews (Bird et al. 1987; Kulicke and Clasen 2004).

The basic notion of solution viscosity is illustrated in Fig. 2.7 of a volume of fluid in a shear field (for instance a film of polymer solution confined between parallel plates where one is stationary and the other is moving in the x-direction at a constant velocity $\mathbf{v}$). Assuming no slippage between the liquid and the plate, the force per unit area applied on the volume, the shear stress τ, results in a rate of deformation or a strain rate $\dot{\gamma}$ where

$$\tau = F/A, \;\; \gamma = \mathrm{d}\mathbf{v}/\mathrm{d}y, \;\; \text{and } \dot{\gamma} = \mathrm{d}\gamma/\mathrm{d}t = \mathrm{d}\mathbf{v}/\mathrm{d}y. \tag{2.32}$$

The viscosity η of the fluid and the quantities τ and $\dot{\gamma}$ are related as follows:

$$\tau = \eta\dot{\gamma}, \tag{2.33}$$

where η is expressed in poise (P) $= \mathrm{g/(cm \cdot s)}$.

The viscosity of water at 20°C is about 1.00 cP whereas that of olive oil is about 10,000 cP at the same temperature.

For simple low-molecular-weight liquids, viscosity η is usually independent of the shear rate (i.e., the linear equation (2.33) applies at constant

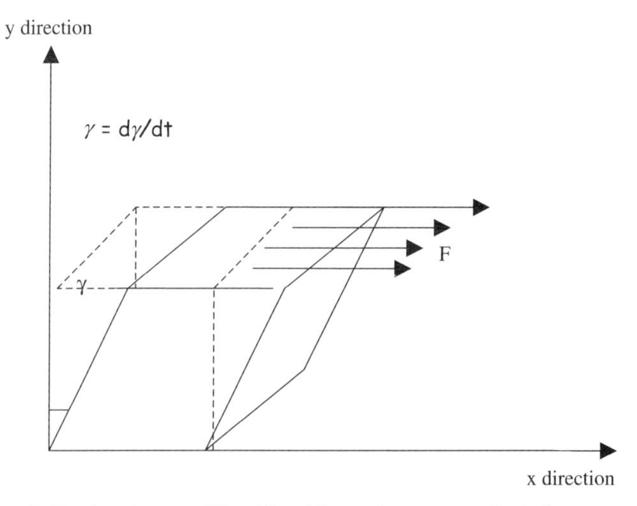

Figure 2.7 A volume of liquid subjected to an applied shear stress.

temperature). Polymer solutions generally do not fall into this category of Newtonian liquids. At moderate shear rates polymer solutions generally show reduced viscosity or undergo shear-thinning. Figure 2.8 compares the shear-rate dependence of viscosity for a Newtonian liquid and a non-Newtonian polymer solution. At very high shear rates, however, a non-Newtonian liquid may revert to Newtonian behavior.

The viscosity η of a polymer solution (at a concentration c) in a solvent of viscosity η_o is conveniently expressed by its relative viscosity η_r and several other common measures:

Relative viscosity (or the viscosity ratio)

$$\eta_r = \eta/\eta_o$$

Specific viscosity

$$\eta_{sp} = \frac{\eta - \eta_o}{c}$$

Reduced viscosity (or the viscosity number)

$$\eta_{red} = \frac{1}{c}\left(\frac{\eta}{\eta_o} - 1\right)$$

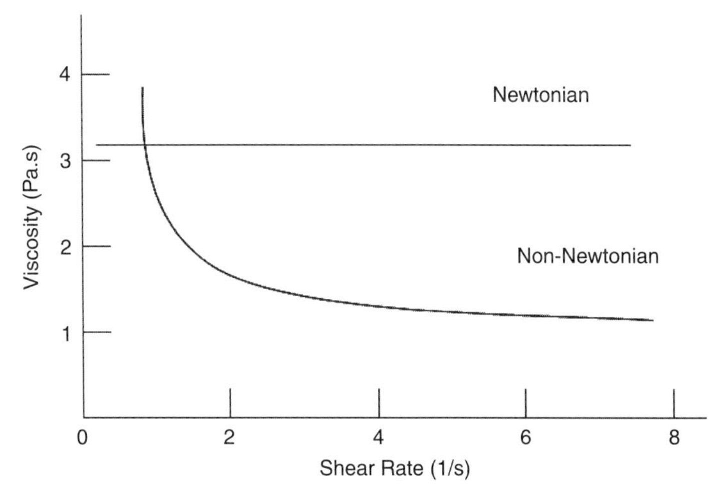

Figure 2.8 Dependence of viscosity on the shear rate for a Newtonian liquid and a non-Newtonian pseudoplastic polymer solution.

Inherent viscosity (or the logarithmic viscosity number)

$$\eta_{\mathrm{I}} = \frac{1}{c} \ln \eta_{\mathrm{r}}.$$

Intrinsic viscosity (or the limiting viscosity number)

$$[\eta] = \lim_{c \to 0} (\eta_{\mathrm{sp}}/c).$$

Note that η_{sp} is essentially the incremental increase in η due to the dissolved solute (polymer). Both η_{sp} or η_{I} are concentration dependent, and extrapolating the linear plots of either of these versus the concentration to zero concentration yields the intrinsic viscosity $[\eta]$, which is related to the dimensions of an isolated polymer chain in solution.

For dilute polymer solutions a simple approximate expression for specific viscosity can be derived from Einstein's equation for a dilute suspension of hard (incompressible) spheres in a liquid. The viscosity of a suspension of N spheres, each with a hydrodynamic volume V_{e} in a total volume V of a liquid, is given by

$$\eta_{\mathrm{sp}} = 2.5\, N(V_{\mathrm{e}}/V) \tag{2.34}$$

and

$$[\eta] = 2.5\, N\, V_{\mathrm{e}}/\overline{M_{\mathrm{v}}}.$$

These equations (which strictly do not hold for nanoscale particles) can be used to obtain the following expression for a monodisperse polymer of molecular weight M_{v} dissolved in a solution of concentration c (where n moles of polymer is dissolved in a volume V), as follows:

$$\eta_{\mathrm{sp}} = \frac{2.5\, nN_{\mathrm{A}} V_{\mathrm{e}}}{V} = \frac{2.5\, cN_{\mathrm{A}} V_{\mathrm{e}}}{M_{\mathrm{v}}}, \tag{2.35}$$

where V_{e} is the macromolecular volume of the random coil and N_{A} is Avogadro's number. Substituting for V_{e} in equation (2.34), the average molecular weight can be introduced into the equation.

Equation (2.35) indicates that, ideally, η_{sp}/c is independent of concentration. Assuming a spherical geometry for the polymer molecules, and that $V_{\mathrm{e}} \infty \langle r^2 \rangle^{3/2}$, the above expression can also be written as follows, where the various constants are combined into a single constant, Φ. This yields the Flory–Fox equation:

$$[\eta] = \Phi\{\langle r^2 \rangle/M\}^{3/2} M^{1/2} = \Phi\{\langle r^2 \rangle_{\mathrm{o}}/M\}^{3/2} M^{1/2} \alpha^3. \tag{2.36}$$

In Θ-solvents $\alpha^3 = 1$ and since $(\langle r^2 \rangle/M)$ is a constant,

$$[\eta]_{\Theta} = \Phi\{\langle r^2 \rangle\,/M\}^{3/2} M^{1/2} = KM^{1/2}. \tag{2.37}$$

The constant K combines quantities such as N_A, molecular mass of the repeat unit m_o, bond length, and the characteristic ratio. Also $\langle r^2 \rangle$ is proportional to (but not equal to) the frictional radius of the macromolecular sphere; an additional constant to take this into account is also included in it.

In non Θ-solvents the exponent is not $1/2$ and a more general form of the above expression (referred to as the Mark–Houwink equation) is obtained. The constant Φ has a value of $\sim 2 \times 10^{23}$, provided $\langle r^2 \rangle$ is expressed in cm and $[\eta]$ in dL per g. For non-Θ conditions the above equation is generalized into

$$[\eta] = KM_v^a, \tag{2.38}$$

where K and a are constants and M_v is the viscosity-average molecular weight. The relationship is called the Mark–Houwink–Staudinger–Sakurada (MHSS) equation. The MHSS constants can be experimentally evaluated from a log–log plot of $[\eta]$ vs M_v and can be useful in understanding the average conformational state of the polymer in solution. Generally, $M_n \leq M_v \leq M_w$, but when $a = 1$, $M_v = M_w$. Value of the temperature-dependent MHSS constant a depends on the polymer/solvent pair and generally ranges from 0.5 (for near-Θ conditions) to about 0.8 for flexible linear polymers. Less flexible polymers generally have a value of $a > 0.8$.

Recalling the expression for the characteristic ratio C_∞ in equations (2.27) and (2.36), equation (2.38) can also be expressed as follows:

$$C_\infty = (K/\Phi)^{3/2}(M_o/l^2), \tag{2.39}$$

where M_o is the mean molecular weight of a repeat unit in the polymer chain, l is the average bond length per skeletal bond of the chain, and C_∞ is the characteristic ratio in the limit of infinite molecular weight. The approach has significant limitations. The treatment, based on a theoretical model for solid impenetrable spheres, does not take into account any thermodynamic interactions between the polymer and solvent. Also, the solutions are assumed to be dilute and the intrinsic viscosity relations employed apply only at zero shear rate.

The Huggins equation relates (η_{sp}/c) to the concentration c in dilute polymer solutions ($[\eta]c \ll 1$) and can also be used in viscometric determination of the molecular weight of polymers:

$$(\eta_{sp}/c) = [\eta] + k'[\eta]^2 c. \tag{2.40}$$

The Huggins constant k' is independent of the molecular weight but depends on the specific polymer/solvent pair and temperature. The value of k' is about

$1/3$ for common polymers in good solvents. The higher terms of this expansion are usually left out as most polymers show (η_{sp}/c) to be a linear function of c at low concentrations.

2.6 CONCENTRATED POLYMER SOLUTIONS

Polymer molecules under Θ-conditions can be represented by an approximately spherical domain with the chain conformations fitted snugly within it. With good solvents these domains will tend to be larger (or "swollen") and relatively more deformable (or "squishy"). In ideal dilute solutions, contact between the spheres, and therefore the intermolecular interactions, tend to be short-lived and minimal. However, as the concentration is increased, such interactions become increasingly important and the rheological properties of polymer solutions deviate from the semi-quantitative descriptions of the low-concentration regime discussed above. Electrospinning generally involves fairly viscous solutions of high polymer concentration and viscosity in nondilute systems are therefore of particular interest.

A requirement for topological interaction between polymer chains in concentrated solution is that the average chain length of the polymer exceeds a certain threshold value. The corresponding molecular weight, M_c, is referred to as the critical molecular weight or the entanglement molecular weight. Experimental observations of the dependence of zero-shear melt viscosity η_o of a polymer on its weight-average molecular weight $\overline{M}_w$ illustrate the dependence of chain entanglement on molecular weight. At about $\overline{M}_w = M_c$, chain entanglements become a significant contributor to melt and solution viscosity. At $\overline{M}_w < M_c$, η_o varies with $M_w^{1.0}$, but at $\overline{M}_w > M_c$, the dependence changes somewhat abruptly to $M_w^{3.4}$. This enhanced viscosity is attributed to interchain overlap. As the molecular weight exceeds M_c, the

TABLE 2.3 Values of M_c and M_{ent} for selected polymers

Polymer	M_c	M_{ent}	M_c/M_{ent}
Polyethylene	3480	980	3.5
Poly(ethylene oxide)	2000	5870	2.9
Poly(vinyl acetate) (atactic)	9100	24,500	2.7
Poly(methyl methacrylate) (atactic)	13,600	29,500	2.2
Polystyrene (atactic)	18,100	31,200	1.7
Poly(dimethylsiloxane)	12,000	24,500	2.0
Poly(isobutylene)	10,500	17,000	1.6

Source: Fetters et al. 2007.

melt will include increasing amounts of topological constraints or entanglements that tend to act very similarly to chemical crosslinks in their ability to constrain bulk deformation of the material. The average molecular weight between a pair of such entanglements is M_{ent}, the "entanglement molecular weight" of the polymer. This again can be experimentally obtained, for instance from the plateau modulus and the density of polymer at a given temperature. At the onset of deviation in the η_o vs M_w curve, the value of $M_c/M_{ent} \approx 2-3$ for most polymers as seen from Table 2.3. Concentrated solutions also exhibit parallel behavior to melts, with a corresponding $(M_{ent})_{soln}$ and an $(M_c)_{soln}$ value (McKee et al. 2004b, 2006a). The experimentally determined ratio $(M_{ent})_{soln}/(M_c)_{soln} \approx 2$. Table 2.3 gives typical values of M_c and M_{ent} for common polymers.

Graessley identified five concentration regimes for polymers dissolved in thermodynamically good solvents (Graessley et al. 1967):

A Ideal dilute particle solution
B Semi-dilute particle solution {concentration regime $c < c^*$}
C Semi-dilute network solution {concentration regime $c > c^*$}
D Concentrated particle solution
E Concentrated network solution

Solutions in regime A and B are of little interest in electrospinning. The change over from solution B to C occurs at a critical concentration c^* that depends on both the average molecular weight of polymer and the nature of the solvent. On further increasing concentration, a second critical concentration (which is independent of the molecular weight of the polymer) is reached when the solution transitions from regime C to D. The value of c^* is more relevant to electrospinning and is readily determined by viscometry or by light-scattering studies.

Recently, de Gennes recognized the analogy between the behavior of polymer solutions and magnetic critical phenomena,[12] allowing the latter analysis to be applied to model concentrated solutions. After de Gennes, three concentration regimes can be identified for high-polymer solutions:

- Dilute ($c < c^*$)
- Semi-dilute ($c > c^*$)
- Concentrated solutions ($c > c''$)

[12]For studies on magnetic critical and tricritical phenomena and the renormalization group theory adopted in the de Gennes approach, see Wilson (1971a, 1971b).

At $c \approx c^*$, crossover phenomena between the domains result in initial overlap of polymer chains and introduces a degree of order into the solution. Essentially the polymer molecules in solution begin to touch at their boundaries. Above the critical concentration, c^*, macromolecules in solution begin to display overlap with adjacent macromolecular domains.[13] Above this concentration the polymer chains are no longer distributed completely randomly in solution. The loose-chain network resulting from this entanglement is described by a mesh parameter ξ (this refers to the average distance between two entanglement points in the loose network). Previously separated randomly arranged spherical chain domains at $c < c^*$ are now arranged as chains of connected, yet individually distinguishable domains. The high-concentration regime c'' is reached when $\xi < 2r$, usually at very high concentrations the mesh parameter approaches unity or the dimensions of monomer units.

The crossover concentration c^* scales with the total number of monomer units N in the system as the individual polymer chains are indistinguishable from each other at these higher concentrations. Thus, measures such as $\langle r^2 \rangle$ and even M_n become irrelevant for the system and are in effect replaced by the mesh parameter ξ:

$$c^* \sim N^{1-vd} \approx 1/[\eta], \tag{2.41}$$

where v is the excluded volume exponent and d is dimensionality. In this concentration regime ξ scales with $c^{-3/4}$ (Wiltzius et al., 1983). Unlike in the dilute regime where solution viscosity scales with concentration ($\eta \approx c$), when $c > c^*$, a power law dependence of the following form is expected (de Gennes 1979):

$$\eta = \eta_o (c/c^*)^{3/(3v-1)}, \tag{2.42}$$

where η_o is the solvent viscosity and the exponent $v = 0.6$ for good solvents. The concentration exponent using single-parameter scaling is about 3.75, and that based on two parameters predicted $\eta \approx c^{4.5}$, closer to the experimentally observed values (Colby et al. 1991).

Recent work has suggested the existence of polymer solutions where $c > c^*$ and therefore have finite overlap, but as yet no significant entanglement (Bordi et al. 2002; Colby and Rubinstein 1990; Krause et al. 1999). They assume a gradual statistical chain entanglement process yielding significant entanglements only at a concentration c_e where $c > c_e > c^*$. This allows for two concentrated solution regimes commonly identified as where polymer chains "overlap with no entanglement" and "overlap with entanglement."

[13]Depending on how c^* may be defined (e.g., $c^* = 3N/4\pi\langle\gamma^2\rangle^{3/2}$ or $c^* = N/\langle r^2\rangle^{1/2}$) it can have different values even at the same molecular weight.

The Power Law equation (η_{sp}/c) relates to the concentration c in dilute polymer solutions and can be used in viscometric determination of the molecular weight of polymers:

$$\eta_{sp}/c = [\eta] + k'[\eta]c + k''[\eta]^3c^2 + \cdots \tag{2.43}$$

Commonly used equations such as the Huggins equation (eqn 2.40) and the Kramer equation can be derived from (2.43). The Huggins constant k' is independent of the molecular weight but depends on the polymer/solvent pair and the temperature. Higher-order terms in the series are generally ignored, as (η_{sp}/c) vs c for moderately concentrated polymer solutions shows a high degree of linearity. The value of k' is about 1/3 for common polymers in good solvents but can be as high as 0.5 for polymers in poor solvents. Equation (2.44) (Martin 1951), which can be derived from the Huggins equation to describe the dependence of viscosity $[\eta]$ on concentration for concentrated polymer solutions is as follows

$$\ln(\eta_{sp}/c) = \ln[\eta] + K_m c[\eta]. \tag{2.44}$$

The magnitude of K_m, the Martin's constant,[14] is indicative of the relative level of interaction between polymer and solvent, with smaller values suggesting a more "open" conformation in solution. Son et al. (2004d) used the constant K_m to interpret their data on electrospinning poly(ethylene oxide) (PEO) from several solvents.

This rather limited introductory discussion into polymer solutions is intended as a prelude to the particular discussion of solution characteristics that influence electrospinning to be discussed in the following chapters. The concept of solvent quality leading to extended average conformations of dissolved macromolecules and the critical concentrations above which such molecules tend to overlap or entangle, determine the electrospinnability of a solution. This results in electrospinning being typically associated with specific concentration regimes, usually the concentrated solutions. However, as discussed in Chapter 4, properties unrelated to solubility of polymers such as solvent dielectric properties, need to be also taken into account in designing polymer solutions for electrospinning.

[14] Transforming equation (2.43) into the logarithmic form and using the approximation $\ln(1+x) = x$ yields the version of Martin's equation shown in (2.44).

3

ELECTROSPINNING BASICS

Based on the discussion of the physics of jet formation (in Chapter 1), some basic requirements for electrostatic spinning might be readily anticipated. A primary such requirement is linear macromolecularity — in other words, a polymer having both a sufficiently high average molecular weight as well as an open chain-like geometry[1] in solution to allow the development of a degree of jet elasticity via chain entanglement. Protein polymers such as casein, for instance, have a high enough molecular weight (M $\sim$ 25,000 g/mol) and good solubility, but are difficult to electrospin because of their globular compact molecular structure due to strong intramolecular hydrogen bonding (Xie and Hsieh 2003). Similarly, dendritic or hyperbranched polymers of higher generations (with high enough M_w values) do not electrospin even from concentrated solutions. Although the present discussion is limited to polymer solutions, as already alluded to elsewhere, polymers can also be electrospun from melts to obtain uniform nanofibers. Common polyolefins, nylon-12, poly(ethylene terephthalate) (PET) and poly(ethylene naphthenate) (PEN) do not dissolve in common solvents at ambient temperature but can be electrospun from melt (Larrondo and St. John Manley 1981a, 1981c). Electrospinning polymer melts might potentially be scaled up using a modification of conventional processing equipment as recently illustrated by the

[1]Most polymers have chain-like molecular geometry and are referred to as linear polymers. However, other geometries such as highly branched polymers, comb-like polymers, cross-linked polymers, and dendritic polymers also exist.

Science and Technology of Polymer Nanofibers. By Anthony L. Andrady

use of a Brabender extruder to electrospin micrometer-scale polypropylene fibers. The required electric field of 6–15 kV/cm was about an order of magnitude higher than that used in solution electrospinning (Lyons et al. 2004).

Experimentally, an electrospinning apparatus consists of three basic components:

- A polymer of adequate average molecular weight dissolved at a high enough concentration in a good solvent having suitable conductivity, surface tension, and vapor pressure.
- A device for electrically charging the polymer solution to obtain a stable jet.
- A gap between the capillary tip (or the charged droplet) and a grounded collecting surface that is set at a suitable distance from the tip carrying the polymer solution.

The principle variables that govern nanofiber quality (or determine if electrospinning will occur at all) are the average molecular weight of the polymer, the nature of the solvent, and the magnitude of the electric field used to induce electrospinning.

3.1 MOLECULAR WEIGHT EFFECTS

Generally, electrospinning can only occur with moderately concentrated solutions, as the process of jet formation relies on the entanglement of polymer chains (McKee et al. 2004b, 2005; Shenoy et al. 2005b). Conventional measures of the extent of chain overlap and entanglement in solution can therefore be useful metrics in describing the amenability of a given polymer solution to electrospinning. These measures usually involve the quantity $\overline{M}_c$, the critical molecular weight at which chain entanglements in solution become significant (see Chapter 2). Two approaches that address the relationship between the concentration regime and electrospinnability of polymer solutions have been proposed. Both can potentially identify concentration regimes where defect-free continuous nanofibers can be electrospun.

3.1.1 The Simha–Frisch Parameter, $[\eta]c$

One of the approaches uses the first term of the Huggins expression for concentration dependence of viscosity in polymer solutions:

$$\eta_{sp} = [\eta]c + k'[\eta]^2c^2 + \cdots \tag{3.1}$$

In Simha's early model (Simha and Zakin 1962), transition from a dilute to a concentrated polymer solution was envisioned as being due to interpenetration of polymer chains that occurs when concentration lies somewhere in the region $1 \leq [\eta]c \leq 10$. This transition is evident from the change in the concentration dependence of viscosity in polymer solutions. The quantity $[\eta]c$, the Simha–Frisch parameter (Frisch and Simha 1956), also sometimes called the Berry number (Gupta et al. 2005), is therefore a reasonable measure of chain overlap in solution. As Shenoy et al. (2005b), however, correctly point out, the dependency, being ultimately based on the equivalent hard sphere hydrodynamic model, is strictly applicable only at low polymer concentrations.

It is convenient to visualize the structure of polymer solutions in terms of critical or crossover concentrations. In dilute solutions with no significant chain overlap or interaction and $[\eta]c < 1$, the chains exist as separate entities or "blobs" in solution. With increasing concentration, however, interaction between the polymer chains also needs to be taken into account. As illustrated schematically in Fig. 3.1, at a critical crossover concentration of $c = c^*$, chains begin to overlap and the blobs begin to touch each other at their boundaries. At this concentration the solution is homogeneous at the molecular level (i.e., the chain concentration inside a blob is the same as that of the bulk solution). At higher concentrations ($c > c^*$), the number of chain entanglements will be proportional to c. The value of c^* will depend on the average molecular weight, polymer chain flexibility (or the chemical structure of the polymer), temperature, and the nature of the solvent. Also, recalling the discussion in Chapter 2, $1/[\eta]$ and therefore c^* can also be calculated from the root-mean-square end-to-end distance $\langle r^2 \rangle$ and the average molecular weight $\overline{M}$ of the polymer using Avogadro's number N_A (for solutions in good

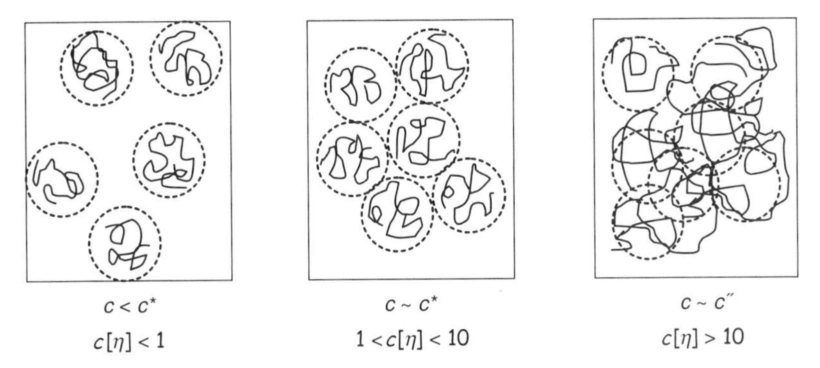

Figure 3.1 Diagrammatic representation of a polymer solution in different concentration regimes.

solvents, the radius of gyration, $\langle s^2 \rangle$ or if available, the hydrodynamic radius might be substituted in place of $\langle r^2 \rangle$ (Gupta et al. 2005)).

$$c^* = \frac{3M}{4\pi \langle r^2 \rangle^{3/2} N_A} \tag{3.2}$$

It is convenient to also express c^* in terms of intrinsic viscosity rather than the chain dimension, using the relationship

$$[\eta] = \Phi \langle r^3 \rangle / M, \tag{3.3}$$

where $\Phi = 2.5 \times 10^{23}$ (c.g.s. units). This yields the relationship

$$c^* \sim 1/[\eta], \tag{3.4}$$

which allows c^* to be conveniently determined experimentally from intrinsic viscosity measurements. Solution viscosity expressed as a function of the normalized concentration (c/c^*) is expected to show different dependencies in different concentration regimes. Note that c^* is directly proportional to M; the longer the chain lengths, the lower will be the concentration needed to effect chain overlap. Chain overlap is the dominant interaction mechanism in the concentration regime $1 < c[\eta] < 10$. Figure 3.1 schematically represents the behavior of solvated chains in solution in the three different concentration regimes.

Increasing the concentration beyond $c > c^*$ further increases the degree of chain overlap and finally, at a concentration $c = c''$ an entanglement regime where the chains begin to topologically constrain each other is reached. Essentially the "open" conformations of the polymer chains in solution becomes increasingly compact at these higher concentrations; this contraction of polymer chains in solution can theoretically proceed until the unperturbed dimension for the chain $\langle r^2 \rangle_\Theta$ is reached at about $c = c''$ (Graessley 1980). Above this concentration the blob structure ceases to exist and individual chains can no longer be discerned in the bulk of the solution. The dependence of viscosity on concentration again changes in this regime. At $c[\eta] \approx 10$, the expansion ratio $\alpha^2 = 1$ and entanglements dominate the solution properties:

$$c'' = c^* \{ \langle r^2 \rangle / \langle r^2 \rangle_\Theta \}^4. \tag{3.5}$$

Therefore c'' will be essentially independent of molecular weight.

Figure 3.2 from Gupta et al. (2005) shows the combined zero-shear viscosity for seven poly(methyl methacrylate) (PMMA) samples ranging in $\overline{M}_w$ from 12,470 to 365,700 (g/mol) dissolved in dimethylformamide (DMF) illustrating this dependence. A change in the gradient of the plot occurs at a

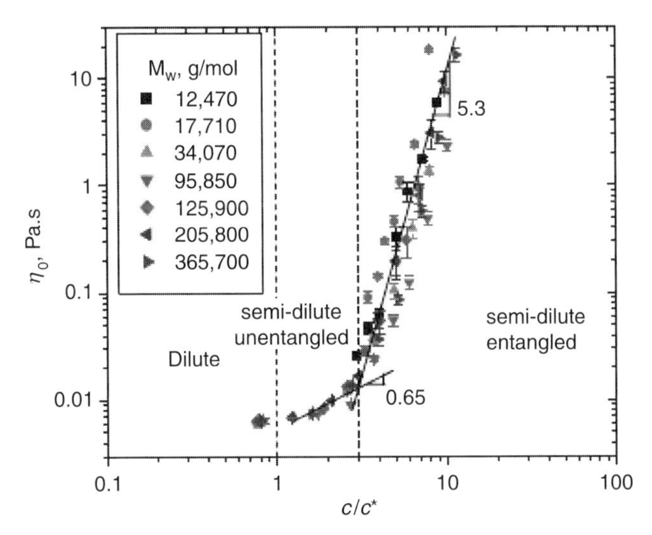

Figure 3.2 A plot of zero-shear viscosity as a function of reduced concentration (c/c^*) for PMMA samples of different average molecular weights. Reprinted with permission from Gupta et al. (2005). Copyright 2005. Elsevier.

concentration where the system changes from semi-dilute $\{1 < (c^*/c) < 3\}$ to the entangled $\{(c^*/c) > 3\}$ regime. The experimental values of the exponents are, however, somewhat different from the theoretical expectations (Colby et al. 2001)

$\{1 < (c/c^*) < 3\}$	$\{(c/c^*) > 3\}$
expected exponent = 1.25	expected exponent = 4.25 – 4.5
experimental value = 0.65	experimental value = 5.3

Colby and colleagues identified four concentration regimes (instead of the three by de Gennes (1979)), subdividing the semi-dilute regime into "semi-dilute unentangled" and "semi-dilute entangled" regimes, defining an entanglement concentration c_e as the boundary concentration[2] between the regimes (Colby et al. 2001). Significantly, this concentration was also shown to be the minimum concentration at which continuous, bead-free nanofibers could be electrospun (McKee et al. 2004b, 2005).

As chain overlap and/or entanglements in solution appear to be critical to achieve bead-free nanofibers (McKee et al. 2004b, 2005; Shenoy et al. 2005b)

[2]McKee et al. (2004b) reported that the fiber diameter d for electrospinning of poly(ethylene terephthalate-*co*-ethylene isophthalate) copolymers from $CHCl_3$/DMF (70/30) scaled with the ratio c/c_e ($d \approx (c/c_e)^{2.6}$).

these different concentration regimes must also be reflected in the electrospinning behavior of the polymer solutions. Data reported by Gupta et al. (2005) for PMMA solutions in DMF illustrate this relationship (see Table 3.1) and can be conveniently interpreted in terms of normalized concentration (c/c^*) or values of $c[\eta]$. At lower concentrations, polymers of different average molecular weights yielded either only droplets or a mix of droplets and short pieces of poorly formed nanofibers in the electrospinning experiments. Bead-free nanofibers were obtained when $(c/c^*) > 6$ or only in the so-called "semidilute entangled" regime (values of c^* for these polymers were estimated from published Mark–Houwink coefficients, and these agreed well with those obtained experimentally from light-scattering measurements). At $(c/c^*) \approx 3.9–4.0$, continuous fibers (but still with some beads) were obtained, then too only with polymer samples of higher average molecular weight. Also, the yield of defect-free nanofibers in electrospinning was also found to be influenced by the polydispersity P of the samples $(P = \overline{M}_w / \overline{M}_n)$. Whereas all samples with narrow molecular weight distributions $(P \approx 1.03–1.35)$ yielded uniform fibers at $(c/c^*) \approx 6$, with samples having a broader distribution (where $P \approx 1.62$ and 2.12), uniform bead-free fibers were obtained in quantity only at the much higher concentration of $(c/c^*) = 9.7$ and 10.1, respectively.

Even within the concentration regime that obtains electrospinning, the quality of nanofibers electrospun can be quite variable. Koski et al. (2004), electrospinning poly(vinyl alcohol) (PVA) in aqueous solution at

TABLE 3.1 Electrospinning behavior of poly(methyl methacrylates) at different (c/c^*) values

Molecular Weight (g/mol)	c^*(exp.)[a]	c/c^*	Electrospun Product[b]
12,470	10.2	2.9	Droplets and beads
		3.9	Beads and a few fibers
		7.2	Continuous fiber
34,070	6.4	0.8	Droplets and beads
125,900	3.3	0.8	Droplets and beads
		2.6	Beads and small fiber
		4.0	Continuous fiber
		6.8	Continuous fiber
205,800	2.5	0.8	Droplets and beads
		2.9	Fibers with numerous beads
		3.9	Continuous nanofiber
		6.8	Continuous nanofiber
365,700	1.9	0.8	Droplets and beads with fibrils

[a]From light-scattering measurements of hydrodynamic size (expressed in wt%).
[b]Descriptions based on scanning electron microscopy (SEM) images of electrospun material.
Source: Gupta et al. 2005.

$[\eta]c \approx 7.5$ (well above the limiting concentration of $[\eta]c \approx 5$), obtained continuous nanofibers from samples of PVA of different average molecular weights (range of $\overline{M}_w = 9000-186{,}000$ g/mol). They varied $[\eta]$ and c of the solutions independently, maintaining a constant value of $[\eta]c \approx 7.5$ by using different molecular weights of the polymer. Although all solutions yielded continuous nanofibers (as expected, since $[\eta]c > 5$), the quality of the nanofibers varied widely with the composition of solutions. As the concentration was decreased from 35% to 6% (switching to progressively higher molecular weight polymer to maintain $[\eta]c \approx 7.5$ constant for all solutions), the fiber quality deteriorated progressively, resulting also in a broader distribution of fiber diameters.

In recent research on electrospinning polystyrene (PS) from tetrahydrofuran (THF) solutions, however, continuous nanofibers were obtained only at a value of $[\eta]c > 13$ (Casper et al. 2006). The very low molecular weight PS ($M_w \approx 36{,}000$ g/mol) could not be electrospun from solutions at normalized concentrations as high as $c/c^* \approx 7$ (of 35 wt% in THF). However, uniform (relatively thicker) fibers could be electrospun at the even higher concentration of $[\eta]c = 16$, corresponding to 80 wt% of PS in solution. In electrospinning poly(ε-caprolactone) (PCL; $M_w = 40{,}000$ g/mol) solutions in $CHCl_3$, nanofibers were obtained at concentrations exceeding 4 wt%, corresponding to a dimensionless viscosity of $[\eta]c \approx 3-4$. However, the best fibers were still obtained only at $[\eta]c \approx 4.5$ (Hsu and Shivkumar 2004a). Although a useful guideline to electrospinnability, the critical values of $[\eta]c$ tend to vary widely with the polymer/solvent system considered, limiting its predictive value. In general, a minimum average molecular weight needs to be exceeded even at high enough $[\eta]c$ values to obtain quality continuous nanofibers.

This observed system-dependence of the value of $[\eta]c$ is to be expected, because the average chain conformations, and therefore the entanglement of chains, depends on solvent quality. Poly(ethylene glycol) (PEG) electrospun from solutions of $[\eta]c \approx 10$ but made up in different solvents yielded very different average fiber diameters, partly because the solvent characteristics also play a role in determining fiber quality (Son et al. 2004d).

The approach does not take into account any polymer–polymer interactions in solution. In poly(methyl methacrylate-*co*-methacrylic acid) copolymers [P(MA-*co*-MAc), 5% MAc], the pendent carboxylic acid group in methacrylic acid (MA) entities may undergo hydrogen-bonded interactions. Yet, the electrospinning behavior of these polymers was found to be similar to that of the higher molecular weight unfunctionalized PMMA (McKee et al. 2004a). The value of the critical concentration for entanglement (~6 wt%) was independent of solvent composition and the nanofiber diameter d was reported to vary with normalized concentration

($d = 0.18(c/c^*)^{2.7}$) — the same as for nonassociating fibers. The hydrogen-bonded interactions in the copolymer, in this instance, appear to be too weak to influence electrospinning behavior. However, when a self-complimentary multiple hydrogen-bonding side group was introduced (in ~5% of the repeat units) into the MA units, the solution viscosity dramatically increased and the value of c^* decreased[3]. Evidently, copious hydrogen-bonded interactions in this system act very much like chain entanglements, lowering the critical concentration of the polymer needed to allow electrospinning. Polyacrylamide (molecular weight 9×10^6 g/mol) also undergoes hydrogen bonding in aqueous solutions. Y. Y. Zhao et al. (2005) studied the electrospun nanofiber morphology of this polymer in the concentration range of 0.3–3.0 wt% in water. Bead-free continuous nanofibers were obtained at concentrations as low as $c/c^* > \sim 2.5$. This value is low compared to that for other polymer solutions. Hydrogen bonding likely provided the additional chain interactions to facilitate electrospinning in this system.

Branched polymers in solution also show concentration-dependent transitions in viscosity. Copolymers of poly(ethylene terephthalate-*co*-ethylene isophthalate) (P(ET-*co*-EI)) were prepared by the polycondensation reaction of an equimolar mixture of dimethyl terephthalate (DMT) and dimethyl isophthalate (DMI) in the presence of a 100% excess of ethylene glycol (EG) (Scheme 3.1). Chain branching was introduced into the polymer using a trifunctional anhydride or a tricarboxylate at a level of 1–1.5 molar percent. The concentration dependence of viscosity for these polymers were $\eta_{sp} \approx c^{1.39}$, $\eta_{sp} \approx c^{2.73}$, and $\eta_{sp} \approx c^{3.7}$ for the semi-dilute unentangled, semi-dilute entangled, and concentrated regimes, respectively (McKee et al. 2004b, 2005). Again, these values differ from the theoretically expected values, in this case primarily because of chain branching. Successful electrospinning of branched copolymer from mixed solvents in $CHCl_3$/DMF (70/30 w/w) was possible at unexpectedly low values of $(c/c^*) > 1$ to obtain bead-free continuous fibers at $(c/c^*) > 2–2.5$. Fiber diameters scaled with the

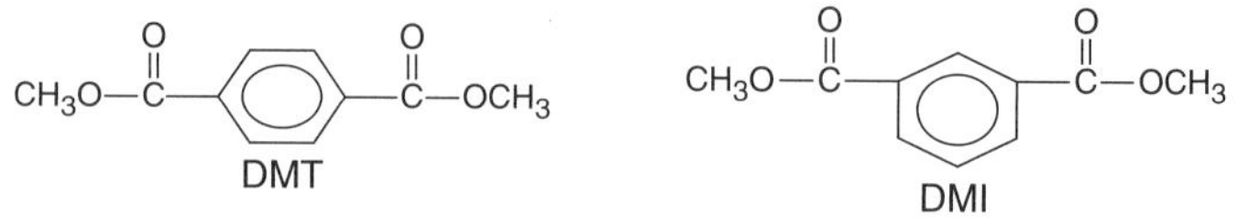

$CH_3O-C(=O)-C_6H_4-C(=O)-OCH_3$ (DMT) $\quad$ $CH_3O-C(=O)-C_6H_4-C(=O)-OCH_3$ (DMI)

Scheme 3.1

[3]Chain entanglement concentration, C_e for the polymer decreased by about 33 percent in DMF/$CHCl_3$ (80/20 v/v) solution and about 25 percent in DMF/$CHCl_3$ (60/40 v/v) solution, relative to that for the PMMA-*co*-PMAA polymer (McKee et al. 2004a).

normalized concentration to the 2.6th power. In highly branched polymers the potential for chain entanglement appears to be higher than that for a linear chain of the same average molecular weight.

Nonpolymeric materials that can form aggregates where physical interactions effectively substitute for chain entanglements can sometimes be electrospun. Lecithin is not a high polymer, but due to its amphiphilic molecular structure, it forms reverse micelles in nonaqueous solutions. Micellar geometry is concentration dependent and assumes a cylindrical, worm-like shape at high enough concentrations. Above the critical concentration of $c^* \approx 35\%$, these worm-like domains yielded entangled associated regimes and the system can be electrospun (from chloroform/DMF 70 : 30 solution) into nanofibers (McKee et al. 2004b).

In polyelectrolyte solutions, such as in poly(2-(dimethylamino)ethyl methacrylate hydrochloride) [P(DMAEMA HCl)], electrospinnability can be achieved without changing the solution concentration, but by altering the ionic strength of the solution. In P(DMAEMA HCl) dissolved in 80/20 w/w% in water/methanol mixtures, the concentration of chain entanglements increased with NaCl concentration. The minimum concentration required for fiber formation therefore decreased as the level of NaCl is increased, due to ionic screening of the repulsive electrostatic interactions between charged repeating units that serve to stabilize the jet (McKee et al. 2006a). As the salt concentration increased from 0 to 50%, the minimum concentration that yielded continuous bead-free fibers decreased by about 66% for the aqueous P(DMAEMA HCl) system.

3.1.2 Solution Entanglement Number n_e

The zero-shear melt viscosity vs the average molecular weight curve abruptly changes its gradient (from M to $M^{3.4}$) at the critical molecular weight $\overline{M}_c$ for the onset of chain entanglement. Provided $\overline{M} > \overline{M}_c$, the average molecular weight between entanglements in the melt, $\overline{M}_e$ is also a good measure of the extent of such entanglements. The ratio $(n_e)_{melt} = \{\overline{M}_c/\overline{M}_e\}$ corresponds to the number of entanglements and has an expected value of $(n_e)_{melt} \approx 2$ (experimentally determined values are also close to 2.0). The same might be extended to concentrated polymer solutions where $(\overline{M}_e)_{soln} = \overline{M}_e/\phi$, where ϕ is the volume fraction of polymer in solution. Corresponding to the behavior of melts, polymer solutions similarly show a marked deviation in their viscosity vs molecular weight plots at a value of about $(n_e)_{soln} \approx 2$ (i.e., at a number of entanglements per chain of $\sim$1). At concentrations $c > c^*$, the following expression applies

$$(n_e)_{soln} = (\phi \overline{M}_w)/\overline{M}_e. \tag{3.7}$$

As values of $\overline{M}_e$ for different polymers are well known, the expression allows the calculation of $(n_e)_{soln}$ values for polymer solutions at different concentrations. As the value of ϕ increases, $(\overline{M}_e)_{soln}$ will decrease but $(n_e)_{soln}$ will increase, according to equation (3.7).

Shenoy et al. (2005b) used this quantity as a semi-empirical measure of chain overlap and entanglement in polymer solutions used in electrospinning. Figure 3.3 (taken from their paper) shows a plot of calculated values of $(n_e)_{soln}$ vs concentration for PS solutions for several values of $\overline{M}_w$. For the sample of molecular weight 190,000 g/mol, the dotted line indicates the value of $(n_e)_{soln} = 2$ for transition from beads to (fibers + beads) in electrospinning. The dashed line indicates $(n_e)_{soln} = 3.5$ for the transition to continuous fibers. The transitions in practice might not be as sharply defined as indicated in the diagram, but concentration dependence in electrospinning behavior is evident in the data. As with values of c^*, those of $\overline{M}_e$ can be theoretically estimated using the proportionality between $\overline{M}_e$ and $(C_\infty)^{-3}$. Their study

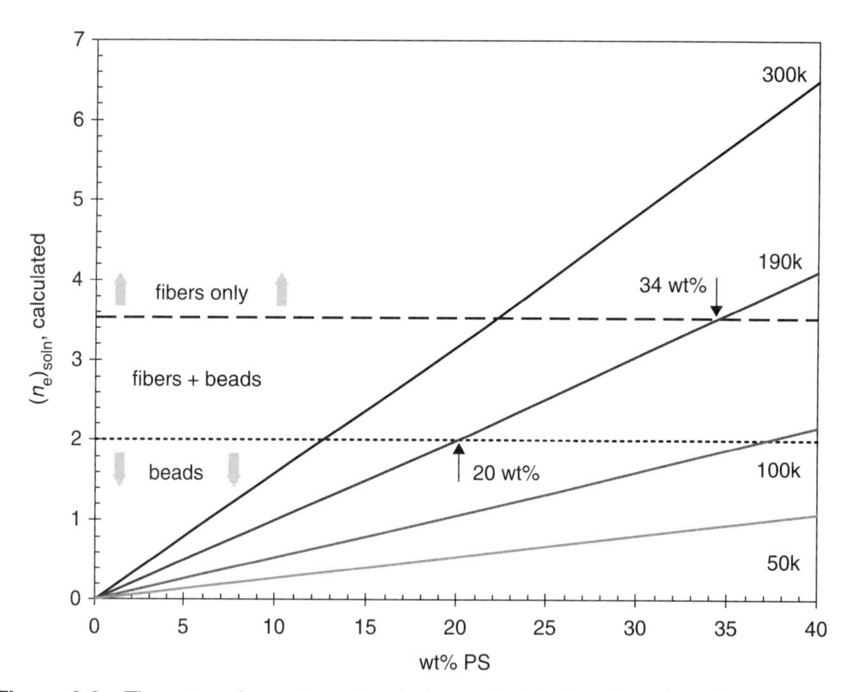

Figure 3.3 The entanglement number $(n_e)_{soln}$ calculated as a function of concentration for the PS/THF system. The dotted lines show the entanglement regime that allows electrospinning of continuous nanofibers. Each solid line is calculated for a different value of average M_w. Reprinted with permission from Shenoy et al. (2005b). Copyright 2005. Elsevier.

TABLE 3.2 Expected and experimentally-observed concentrations at which fiber initiation [at $(n_e)_{soln} \approx 2$] and continuous fiber formation [at $(n_e)_{soln} \approx 3.5$] occur in several polymer/solvent systems

		Initiation of Fiber/Beads		Continuous Fiber	
System	M_w (g/mol)	Expected (wt%)	Actual (wt%)	Expected (wt%)	Actual (wt%)
PS/THF	190,000	20	18	34	30–35
PDLA/DMF	109,000	18.5	<20	32	30–35
PLLA/$C_2H_2Cl_4$	670,000	1.9	<3	3.4	>4
PLLA/CH_2Cl_2	670,000	2.3	<1	4	3
PVP/EtOH	1,300,000	4	3	7.5	7.9

[a]PDLA = poly(D,L-lactide). PLLA = poly(L-lactide). PVP = poly(vinyl pyridine).
Source: Adapted from Shenoy et al. 2005b.

included other polymer/solvent systems (e.g., polylactide (PLA)/CH_2Cl_2, PLA/DMF, PEO/H_2O, and poly(vinyl pyridine) (PVP)/EtOH) (Shenoy et al. 2005b). The utility of this methodology to predict fiber formation *a priori* in electrospinning the solutions (in the absence of other intermolecular interactions) has been studied. Table 3.2 includes some of the data from Shenoy and colleagues to illustrate the relationship between the value of $(n_e)_{soln}$ and electrospinnability.

Data on electrospinning solutions of PS in several different solvents, reported by Megelski et al. (2002), might be used to further illustrate this approach. Using PS of $M_w = 190{,}000$ g/mol, attempts at electrospinning obtained only bead formation at $(n_e)_{soln} < 2$. Short nanofibers intermixed with beads started appearing at $(n_e)_{soln} \approx 2–3$ (corresponding to about 20–25 wt% concentration of PS). Fully formed bead-free fibers were obtained only with solutions where $(n_e)_{soln} > 3$ (corresponding to 30 wt% concentration of the polymer). The transition from beads to fibers + beads or to fibers occurs gradually.

Based on experimental observations, concentrations where $\phi M_w < 2M_e$ are anticipated to yield only droplets and short fibers, and concentrations where $\phi M_w > (3–4)M_e$ will generally yield well-defined bead-free nanofibers. The treatment, however, assumes negligible polymer–polymer interactions in solution and applies to only good solvents where entanglements provide the primary mechanism for stabilization of an electrospinning jet. Specific polymer–solvent interactions were also not considered in arriving at equation (3.7), in effect assuming that the concentrations of interest are very high ($c \gg c^*$). This discussion also assumes the upturn in the viscosity vs concentration curve to be solely due to chain entanglement effects (Shenoy et al. 2005b).

Other interactions, as already pointed out, change these relationships significantly. The presence of hydrogen bonding in the polymer (as in the case of electrospinning PVA (Koski et al. 2004), where fibers were obtained even at $(n_e)_{soln} < 1$), phase separation (either solid–liquid or liquid–liquid) in the system (Kenawy et al. 2003), and gelation due to the formation of thermoreversible junctions (Shenoy et al. 2005a) can all result in very significant deviations from the expected electrospinning behavior. For instance, Shenoy et al. (2005a) found thermoreversible gelation to be the dominant stabilizing mechanism in PVA solutions at 80°C, where fiber formation is obtained at $(n_e)_{soln} \geq 1$. At the higher dissolution temperature of >92°C, they obtained a similar result. Also, polymer solutions that meet the concentration criteria suggested, cannot always be electrospun into continuous fiber because of other factors.

Electrospinning blends of high-molecular-weight poly(ethylene oxide) (PEO) mixed with PEG (molecular weight of about 10,000 g/mol) in aqueous solution have been studied by Yu et al. (2006). These solutions were selected to have the same polymer concentration, zero-shear viscosity, and surface tension. They are non-Newtonian liquids, well below the entanglement threshold of $[\eta]c < 1$ and are therefore essentially unentangled. Yet some nanofiber formation was obtained. The high degree of hydrogen bonding via the terminal hydroxyl groups of PEG likely facilitated electrospinning in this case.

The crux of the above discussion is that a semi-quantitative relationship exists between electrospinnability of a given polymer solution and its concentration as well as the average M_w of the polymer. With polymer solutions a number of generic concentration regimes based on their rheological behavior can be identified and have been proposed by Graessley (Graessley 1980) in the form of a concentration/molecular weight diagram. Figure 3.4 shows the general features of the diagram based on the solution rheology of PS/DMF systems. Polymer concentration (g/dL) can also be conveniently expressed in terms of the volume fraction of polymer ϕ (Wang, C., et al. 2006). The relationships in equations (3.3) and (3.4) can also be written approximately as follows (the value of the Flory constant is assumed to be $\Phi \sim 2.5 \times 10^{23}$ c.g.s units) (Graessley 1980)

$$c^* = (K\Phi/N_A)/[\eta] = 0.77/[\eta]. \tag{3.8}$$

The values of $[\eta]$ can be converted to average molecular weight using the Mark–Houwink constants, K and a, allowing the values of c^* to be plotted at different molecular weights. The boundary line between the "dilute" and "semi-dilute" regions is indicated by the lower solid line in Fig. 3.4. The

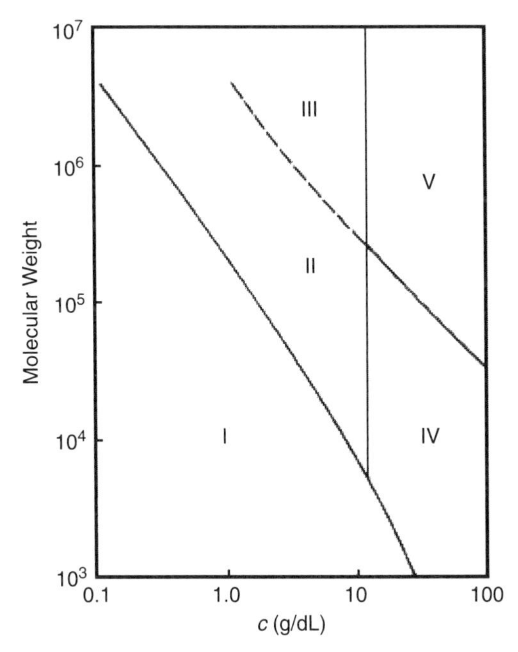

Figure 3.4 Concentration regimes calculated for PS dissolved in a good solvent. Regimes: I, dilute; II, semi-dilute unentangled; III, semi-dilute entangled; IV, concentrated unentangled and V, concentrated entangled. Reprinted with permission of Graessley (1980). Copyright 1980. Elsevier.

vertical line that separates the "semi-dilute" and "concentrated" regimes varies with the system and was calculated as

$$c'' = 0.77/[\eta]'' \approx 11\,\text{g/dL} \tag{3.9}$$

for the PS/toluene system, where $[\eta]'' = 0.07\,\text{dL/g}$. The solid line in the figure at values of $c > 11\,\text{dL/g}$ was determined using the relationship

$$cM > \rho M_c \tag{3.10}$$

for chain entanglement substituting $M_c = 31{,}000$ g/mol and density = 1.07 g/cm^3 for polystyrene. The broken and solid lines as well as the vertical line divide the plot area into five regimes. However, the vertical line separating the "semi-dilute" and "concentrated" regimes (calculated here using equation (3.9)) is invariably based on the specific overlap criterion used or on the definition of c^*. The value of c^* corresponds approximately to the concentration at which the average separation between chains is twice their radius of gyration, leading to the equation $\left(\langle s\rangle^2 = 6\langle r^2\rangle\right)$

$$c^* = \frac{1.84M}{N_A\langle r^3\rangle_0^2} \tag{3.11}$$

As there is no strict definition of c^* for polymer solutions, alternative criteria might also be used, changing equation (3.11) and therefore also changing where on the concentration axis the vertical line demarcating the beginning of entangled regimes is located. In any event, the transition from unentangled to entangled state will not be an abrupt one as is shown in the figure, especially for polymers that are polydisperse.

Figure 3.4 can potentially be useful in establishing the electrospinnability of polymer solutions (Wang, C., et al. 2006). As indicated in the figure, polymer solutions in regimes I, II, and IV include concentrations where the polymer chains that are not entangled sufficiently to allow electrospinning (although electrospraying will of course be possible). Solutions in regimes III and V have adequate chain entanglement (Wang, C., et al. 2006). But the regime III is a semi-dilute regime, and it is the 'concentrated entangled' regime V that is more likely to allow electrospinning. At very high concentrations the solutions of very high molecular weight polymers may, however, be too viscous for the purpose. This representation of concentration regimes can be particularly useful if it can be generally applied to all polymers to assess their amenability to electrospinning. One difficulty is, however, that the nature of solvent is ignored in the analysis. The chain expansion factor for polymers that influences entanglement is dependent on the solvent. In comparing different polymer/solvent systems it is M^a rather than M that is pertinent, at least at moderate concentrations. If electrospinning is consistently obtained only in regimes V, however, the Mark–Houwink parameter a has limited relevance because of the high polymer concentrations in these regimes. Also, experimental data show that other properties of the solvent (surface tension or dielectric constant) also determine electrospinnability, quite apart from polymer molecular weight considerations.

3.2 ELECTRICAL CHARGE

In electrospinning, the liquid jet travels across the gap distance from the highly charged tip to the grounded collector plate. It is the presence of a surface charge that is responsible for the acceleration of the initial jet towards the grounded collector, extending it as by as much as a million times during the short span of travel. In the process, along with mass transfer there is a corresponding charge transfer across the gap. The current flow due to this transfer can be measured and is generally found to increase smoothly with the applied voltage. Accumulation of charge on the collector is experimentally observed to be quite uneven over time, with spikes of charge corresponding to the incidence of larger volumes such as beads (or droplets) being transferred across the gap (Samatham and Kim 2006). An example of

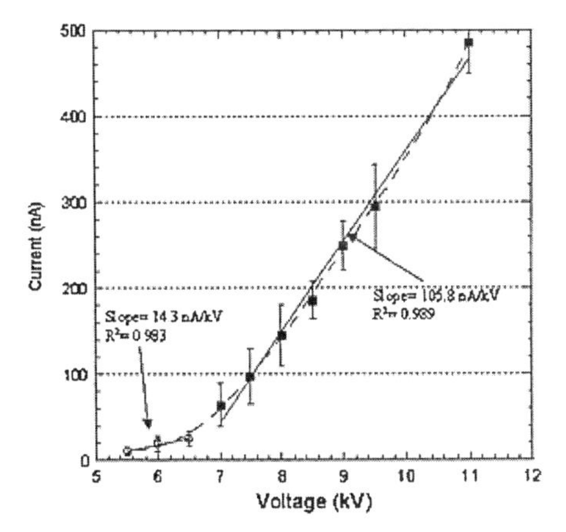

Figure 3.5 Electrospinning current I as a function of the spinning voltage. Reprinted with permission from Deitzel et al. (2001a). Copyright 2001. Elsevier.

an I vs V plot for aqueous solutions of PEO shown in Fig. 3.5 illustrates the relationship between the voltage and current in the electrospinning circuit. Deitzel et al. (2001a) reported that the change in gradient of the plot around 7 kV corresponded to the transition from continuous well-formed nanofibers at the lower voltage to highly beaded fibers at the higher voltage. As mass transfer increases with the appearance of fibers (and beads), consequent higher current flow is anticipated.

The charge on the electrospinning liquid jet primarily resides on its surface (Feng 2002; Hohman et al. 2001a; Spivak and Dzenis 1999). Typical surface charge densities can be calculated from the feed rate of solution, polymer concentration, and the current flow I across the gap.

Where the collector plate is grounded through an ohmic resistance, the current can be estimated from the measured potential drop across the resistance. If the volumetric feed rate of polymer solution is Q, then conservation of mass and charge (assuming no adventitious losses in charge from the jet) require

$$\pi r^2 \rho u = Q \tag{3.12}$$

and

$$2\pi r u q_s + k\pi r^2 E = I, \tag{3.13}$$

where u is the axial velocity of the jet, r is its radius, ρ is the density, q_s is the surface charge density, k is the dimensionless conductivity of the liquid, E is the applied field, and I is the current passing through the jet (Feng 2002, 2003).

Assuming constant acceleration of a jet of uniform diameter He, J.-H., et al. (2007c) obtained the following scaling relationship for the radius r of the jet:

$$g \sim r^{\alpha} \tag{3.14a}$$

$$g \sim c^{\beta} \tag{3.14b}$$

where g ($I = g \cdot E$) is the conductance of the jet and α and β are the scaling exponent. The current balance in the jet was expressed as

$$2\pi r q_s + K c^{\beta} r^{\alpha} E = I \tag{3.14c}$$

where K is a constant.

As the jet extends and reduces in diameter, the surface charge density, as well as the force with which the jet is attracted by the collector, also increase.

Theron and colleagues (Theron et al. 2001) expressed the volume and surface charge densities in terms of I, Q, and d as follows:

Volume charge density

$$q_V = I/Q. \tag{3.15}$$

Surface charge density

$$q_S = q_V(0.25d) \times 10^{-7}, \tag{3.16}$$

where d (μm) is the diameter of the jet measured just below the Taylor's cone.

Naturally, the charge density will be particularly sensitive to the solvent used. In electrospinning bisphenol A polycarbonate (PC) from a DMF/THF mixed solvent, the charge per unit mass of fiber could be changed by as much as 80% by changing the ratio of the solvents from 40/60 to 20/80 (Shawon and Sung 2004). However, nanofiber diameters also change with the solvent composition. Evaporation of the solvent from the spinning jet does not in general dissipate the electric charge on its surface (Kalayci et al. 2005). The solvent content in the fiber will change as the solvent evaporates off the jet.

Theron and colleagues (Theron et al. 2001) found a power-law dependence of q_V on the variables of applied potential V, the feed rate Q, the polymer concentration C, the molecular weight of the polymer, and the gap distance, in electrospinning PEO solutions. Values of q_S measured at a point just below the Taylor's cone in electrospinning PCL solutions showed a similar dependence on V and on the feed rate. The mass charge density on the dry nanofibers can be conveniently measured by collecting a quantity of the spun fiber in a Faraday cup (Kalayci et al. 2005). As the magnitude of the charge is expected to be low, a nanocoulomb meter that has a range up to a few hundred nanocoulombs is best suited for the purpose. Rutledge and Warner (2003) reported the charge on PCL nanofibers electrospun from 3 : 1 $CHCl_3$: MeOH solvent mixture to be 30–60 nC/mg of nanofiber. The q_s (dry) can be calculated from microscopically established distribution functions for fiber diameter, d. For polyacrylonitrile (PAN) nanofibers spun from DMF, the surface charge estimated using a value of $d = 900$ nm, was found to be in the range of 1.05–1.10 (nC/cm^2) (Kalayci et al. 2005). Interestingly, a modified version of an equation used to estimate the mass charge density of dry powders was used with nanofibers in this work and yielded values that were close to (or within an order of magnitude) of that experimentally obtained.

3.3 BEAD FORMATION IN ELECTROSPINNING

Bead formation is the most common type of defect encountered in electrospun nanofibers (Lee, K. H., et al. 2003a; Tomczak et al. 2005) and occurs primarily as a result of the instability of the jet under different process conditions (Entov and Shmarayan 1997). Qualitatively, beads may be expected at times during electrospinning whenever the surface tension forces tend to overcome the forces (such as charge repulsion and viscoelastic forces) that favor the elongation of a continuous jet (Fong et al. 1999). This occurs intermittently, as fiber formation still remains the dominant process and consequently leads to the typical "beads on a string" morphology described for a variety of different polymer/solvent systems (Fong et al. 1999; Gupta et al. 2005; Lee, K. H., et al. 2003a; Wannatong et al. 2004). Figure 3.6 shows an example of beaded nanofibers of PAN electrospun from 10 wt% solutions in DMF.

This is consistent with the observation that electrospinning dilute solutions of polymers (Deitzel et al. 2001a; Dong et al. 2004; Zuo et al. 2005) or low molecular weight polymers, where chain entanglement is limited (Morota et al. 2004), tends to result in bead formation. With most polymer solutions electrospun at a fixed electric field, some beading is invariably obtained in the critical concentration regime where transition from electrospraying of droplets into electrospinning occurs. This is related to incipient chain entanglement gradually overcoming the surface tension forces at higher polymer concentrations. The slow transition from beads only (essentially electrospraying) to beaded nanofibers and eventually to bead-free continuous nanofibers with increasing polymer concentration is shown in Fig. 3.7 for electrospinning of PEO. In the bead formation regime, increasing the polymer

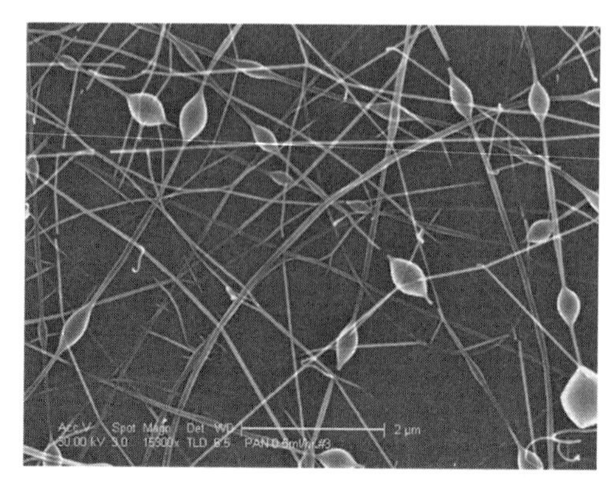

Figure 3.6 Polyacrylonitrile (PAN) electrospun from DMF (10 wt%) showing slightly elongated beads. Feed rate = 0.5 mL/h; applied voltage of 25 kV; gap distance of 8 in. (Courtesy of RTI International.)

concentration can result in increasing the aspect ratio of beads, leading to spindle-like shapes (Lee, K. H., et al. 2003a) and also larger bead sizes (Hsu and Shivkumar 2004b). Zuo et al. (2005) recently reported optical microscopic images of unstable jets at different distances from the capillary tip, showing instability and incipient bead formation in the electrospinning of poly(3-hydroxybutyrate-*co*-3-hydroxyvalerate) (PHBV) in $CHCl_3$.

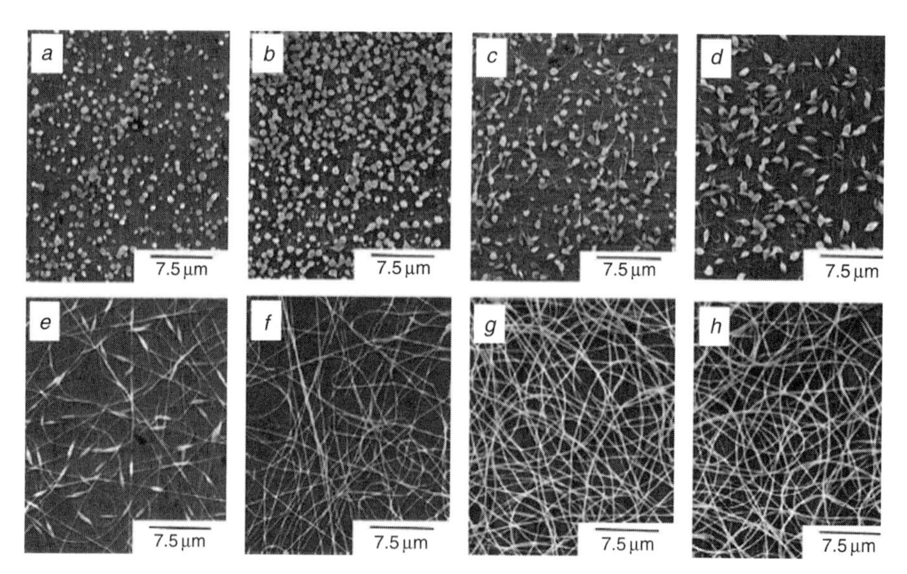

Figure 3.7 Electrospraying and electrospinning of aqueous PEO solutions (M_w = 500,000) under a constant applied voltage of 4.5 kV and a gap distance of 7 cm: (*a*) 5, (*b*) 10, (*c*) 20, (*d*) 30, (*e*) 40, (*f*) 50, (*g*) 60, (*h*) 70 g/L. Reprinted with permission from Morota et al. (2004). Copyright 2004. Elsevier.

Beads need not be always spherical or spindle like; collapsed shapes and "prune-like" beads (Shawon and Sung 2004) have been described. When a particularly volatile solvent such as THF is used to electrospin PS from 13 wt% solution, cup-shaped beads, likely resulting from the collapse of thin-walled spherical beads, were observed. Even more complex bead morphologies, such as porous cups with PMMA/acrylonitrile (8 wt%) (Liu and Kumar 2005) or "raisin-like" morphologies with polycarbonate/(THF/$CHCl_3$ 1 : 1) (14–15 wt%) (Krishnappa et al. 2003), have been reported. Several complex shapes of beads encountered in electrospinning are illustrated in Fig. 3.8.

Theoretical analyses of the mechanisms of bead formation have been attempted by several groups. The rapidly elongating jet undergoes several different modes of instability. Analysis by Hohman et al. (2001a, 2001b) and Y. M. Shin et al. (2001b), for instance, predicted three modes of instability that can develop in the extending jet, of which two are axisymmetric (Fig. 3.9). One of these is Rayleigh instability, which is primarily governed by surface tension, and the other is a conducting instability governed mainly by the electrical conductivity of the fluid. In axisymmetric instability the axis of the fiber remains undisturbed but its radius is modulated, yielding wave-like deformations of the fiber that are the precursors of beads. (Fong et al. 1999; Fong and Reneker 1999; Zuo et al. 2005). Therefore, processing conditions that favor axisymmetric instabilities also favor bead formation whereas increased whipping instability discourages bead formation. Higher surface charge density that favor whipping instability over axisymmetric modes therefore generally suppresses bead formation (Fong et al. 1999).

Reneker and colleagues, however, suggested factors such as surface tension, viscosity, and charge density of the jet to be the primary factors

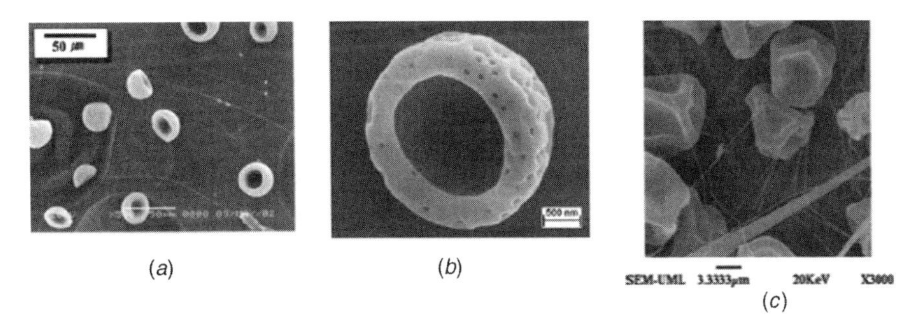

Figure 3.8 (*a*) Polystyrene electrospun from 13 wt% solution in THF, showing cup-shaped beads. Reproduced with permission from K. H. Lee et al. (2003a). Copyright 2003. Elsevier. (*b*) Porous cup-shaped beads of PMMA electrospun from nitromethane. Reproduced with permission from Liu and Kumar (2005). Copyright 2005. Elsevier. (*c*) Prune-shaped beads of PC electrospun from THF. Reproduced with permission from Shawon and Sung (2004). Copyright 2004. Elsevier.

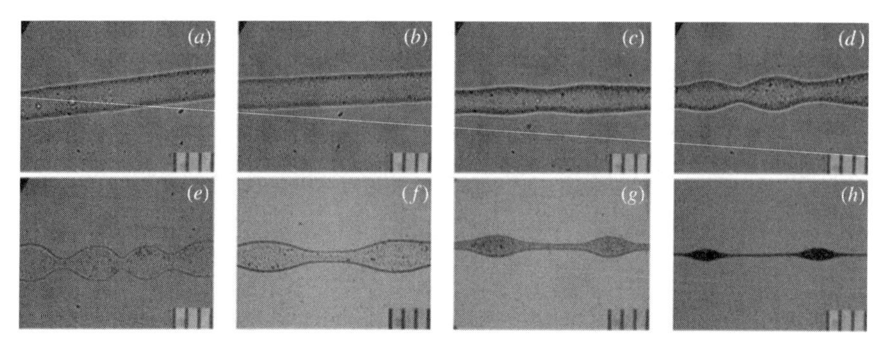

Figure 3.9 An electrospinning jet of 4% solution of PHBV in chloroform imaged at different distances from the tip showing the development of axisymmetric instabilities (applied voltage 20 kV and feed rate 4 mL/h). Images (*a*) through (*h*) correspond to distances of 1, 3, 5, 7, 9, 12, 15, and 30 cm from the capillary tip. Reprinted with permission from Zuo et al. (2005). Copyright 2005. John Wiley & Sons.

that govern bead formation, also pointing out the possible role of solvents in the process (Fong et al. 1999). For instance, an incompletely dry section of jet may lose its charge on contact with the collector and contract to form a bead. This would depend on the volatility of the solvent — the influence of solvent volatility on bead formation is well known (Wannatong et al. 2004). Highly conductive solvents or those with high dielectric constants encourage splaying of the jet (Hsu and Shivkumar 2004b). As already alluded to in Chapter 1, the smaller branch jets do not spin well and invariably disintegrate into droplets or beads (Hsu and Shivkumar 2004a).

Several approaches are available to control bead formation in electrospinning of polymer nanofibers:

1. For continuous defect-free nanofibers to be obtained, ideally the rate at which the polymer solution is pumped into the capillary tip needs to be approximately equal to the rate at which nanofibers are spun out of it. Other factors being constant, increasing the solution feed rate in such a system will therefore lead to increasing amounts of bead formation (Lin et al. 2004; Zuo et al. 2005). This is qualitatively understood in terms of the available electric field being inadequate to generate a high enough surface charge to stretch the jet as throughput from the tip is increased, leading to volumes that are not stretched sufficiently (or beads) (Deitzel et al. 2001a). Decreasing the feed rate will control bead formation under such operating conditions.
2. Increasing the applied voltage increases the surface charge of the jet and helps to reduce the frequency of occurrence of beads. However, this will generally suppress bead formation only at adequate feed rates

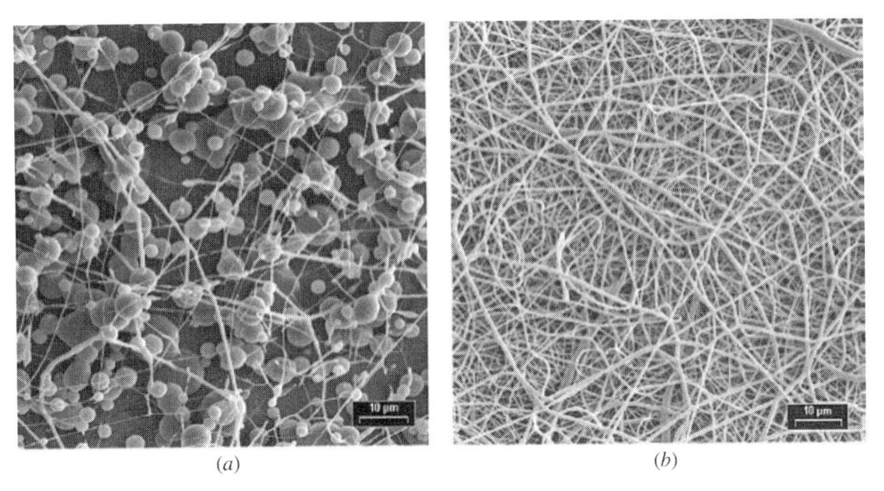

(*a*) (*b*)

Figure 3.10 Poly(ε-caprolactone) (PCL) nanofibers electrospun from 5 wt% solution in $CHCl_3$ at a gap distance of 7.5 cm. The applied voltages in the two panels are (*a*) 20 kV and (*b*) 25 kV. Reprinted with permission from Hsu and Shivkumar (2004a). Copyright 2004. Springer.

(Krishnappa et al. 2003; Zong, X. H., et al. 2002). As seen in Fig. 3.10, for PCL electrospun from chloroform, a relatively modest change in the applied voltage can dramatically reduce the incidence of beads in the nanofiber mat. If the applied voltage is too high, however, the increased mass flow can lead to increased bead formation. Also, multiple jets may emanate from a single droplet or branching off may occur, both resulting in the charge density per each jet being reduced, again encouraging bead formation depending on the solvent used (Shukla et al. 2005). Electrospinning DMF solutions of PDLA, X. H. Zong et al. (2002) observed increased bead formation with increasing voltage while maintaining the same feed rate of 20 μL/min. The feed rate, however, was inadequate to sustain a stable Taylor's cone at the capillary tip when the voltage and therefore mass flow was increased. Fibers spun off the unstable, oscillating, asymmetrical Taylor's cone from a very small droplet increased bead formation, with the shape of beads changing with increasing voltage. Similar observations have been made by others (Deitzel et al. 2001a; Demir et al. 2002).

3. Increasing the surface charge on the jet using additives to increase the conductivity of the polymer solution can help avoid beads at a given applied voltage (Hsu and Shivkumar 2004a; Jun et al. 2005; Lin et al. 2004; Zuo et al. 2005). Increasing the applied voltage may be undesirable in some instances, as it may also change fiber dimensions. In the PS/(DMF : THF 1 : 1) system, bead formation was effectively

suppressed by as little as 10^{-6} mol/L of organic salts [such as dodecyl trimethyl ammonium bromide (DTAB) or tetrabutylammonium chloride (TBAC)] in the spinning solution (Lin et al. 2004). (DTAB is also a cationic surfactant and would have lowered the surface tension of the solution as well.) In electrospinning PEO, NaCl was similarly used successfully (Arayanarakul et al. 2006). However, increasing solution conductivity will also increase the throughput from the capillary tip (Lee, K. H., et al. 2003a). Conversely, partial charge neutralization can promote bead formation (Fong et al. 1999).

4. Using additives to decrease the surface tension of solutions while leaving other parameters unchanged should also reduce bead formation. Adding a surfactant such as Triton X-100 or sodium dodecyl sulfonate (SDS) suppressed beads in electrospun PVA (Yao et al. 2003) or PEO (Arayanarakul et al. 2006); the occasional beads that still occurred could be controlled by adding a small amount of acetic acid along with the surfactant (Yao et al. 2003). With the PEO/water system, however, adding hexanol (up to 0.5%) to the solution to lower its surface tension did not result in a corresponding change in jet morphology (Morota et al. 2004), illustrating the complex interplay of variables involved in electrospinning.

3.4 INTRODUCTION TO ELECTROSPINNING PRACTICE

From the time Zeleny first developed the technique, using a needle or an open nonconducting capillary tip[4] carrying an applied voltage placed at a fixed distance from a grounded collector has been the popular design for laboratory electrospinning equipment. Although generally used vertically, this "point-plate" design can also be operated horizontally (with the advantage that any dripping from the tip does not damage the fiber mat). Its popularity arises at least partly because the equipment can be conveniently assembled using off-the-shelf components such as a syringe or a glass pipette (drawn to a fine point), hypodermic syringe needles, laboratory pump, metal plate, and flexible plastic tubing. A great majority of the reported data on electrospinning is based on either the point-plate geometry or its modification by substituting a rotating cylindrical drum collector in place of the plate. Although it has worked well with a wide variety of polymers, the electric

[4]Although a great majority of electrospinning experiments use a conductive capillary tip or an electrode in contact with solution in a non-conducting tip, an adequate electric field for the purpose can also be created via induction. For instance, PAN solutions in DMF have been electrospun successfully using a ring electrode placed around the tip — although the ring electrode was connected to a power supply the capillary tip was not (Kalayci et al. 2005).

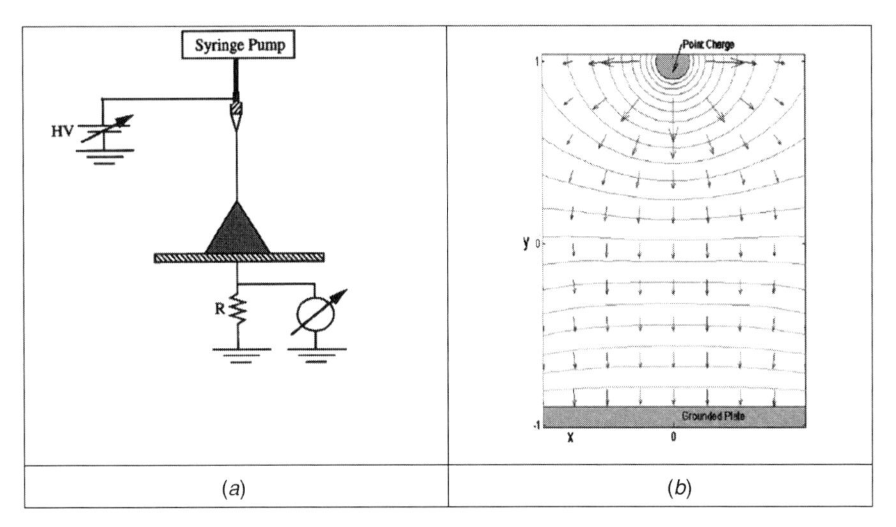

Figure 3.11 (*a*) Electrode arrangement for point-plate electrospinning. Modified from Y. M. Shin et al. (2001b). (*b*) Lines of forces calculated for the electric field. Reprinted with permission from Theron et al. (2001). Copyright 2001. Elsevier.

field in the point-plate arrangement is inherently nonuniform (see Fig. 3.11b). Small samples of nanofibers can be obtained using a single pendent drop of solution on the tip electrode or using a few drops of the solution in a glass pipette carrying an electrode. However, for quality nanofiber production a pulse-free pump that operates at a low feed rate is generally necessary. This allows the feed rate to be adequately matched to the applied voltage. A high-voltage DC power supply unit with a stable output is critical for good electrospinning (Matsuda et al. 2005; Zhang, Y. Z., et al. 2006c).

The electric field (lines of forces) resulting from the point-plate arrangement described above is illustrated in Fig. 3.11b. To improve the uniformity of the field a parallel plate design that includes a disc electrode affixed above the capillary tip was later introduced (Shin, Y. M., et al. 2001b) (Fig. 3.12). The latter arrangement results in uniform parallel electrical lines of forces and therefore simplifies theoretical studies undertaken to better understand the physics of electrospinning (Yarin et al. 2001b). Also, as the diameter of the source disk at the capillary tip is increased, the divergence of the whipping jet and therefore the area of nanofiber mat on the collector was found to decrease as well (Bunyan et al. 2006). With aqueous PEO solutions the reduced divergence was coupled with smaller average fiber diameters. Tilting the disk by small angles shifts the center point of deposition without affecting the diameter of the mat, allowing a wider spray area to be covered by the tip (Bunyan et al. 2006). However, the parallel plate electrode arrangement has not been extensively used in reported studies.

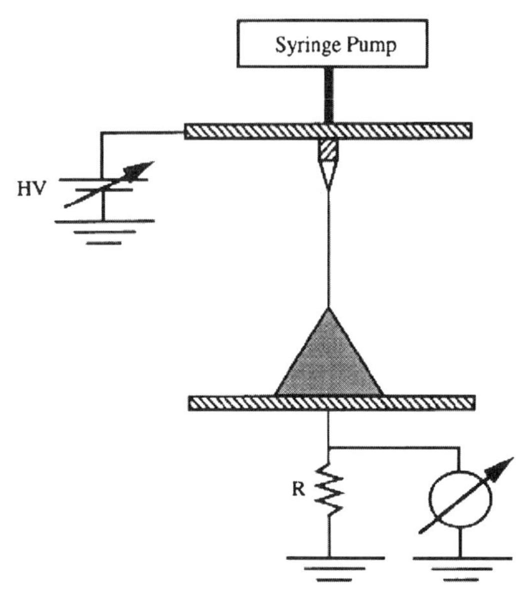

Figure 3.12 Basic electrode arrangement in electrospinning. Reprinted with permission from Y. M. Shin et al. (2001b). Copyright 2001. Elsevier. The shaded area represents the whipping instability region.

Other attempts at changing the electric field have been mainly efforts to control the instability region of the jet to alter the area of fiber mat depositing on the collector. Using secondary electrodes about the whipping jet for this purpose has been explored. Placing parallel planar or tubular electrodes (Kim, G.-H. and Kim 2006; Kim, G.-H. 2006) (at the same polarity as the jet) about the whipping region helps focus the jet. Using ring electrodes of the same polarity as the main electrode, encircling the spinning jet has also been attempted (Deitzel et al. 2001b, 2001c) (Fig. 3.13). This modification dampened the jet even more, but did not completely eliminate the whipping instability region, and nanofibers were deposited over a relatively smaller area on the collector.

Electrospinning from a single capillary tip is a slow low-throughput process. At a typical feed rate of 2 mL/h of a 10% polymer solution being pumped through a tip, for instance, only about 5 g/day of 500 nm nanofiber is generated (Burger et al. 2006). Using multiple needles or an array of tips is a somewhat obvious route to scaling up the process and producing multi-component fiber mats as well (Ding et al. 2004c). Theron and colleagues (Theron et al. 2005) modeled the operation of multiple jets placed adjacent to each other and also studied multiple tip systems experimentally. Each jet develops and undergoes instability as with a single tip, but, as might be expected, the jets are repelled by their nearest neighbors to an extent dictated by the inter-tip distance. With multiple tips, however, the electric field is

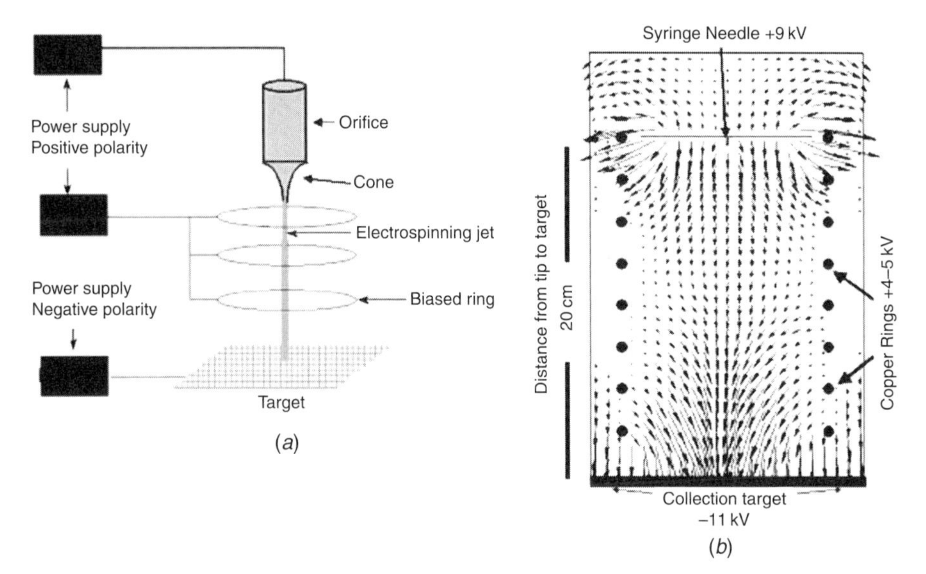

Figure 3.13 Using multiple ring electrodes to modify the electric field. (*a*) Experimental design. (*b*) Calculated lines of forces using a series of copper ring electrodes. Reproduced with permission from Deitzel et al. (2001c). Copyright 2001. Elsevier.

complex, as the jets themselves undergo coulombic interaction with each other. The electric field at a single ith electrode in the array E_i is given by (Fang et al. 2006)

$$E_i = E_i^{\mathrm{o}} + \sum_{j \neq i} E_{ij} + \sum_{j} E_y''(J_j), \tag{3.17}$$

where E^{o} is the unperturbed field strength due to the ith capillary tip electrode, E_{ij} is the field at location i due to electrode j, and $E_y''(J_j)$ is the interference field due to current J of the jet j.

Screening the electrodes (or shaping them) can partially control the interference due to adjacent electrodes. For instance, the use of ring-shaped secondary electrodes with appropriate voltage about the primary electrodes has been proposed as a means of controlling interference (Chu et al., U.S. patent # 6713011 (2004); Fang et al. 2006). The patent describes the translational platform for a multiple-tip arrangement allowing the assembly to be moved back and forth (as well as side to side) over a moving belt collector to obtain an even deposition of the nanofiber. With arrays of jets, however, clogging of a few of the capillary tips can easily affect the uniformity of the mat. Circular, elliptic, and other geometric arrangements of the multiple capillary tips in arrays have also been proposed (Tomaszewski and Szadkowski 2005). Stability analysis carried out for arrays with auxillary

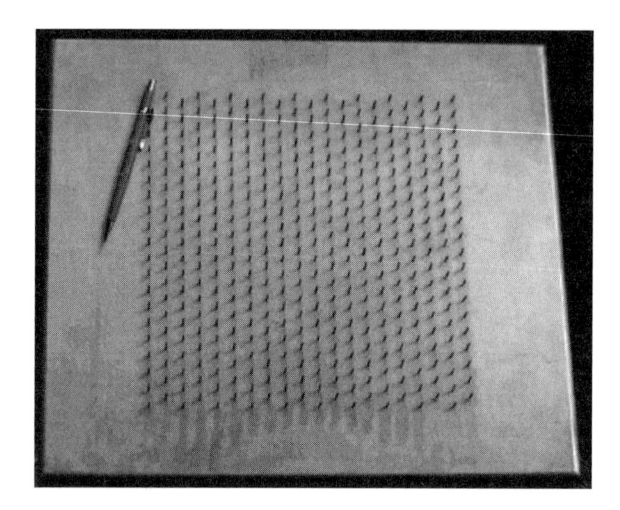

Figure 3.14 A 400-jet modular spin head available for scaling up electrospinning. Line speeds of 20–1000 fpm are claimed. (Courtesy of Nanostatics Inc., Columbus, OH.)

electrodes suggest that the approach can potentially result in high production rates (Kim, G. H., et al. 2006).

Success in scale-up of electrospinning is evidenced by the large volume of nanofiber products, such as air filters, already available in the market. The scale-up approaches used and the manufacturing rates achieved by the few manufacturers are proprietary. The idea of using arrays of tips, however, has been commercialized in the form of a modular 100–400 tip per square foot array (Fig. 3.14) that might be combined to design manufacturing lines for large-area mats.

A recently commercialized technology (Nanospider™), uses a thin film of polymer solution or melt on a cylindrical surface that is subjected to a high electric field to obtain tip-less electrospinning. The cylinder is partially immersed in the polymer solution and, as it rotates, a controlled amount of polymer solution is carried to the top part of the cylinder in the electric field where a series of Taylor cones are created.

4

FACTORS AFFECTING NANOFIBER QUALITY

Given the complexity of the process itself and the numerous variables that seem to affect it, developing comprehensive models of electrospinning is a challenge. The complex interplay between even the key interrelated variables that govern the process is not understood at a level that allows practical predictive models to be developed. Most of the research literature devoted to the study of such variables is focused on the major well-known process parameters. Only a few works touch upon minor variables such as the polarity of the capillary used in spinning (Supaphol et al. 2005a, 2005b), the nature of the gaseous species in the environment (Casper et al. 2004; Larsen et al. 2004b), temperature (Mit-uppatham et al. 2004a), the dielectric properties of the collector (Mitchell and Sanders 2006), or the pH of the aqueous solutions of polymers being electrospun (Son et al. 2005). These can nevertheless be very important in determining fiber quality and yield. The rapidly growing database on electrospinning practice, however, allows some pertinent general observations to be made based on experimental data.

Two classes of variables that affect electrospinning can be identified from the literature: the materials variables relating to polymer and solvent characteristics and the process variables pertaining to either the choice of equipment (e.g., the collector plate material or the spinning environment) or the operating parameters. For convenience of description these are categorized as shown in Scheme 4.1 as the materials variables (A, B, and C) and process variables (D, E, F, and G). However, because these are interrelated, a small

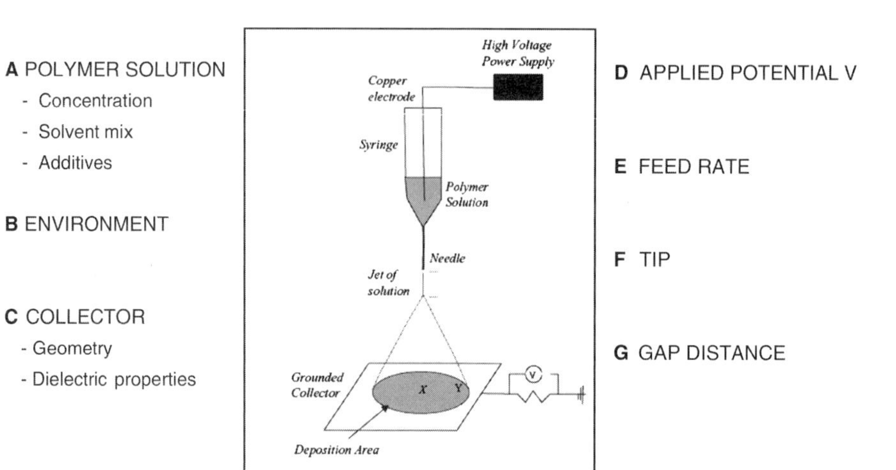

Scheme 4.1 The basic materials and process variables involved in electrospinning of polymer nanofibers. Electrospinning schematic reproduced here with permission from Hsu and Shivkumar (2004b). Copyright 2004. Springer Science and Business Media.

change in any one of these in a smoothly operating electrospinning process can significantly impact the others, often dramatically affecting nanofiber quality or even halting the spinning process altogether.

Only a limited number of studies (Deitzel et al. 2001a; Tan, S.-H., et al. 2005) have attempted a direct comparison of the relative importance of different variables on fiber quality. However, these too have not comprehensively studied all the variables and their combinations. S.-H. Tan et al. (2005), studying a copolymer of L-lactide-*co*-caprolactone (30% caprolactone) [P(L-CL)] in several solvents, concluded polymer concentration, average molecular weight, and the electrical conductivity of the solvent to be the dominant parameters that control nanofiber morphology. Mitchell and colleagues studied melt electrospinning of a thermoplastic polyurethane using a dielectric material as an auxiliary collector that was not grounded (but placed over the grounded primary collector) within the gap to collect the nanofiber mat samples. Interestingly, they found the dielectric strength of the collector material[1] and its surface area to be the dominant variables that influenced fiber diameter and fiber spacing (or porosity) of the mat (Mitchell and Sanders 2006).

[1]Also, the experimental setup used here was different from the conventional in that the grounded electrode was placed in the polymer melt and the charged electrode was placed at the position below it where a collector plate is generally placed. An auxiliary collector plate (ungrounded) was placed in between the two electrodes.

The nature of the collector surface influenced the electric field in its vicinity under these conditions. Although these observations do not necessarily agree with general findings reported in the literature for studies carried out for the most part on polymer solutions, the data underscore the potentially large parameter space that needs to be considered in deriving meaningful generalizations. Scheme 4.1 also shows a diagram of electrospinning equipment and identifies the variables.

4.1 THE POLYMER SOLUTION

4.1.1 Concentration Effects

At concentrations allowing adequate chain entanglement, continuous uniform nanofibers can be electrospun from polymer solutions in a strong enough electric field (Deitzel et al. 2001a; Pornsopone et al. 2005; Subbiah et al. 2005). The concentration of polymer in solution often determines if it will electrospin at all and generally has a dominant effect on the fiber diameter, *d* (nm), as well as fiber morphology (Demir et al. 2002; Zong, X. H., et al. 2002). Higher concentrations generally yield nanofibers of larger average diameter but the quantitative relationship between the solution concentration *c* (wt%) and *d* (nm) appears to be variable. Recent studies have attempted to obtain the scaling exponent for concentration dependence of *d* over a large enough range of average molecular weights of the polymer. Figure 4.1

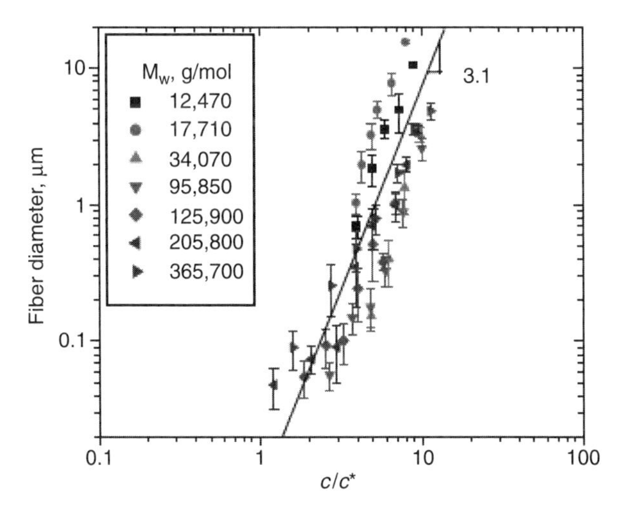

Figure 4.1 Dependence of average fiber diameter *d* (μm) on the reduced concentration of PMMA in solution. Reproduced with permission from Gupta et al. (2005). Copyright 2005. Elsevier.

illustrates the dependency obtained for poly(methyl methacrylate) (PMMA) samples in dimethylformamide (DMF) solutions, as reported by Gupta et al. (2005). Based on this plot (and a similar plot of fiber diameter vs zero shear-rate viscosity that also depends on $\overline{M}_w$), the following scaling relationships were proposed:

$$d \sim (c/c^*)^{3.1} \quad \text{and} \quad d \sim (\eta_o)^{0.72} \tag{4.1}$$

McKee et al. (2004a, 2004b) studied the electrospinning of poly(ethylene terephthalate-*co*-ethylene isophthalate) synthesized either with or without chain branching (branching was introduced using trifunctional co-monomers during the copolymerization) electrospun from $CHCl_3$/DMF (70/30) solutions. The exponent for concentration dependence for their system was ~3.0 and that for viscosity dependence was 0.80, over the range of average M_w values (11,700–106,000 g/mol). The same exponent of 3.0 was also reported for concentration dependence of d (nm) in electrospinning of segmented poly(urethane-urea) from DMF (Demir et al. 2002). Despite such agreement, it is reasonable to expect the scaling exponent to vary with the nature of solvent and even with other process variables such as the feed rate of the solution. Factors such as solvent quality (Supaphol et al. 2005a, 2005b), spinning conditions (Kang et al. 2002) and temperature (Mit-uppatham et al. 2004b) also understandably play a role in determining fiber diameter. However, in general, the relationship between c (wt%) and d (nm) also appears to apply over a range of values of the electric field (Sukigara et al. 2003). This is illustrated in Fig. 4.2 with data on *Bombyx mori* silk electrospun from formic acid solution (Sukigara et al. 2003) under several electric fields of different strength.

Most of the published data relate to specific polymer/solvent systems, and are often limited to a single average molecular weight of polymer. These generally agree that the fiber diameter d (nm) increases with polymer concentration (Huang et al. 2006; Jun et al. 2003; Mit-uppatham et al. 2004b; Supaphol et al. 2005b). For instance, in electrospinning aqueous solutions of poly(ethylene oxide) (PEO), a linear dependence of log d (nm) vs log c (wt%) with a gradient of about 0.5 has been reported (Deitzel et al. 2001a). Baumgarten (1971) also found the logarithm of nanofiber diameter to increase with the concentration of polyacrylonitrile (PAN) in dimethylformamide (DMF) solution.

Solution viscosity is generally identified as the dominant variable that determines fiber diameter (Jun et al. 2003). The minimum viscosity needed corresponds to some value of $c > c^*$ in solution and varies with the molecular weight of the polymer as well as the nature of the solvent used, as already discussed in Chapter 3. Although solution viscosity is primarily adjusted by changing polymer concentration, varying the solvent composition at a

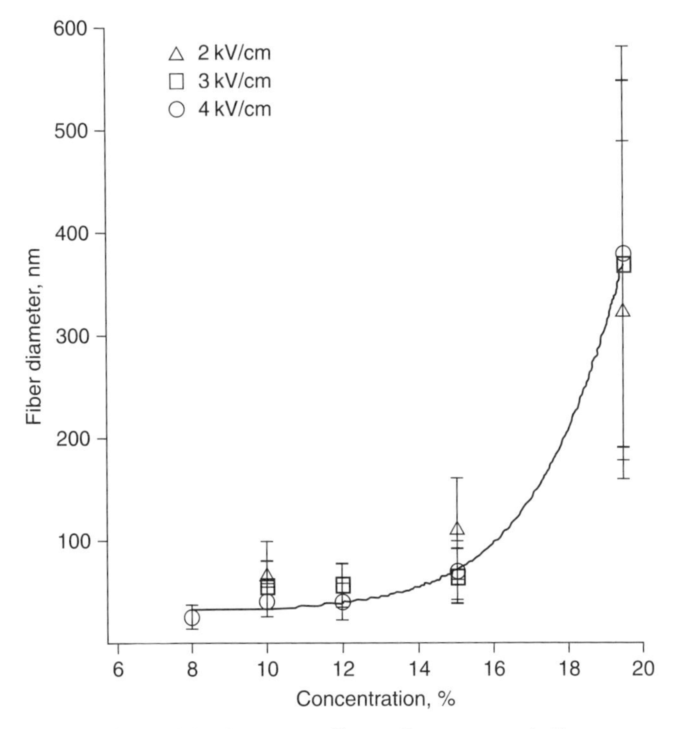

Figure 4.2 The relationship between fiber diameter and the concentration at different values of the electric field. Reproduced with permission from Sukigara et al. (2003). Copyright 2003. Elsevier.

constant concentration of polymer can also be used for the purpose. For example, the viscosity of a 13 wt% polystyrene (PS) solution can be changed from 42 cPs to 48 cPs by merely changing the composition of the binary solvent system (tetrahydrofuran (THF) + DMF) (Lee et al. 2003a).

The conjugated polymer poly(*p*-phenylene vinylene) (PPVy) in ethanol/DMF solutions has a higher viscosity than when dissolved in ethanol alone at the same concentration. Although both solutions are electrospinnable (Xin et al. 2006), only the higher concentrations tend to yield the bead-free nanofibers. Concentrations that are too high, however, can lead to practical difficulties (such as tip blockage) in electrospinning (Subbiah et al. 2005; Zong, X. H., et al. 2002). Adding a polyelectrolytes in small quantity, as demonstrated by the addition of prepolymer to a poly(vinyl pyridine) (PVP) polymer solution (Xin et al. 2006), dramatically increased the viscosity, without appreciably increasing the concentration of the solids in solution. Solutions with a high solids content may not only impede the fiber-extension process (in the whipping instability region of the jet), yielding relatively thicker nanofibers, but may also make it difficult to pump the polymer solution because of its high viscosity (Kameoka et al. 2003). Nanofibers spun from such solutions can have an uneven appearance and tend to be deposited over a relatively smaller area on the collector plate.

An innovative approach to reducing solution viscosity is by using low-frequency vibration of the solution during electrospinning (He et al. 2004; Wan, Y.-Q., et al. 2006). Vibration facilitates temporary disentanglement of polymer chains by effectively disrupting the van der Waal's interactions between them, reducing the solution viscosity. The drop in viscosity with frequency is exponential within a range of vibration frequencies; the solution viscosity (Pa/s) of PMMA was shown to reduce by an order of magnitude when a vibration frequency of $\omega = 300$ (rad/s) was imposed on the solution (He et al. 2004). Vibrating the capillary tip ($\sim$400 kHz) during electrospinning of poly(butylene succinate) (PBS)/$CHCl_3$ solutions led to the generation of thinner nanofibers compared to when vibration was not used (Wan, Y.-Q., et al. 2007). At higher concentrations the solutions were electrospinnable only with vibration assistance. Another advantage of this technique might be its use in electrospinning coagulated materials and suspensions that cannot be otherwise electrospun.

4.1.2 Solvent System

The choice of solvent[2] (Table 4.1) primarily determines

- conformation of the dissolved polymer chains
- ease of charging the spinning jet
- cohesion of the solution due to surface tension forces
- rate of solidification of the jet on evaporation of the solvent.

Unlike with droplets of low-molecular-weight liquids or monomers that subdivide into smaller droplets under a strong electric field, polymer solutions undergo a degree of elongational flow and orientation in an electric field. It is the entanglement of the partially oriented, expanded conformations of polymer chains that makes their electrospinning possible in the first place. Solvents that yield open conformations of polymer chains and

[2]Electrospinning can also be carried out in a supercritical CO_2 medium. A pair of electrodes with a gap of 3 mm was used with poly(dimethyl siloxane) (PDMS) applied as a solid on the electrode to which a potential was applied. When the assembly placed in a pressure vessel was charged with CO_2, electrospinning was obtained under supercritical conditions (Levit and Tepper 2004).

TABLE 4.1 Data for solvents commonly used in electrospinning of polymers

	$\rho\,(\mathrm{g/cm^3})$	T_b (°C)	Dielectric Constant	γ (mN/m)[b]	η (mPa.s)
Acetone	0.790	56	20.7	23.46	0.324
$CHCl_3$	1.483	62	4.81	26.67	0.568
THF[a]	0.889	66	7.6	23.97	0.468
Cyclohexane	0.779	81	2.02	24.65	0.979
Water	0.998	100	78.5	71.99	1
Toluene	0.867	111	2.38	27.93	0.59
DMF[a]	0.944	153	36.7	—	0.92
Formic acid	1.213	101	58.5	37.7	1.8
Ethanol	0.785	78	24.6	22.0	1.1
Acetic acid	1.050	118	6.19	26.9	1.1

[a]THF, tetrahydrofuran; DMF, dimethylformamide.
[b]Surface tension. Dielectric constants measured at 25°C.

those solutions with high solids contents are therefore better suited for electrospinning.

The dramatic influence of solvents on electrospinnability is well illustrated by a study that investigated 18 different solvents for PS ($\overline{M}_n = 119{,}000$ g/mol) (see Fig. 4.3). All but one of these dissolved the PS, but only four (DMF, MEK, THF, DCE) yielded electrospinnable solutions (at 10–30% w/v). Both the dipole moment and conductivity of the solutions were identified as key properties that determined electrospinnability (Jarusuwannapoom et al. 2005). Son et al. (2004d) studied PEO dissolved in five solvents at different concentrations selected to obtain $[\eta]c \approx 10$ (i.e., entangled semi-dilute regime). The average nanofiber diameters d (nm) obtained on electrospinning these varied widely with the solvent used and thinner fibers were obtained with solvents with higher dielectric constant. A qualitative indication of the average conformation of polymer in solution is afforded by the value of Martin's constant K_m (see Chapter 2). PEO solutions in

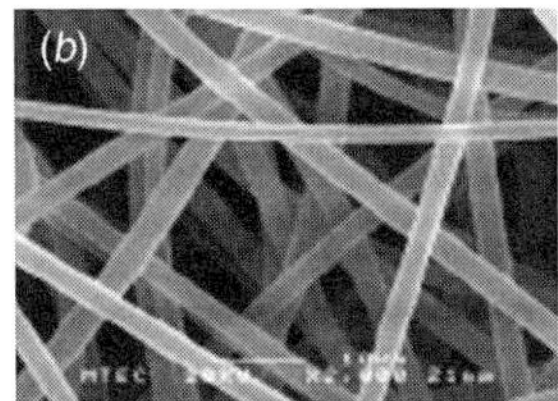

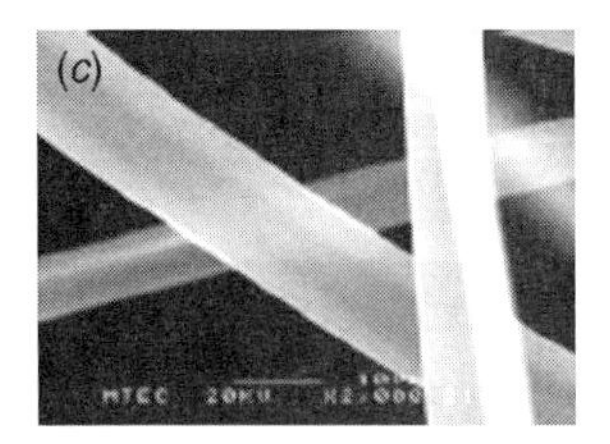

Figure 4.3 Scanning electron micrograph (SEM) images (same magnification) of PS nanofibers electrospun from 30% w/v solutions of (*a*) 1,2-dichloroethane, (*b*) DMF, and (*c*) ethyl acetate solutions at an applied potential of 20 kV and a gap distance of 10 cm, illustrating the widely different fiber diameters. Reproduced with permission from Jarusuwannapoom et al. (2005). Copyright 2005. Elsevier.

chloroform, with a low value of $K_m = 6.51$ (therefore leading to an open conformation of the polymer chains in solution) yielded nanofibers at the lower x concentrations ($\sim$3%). Solutions in DMF with a high value of $K_m = 27.71$ yielded fibers at 7% concentration, and were difficult to electrospin. Shenoy and colleagues also found a strong solvent dependence on the electrospinnability of poly(vinyl alcohol) (PVA) and poly(vinyl chloride) (PVC) (Shenoy et al. 2005a). In this case the differences were attributed to differences in thermoreversible gelation obtained with different solvents.

Sensitivity of the yield and quality of nanofiber produced to the nature of solvent was reported by numerous researchers: polycarbonates (PC) (Krishnappa et al. 2003; Shawon and Sung 2004); nylon-6 (Mit-uppatham et al. 2004b); polystyrene (PS) (Wannatong et al. 2004); ethyl cellulose (Wu, X. H., et al. 2005); and poly(vinyl pyridine) (PVP) (Yang, Q. B., et al. 2004). These, in general, identify four key properties of solvents as being particularly important in electrospinning:

- conductivity
- surface tension
- dielectric properties
- volatility.

Although efforts are being made to model solvent effects in electrospinning (Lu, C., et al. 2006), selecting an ideal solvent system is a complex task and, except for broad guidelines, the selection is mostly based on trial and error. Not only does each of these solvent characteristics have a direct critical influence on electrospinnability and nanofiber quality, but they cannot generally be independently varied to optimize a solution. Varying any one of these solvent properties with an additive, keeping everything else constant, to isolate its effect experimentally is usually difficult, as most additives tend to change several properties simultaneously. For instance, adding an alcohol to change the surface tension of a PEO/water system changes the viscosity and conductivity as well (Morota et al. 2004).

4.1.3 Conductivity

The electrospinning process fundamentally requires the transfer of electric charge from the electrode to the spinning droplet at the terminus of the tip. A minimal electrical conductivity in the solution is therefore essential for electrospinning; solutions of zero conductivity cannot be electrospun. Solvents commonly used in electrospinning have conductivities that are much lower than that of even distilled water[3]; dichloromethane has a value of only 0.03 mS/m. On dissolving a polymer in the solvent, however, the solution conductivity generally increases due to the availability of conducting ionic species (mostly from impurities or additives) from the polymer. With increasing polymer concentration in solution, however, its electrical conductivity may decrease (Jun et al. 2003). Where the polymer itself has ionic functionalities as with polyelectrolytes, however, the solution conductivity will be much higher (relative to those of uncharged polymers) and markedly concentration dependent (McKee et al. 2006a).

Solution conductivity afforded by incidental ionic species in solution might be inadequate to electrospin smooth continuous fibers with some polymer/solvent systems. In these instances an additive might be used at very low concentrations to increase conductivity. These tend to be either inorganic salts such as NaCl (0.01 M) (Kim, S. J., et al. 2005; Lee, C. K., et al. 2005; Wannatong et al. 2004; You et al. 2006a) or ionic organic compounds such as pyridinium formate (PF) (Jun et al. 2003), palladium diacetate (Yu et al. 2004), trialkylbenzyl ammonium chloride in L-polylactide (PLLA) (Zeng, et al. 2003b) or in poly(3-hydroxybutyrate-*co*-3-hydroxyvalerate) (PHBV) solution (Choi, J. S., et al. 2004). Adding ionic species to the solution allows a relatively higher surface charge density to be maintained on the jet and consequently promotes improved fiber extension during the whipping stage. Ionic sizes of the conducting species influence both their mobility as well as their charge density (Zong, X. H., et al. 2002). Conductivity of polymer solutions can also be changed by using solvent mixtures. S. H. Tan and colleagues (2005), used either CH_2Cl_2/DMF or CH_2Cl_2/pyridine blends as the solvent to make up 10 wt% solutions of copolymer of L-lactide with caprolactone (P(LLA-CL)). Solutions of higher conductivity yielded nanofibers smaller in average diameter (d (nm)). Figure 4.4 illustrates the effect of an organic additive on d (nm) of poly(lactide-*co*-glycolide) (50 : 50) PLGA copolymer from $CHCl_3$ solution.

A 2 wt% solution of PLLA ($\overline{M}_w = 670{,}000$ g/mol) in CH_2Cl_2, for instance, yields beads and fibers on attempted electrospinning. But on adding 0.8 wt% (with respect to solvent content) of pyridinium formate (PF), the system yields nanofibers exclusively (Shenoy et al. 2005). The ionic additive cannot increase chain entanglement, but enhances the surface electric charge. A simple interpretation of the observation is afforded in terms of the two primary forces involved: Vq, the electrical energy working to disrupt the droplet, and γR^2, the surface free energy of the droplet (where γ is the surface tension of the solution). It is reasonable to argue

[3]The SI unit for conductivity is siemens per meter, where 1 siemen = 1 ampere/volt = $(\Omega)^{-1}$.

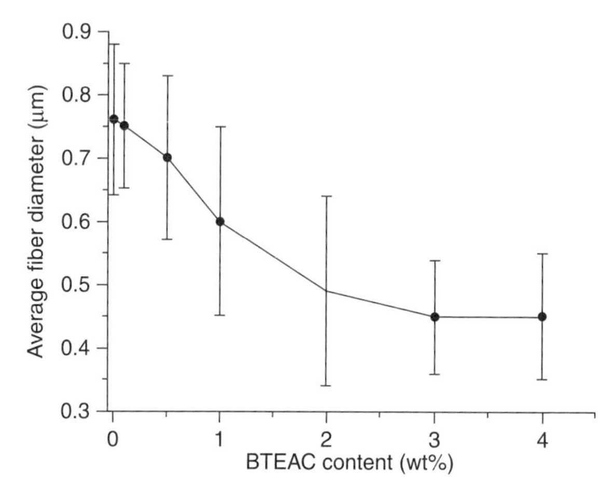

Figure 4.4 Change in average nanofiber diameter of PLGA copolymer (LA : GA, 50 : 50) electrospun from $CHCl_3$ (15 wt%) on adding benzyl triethyl ammonium chloride (BTEAC) to the spinning solution. Reprinted with permission from You et al. (2006a). Copyright 2006. John Wiley & Sons.

that electrospinning occurs only when $Vq/(\gamma R^2) > 1$ (Shenoy et al. 2005). Ionic additives in the spinning solution that at low concentrations can significantly increase Vq without increasing γ will therefore increase the magnitude of this dimensionless ratio, facilitating easier spinning at lower applied voltages. The reverse effect is found when an appropriate radioactive source is brought near an electrospinning setup. Radiation from the source neutralizes the charge on the droplet or jet, reducing Vq (again without affecting surface tension), and halts the electrospinning process. Corona discharges similarly retard or arrest electrospinning.

There is little to be gained by increasing the value of Vq indiscriminately using high levels of the additive. At very high conductivities, maintaining a surface charge on the droplet in fact becomes increasingly difficult, and the characteristic cone geometry described by the jet is affected as well. In aqueous solutions of poly(ethylene oxide) (PEO) (40 g/L), addition of $CaCl_2$ changed the spinning jet from a simple cone-jet mode to a multijet mode (Jaworek and Krupa 1999), and solutions with conductivity exceeding 5 mS/cm could not be electrospun at all (Morota et al. 2004). Similarly, 5 wt% poly(acrylic acid) (PAA) could be electrospun from 0.01 M and 0.1 M aqueous solutions of NaCl, but not from 1 M NaCl (Kim, B., et al. 2005).

The effect due to conductivity is not always easy to isolate or quantify. Changing solvent composition will often change the surface tension and dielectric constant as well, and consequent changes in electrospinning behavior cannot be uniquely attributed to changes in conductivity. Adding

0.5–1.5% of triethyl benzyl ammonium chloride (TABC) to 5 wt% solutions of PLA in CH_2Cl_2 yielded thinner nanofibers on electrospinning (Bognitzki et al. 2000). However, the relative contribution of increased conductivity by four orders of magnitude compared to that due to the near doubling of the surface tension by the additive cannot be ascertained. In the practice of electrospinning, the goal is to obtain defect-free nanofiber mats and, regardless of the mechanism involved, the use of either ionic additives and solvent mixtures does often result in improved fiber morphology.

4.1.4 Surface Tension

Surface tension is the primary force opposing coulomb repulsion and its role in determining electrospinnability cannot be overstated. In the instability region of the jet that obtains fiber extension, electrostatic forces are countered primarily by surface tension forces. It is this balance between surface tension cohesive forces and the surface electrostatic repulsion (i.e., $Vq/\gamma R^2$) that determine the curvature in bending of the jet during whipping instability. Also, bead formation in electrospinning can be induced by changing the surface tension of the solution (Fong et al. 1999; Shawon and Sung 2004). It is often the surface tension and viscosity of the solution that determine the window within which a specific polymer/solvent combination can be electrospun (Deitzel et al. 2001a). The minimum spinning voltage V_c therefore increases (but not necessarily linearly) with the surface tension of the spinning solution. In the Taylor's equation the critical voltage to initiate jet formation, V_c, is proportional to $\gamma^{1/2}$, suggesting that electrospinning is possible at lower voltages when the γ of the solution is reduced. All other factors being equal, lower surface tension is therefore a desirable solvent characteristic (Fridrikh et al. 2003). However, the surface tension of polymer solutions change with concentration (Deitzel et al. 2001a) as well as with the chemical nature of the polymer (Lee et al. 2003b). It also likely changes in time, as the jet becomes progressively concentrated during its passage from the tip to the collector plate. Surface tension is temperature-dependent and is affected by the presence of an electric field, making it one of the more elusive factors to quantify in an electrospinning model.

Surface tension of a polymer solution can be controlled by judicious selection of the solvent. The very different surface tensions of poly(vinyl chloride) (PVC) solution at constant concentration in various DMF/THF mixtures are shown in Fig. 4.5*a*. Note, however, that on changing the solvent composition the viscosity also increases threefold at constant polymer concentration (Lee, K. H., et al. 2002). DMF is a polar solvent with a high dielectric constant (36.7 at 25°C) and high boiling point (153°C), whereas THF has a low dielectric constant (7.6 at 25°C) as well as a low boiling point (65°C).

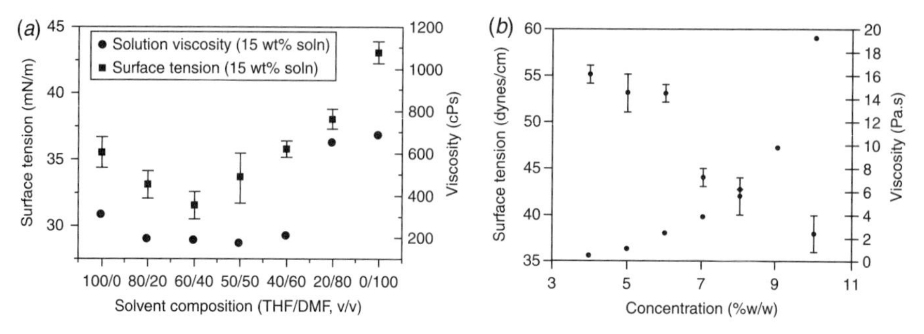

Figure 4.5 (*a*) Effect of solvent composition on the surface tension of PVC solutions of constant polymer concentration. Reproduced with permission from K. H. Lee et al. (2002). Copyright 2002. John Wiley & Sons Inc. (*b*) Effect of PEO concentration in water on the surface tension of the solution. Reproduced with permission from Deitzel et al. (2001). Copyright 2001. Elsevier.

Although a good solvent for PVC, because of its high volatility, THF is not a convenient solvent to use in practice and often results in blockage of the capillary tip. The nanofiber quality obtained is also very different at the various solvent compositions: THF alone yielded a very broad distribution of fiber diameters and DMF resulted in a narrow distribution with an average diameter of about 200 nm. Mixtures in the range of 80/20 to 20/80 (DMF/THF v/v) yielded fibers intermediate in average diameter. In general, increasing polymer concentration will decrease the surface tension of solutions as illustrated by the PEO/water system (see Fig. 4.5*b*). The influence of surface tension of solvents in electrospinning is well discussed in the literature (Fong et al. 1999).

As with electrical conductivity, low concentrations of additives may be used to alter the surface tension of polymer solutions. Surfactants, used at very low concentrations, dramatically decrease the surface tension of solutions (Jung et al. 2005; Lin, T., et al. 2005a; Zeng et al. 2003b) and facilitate electrospinning. Numerous examples are available in the literature. Nanofibers of PS electrospun from DMF/THF mixtures (5–15% w/v) yielded bead-free nanofiber mats only when 0.03–30 mmol/L of the cationic surfactant dodecyl trimethyl ammonium bromide (DTAB) was added to the polymer solution (Lin, T., et al. 2004). Nonionic surfactant Triton X-100 was used in electrospinning fully hydrolyzed PVA from water solutions to yield bead-free nanofibers (Yao et al. 2003). Triethyl benzyl ammonium chloride (TEBAC) and aliphatic PPO–PEO ether (AEO10) were used to improve the fiber quality of PLLA nanofibers electrospun from $CHCl_3$/acetonesolutions (Zeng et al. 2003b). Figure 4.6 compares SEM images of PLLA nanofiber mats spun with and without the added surfactant, illustrating the effect of this key variable on the distribution of fiber diameters (Zeng et al. 2003b).

Figure 4.6 The effect of adding 5% (on the weight of polymer) surfactant on the fiber morphology in electrospinning of PLLA from a $CHCl_3$/acetone (2 : 1) solution: (*a*) no surfactant; (*b*) nonionic surfactant. Reproduced with permission from Zeng et al. (2003b). Copyright 2003. Elsevier.

As a nonionic surfactant was used in this study, any change in conductivity did not significantly contribute to the observed changes in morphology.

The surface tension of polymer solutions can be conveniently measured[4] using the simple Du Nouy ring technique (Geng et al. 2005), where a wire loop of known circumference is dipped in the solution and the maximum force needed to slowly raise it out of the solution is measured. The Wilhelmy plate method (Seoul et al. 2003) essentially substitutes a platinum plate for the ring. With small samples of solution available for the measurement, however, a platinum rod might be substituted for the plate. In practice it is important to minimize the evaporation of solvent during the measurement.

An elegant technique that works particularly well with volatile solutions is the pendant drop method, especially relevant as the electrospinning process starts with a pendant drop of the polymer solution. The droplet can be readily imaged (in the absence of the electric field). Then

$$\gamma = (\Delta\rho)gR^2/\beta, \tag{4.2}$$

where γ is the surface tension, $\Delta\rho$ is the difference in the densities of the fluids at the interface, R is the radius of curvature of the drop at its apex, and β is a shape factor calculated from the imaged shape of the drop. Recent equipment automatically calculate iterative solutions for the Young–Laplace equation to yield values of β. These typically use a digital CCD camera to image the droplet and the error in γ is typically about ± 0.01 mN/m.

4.1.5 Dielectric Constant ε

The dielectric constant ε of a material is essentially a measure of how effectively it concentrates the electrostatic lines of flux when placed in an electric

[4]The SI unit is mN/m, which has the same magnitude as the conventional unit of dynes/cm.

field.[5] In a practical sense it is a measure of how much electrical charge the solvent is capable of holding. Solvents with different values of ε used in electrospinning will interact very differently with the electrostatic field and it is therefore an important material parameter in electrospinning (Wannatong et al. 2004). The Rayleigh equation includes ε as one of the key parameters. With solutions of high dielectric constant, the surface charge density on the jet tends to be more evenly dispersed. This translates into high nanofiber quality and productivity during electrospinning. Wannatong et al. (2004) found the productivity (quantified as the number of webs per unit area per min) in electrospinning of PS to increase with ε of the solvent used.

The effect of solvent dielectric constant on fiber morphology is illustrated by the comparison of the quality of nanofibers of poly(lactide-*co*-glycolide) (PLGA) (LA : GA 50 : 50) electrospun from 15 wt% solutions of either HFP ($\varepsilon \approx 16.7$) or chloroform ($\varepsilon \approx 4.81$) at 25°C. The average fiber diameter *d* (nm) obtained with the less polar chloroform was 760 nm, while that for HFP was 310 nm (Min et al. 2004d). HFP, with the much higher value of ε, yields nanofibers that have the smaller average *d* (nm) and very different morphologies were obtained with the two solvents. Solvents with the higher ε also resulted in smaller average values of *d* (nm) also in electrospinning of PEO carried out in multiple solvents (Son et al. 2004d). Figure 4.7 shows the reported relationship between the value of ε and the average fiber diameter *d* (nm) for the system. However, this must be regarded as a qualitative observation as the concentrations were adjusted to $[\eta]c \approx 10$ in different solvents (i.e. the concentrations were not the same).

The extent to which the choice of solvent affects the nanofiber characteristics is well illustrated in the electrospinning of poly(ε-caprolactone) (PCL) from $CHCl_3$/DMF mixed solvent. As the volume fraction of DMF in the mix is increased from 0 to 10 wt%, the average *d* (nm) (at the same polymer concentration) was shown to decrease from ~450 nm to 150 nm (Hsu and Shivkumar 2004b). This was likely a result of the increased dielectric constant of the solvent due to addition of DMF [$\varepsilon(DMF) \approx 36.7$ and $\varepsilon(CHCl_3) \approx 4.8$]. Furthermore, the distribution of *d* (nm) was also affected by the higher volume fractions of DMF in the mixture. Bimodal distributions of *d* (nm) typically obtained when electrospinning from $CHCl_3$ solutions changed to narrow unimodal distributions with the addition of DMF into the solvent. This was attributed to multiple splaying of the jet obtained with solvent blends containing DMF. Similar data on the effect of solvent composition on

[5]The force of attraction between a pair of point charges of opposing polarity generally varies with the square of the distance *r*. But the value depends on the nature of the intervening medium (air or solvent) in that the force is instead proportional to $[(\varepsilon^{1/2})r]^2$, where ε is the dielectric constant (or the relative permittivity) of the material.

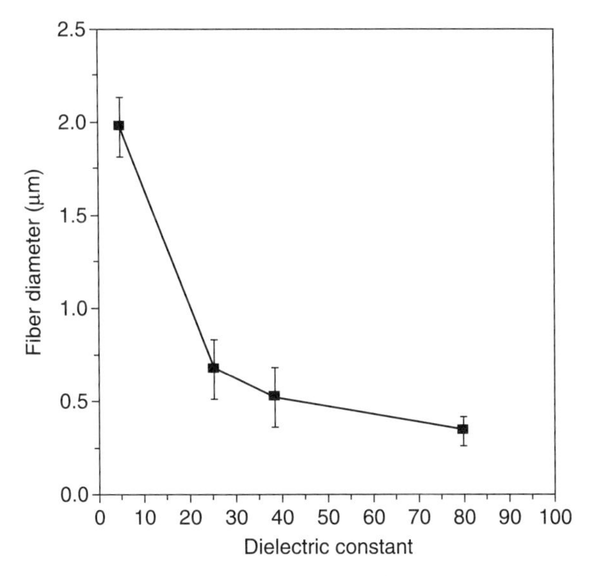

Figure 4.7 Relationship between the average fiber diameter and the dielectric constant of the solvent for electrospun PEO nanofibers. All solutions were at concentrations corresponding to $[\eta]c \approx 10$. Reproduced with permission from Son et al. (2004d). Copyright 2004. Elsevier.

fiber morphology have been reported for PCL electrospun from CH_2Cl_2/DMF solutions (Lee, K. H., et al. 2003b), for PS electrospun from THF/DMF (Lee, K. H., et al. 2003a), and PVC electrospun from THF/DMF (Lee, K. H., et al. 2002).

Dong et al. (2004) found a qualitative correlation between ε of the solvent and the quality of electrospun nanofibers for PMMA (at a concentration of 100 mg/mL) (Table 4.2), illustrating the effectiveness of high-ε solvents (Dong et al. 2004). Similar qualitative observations have been made for other polymers such as PCL (Lee, K. H., et al. 2003b), PMMA

TABLE 4.2 The effect of the dielectric constant ε of the solvent on electrospinning of PMMA at a concentration of 100 mg/mL[a]

Solvent	ε	Average *d* (nm)	Morphology
Toluene	2.38	—	No jet
$CHCl_3$	4.81	—	Beads with tiny fibers
$CHCl_3$/DMF (1 : 1)	—	620	Fibers with beads
THF	7.52	2490	Fibers with beads
THF/DMF (1 : 1)	—	770	Fibers
DMF	38.2	530	Fine fibers
Formic acid	51.1	280	Fine fibers

[a]Applied voltage = 20 kV and gap distance = 25 cm. The tip was tilted at a 10° angle.
Source: Reproduced from Dong et al. 2004.

(Dong et al. 2004), PLGA (You et al. 2006a), and PVC (Lee, K. H., et al. 2002). However, in comparisons such as these it is important to appreciate that changing solvents affects not only ε, but also the conductivity, surface tension, and the chain conformation of the polymer in solution. With hydrogen-bonding methacrylate copolymers, varying ε by changing the solvent composition resulted in changes in apparent entanglement threshold affecting electrospinnability (McKee et al. 2004b). In this case too, however, changing ε was accompanied by a change in ionizability of the pendant —COOH groups in the polymer, and therefore the conductivity of solution. The observation cannot solely be attributed to the dielectric constant of the system.

4.1.6 Volatility

Invariably, it is the evaporation of solvent from the jet that yields a solid polymer nanofiber at the collector plate. Ideally, all traces of solvent must be removed by the time the nanofiber hits the collector. If not, the wet fibers may fuse together to form a melded or reticular mat as shown in Fig. 4.8*b* (Hsu and Shivkumar 2004a). Sometimes a flat ribbon-like nanofibers derived from the fluid–filled, incompletely dry nanofiber due to slow subsequent evaporation of solvent and collapse of the tube, are obtained (Koombhongse et al. 2001). Using volatile solvents avoids this difficulty. However, when using highly volatile solvents the solution may dry on the capillary or needle, causing blockage to flow (Megelski et al. 2002). (Even

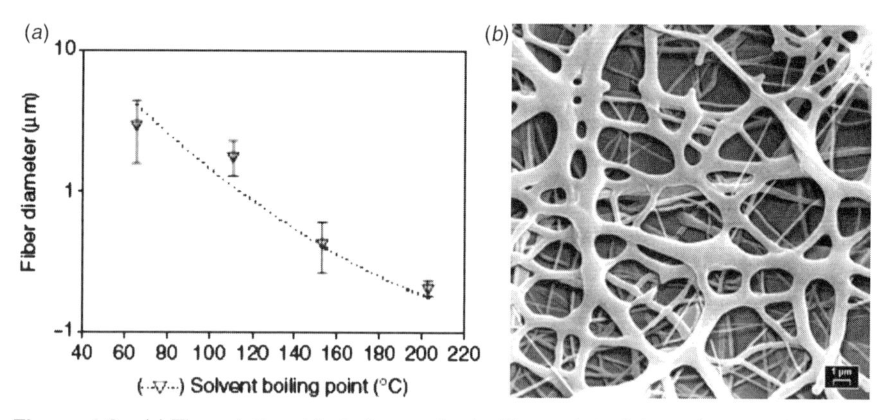

Figure 4.8 (*a*) The relationship between the boiling point of the solvent and average fiber diameter in electrospinning PS from solutions in solvents of different boiling points (polymer concentration 23–27 wt%) (Wannatong et al. 2004). Reprinted with permission from Wannatong et al. (2004). Copyright 2004. John Wiley & Sons Inc. (*b*) Reticular mesh from fused "wet" fibers reaching the collector. PCL, 4 wt% solutions in $CHCl_3$. Reprinted with permission from Hsu and Shivkumar (2004a). Copyright 2004. Springer Science and Business Media.

the earliest patents on the process — for example, U.S. patent # 745,276, Cooley 1903 — refer to this difficulty.) The wrinkled or "raisin-like" surface features observed in polycarbonate (PC) nanofibers electrospun from THF (or a THF : DMF mixed solvent) have also been attributed to rapid drying of the surface of the fiber (Krishnappa et al. 2003).

The smaller fiber diameters in electrospinning result from jet extension and it is critical that a minimal elongational viscosity be maintained in the jet during this stage. Essentially, the relaxation time for the polymer chains in solution needs to be matched to the rate of extensional deformation due to instability. Very rapid drying can therefore hinder the development of smaller diameters in the nanofibers. In electrospinning PS from several solvents, the average fiber diameter d (nm) was reported to decrease with increasing boiling point of the solvent (Wannatong et al. 2004), as shown in Fig. 4.8. The rate of drying is determined primarily by the vapor pressure of the solvent and by the degree of whipping instability that governs the rate at which the surface area to volume ratio[6] of the jet increases during spinning. At a given gap distance, drying rate also depends on temperature in the spinning environment; this may afford some independent control over fiber diameter.

The volatility of solvent also influences the kinetics of phase separation and therefore fiber morphology obtained in electrospinning. In electrospinning blends of polycarbonate (PC) and polybutadiene (PB) from $CHCl_3$ solutions, a fine phase morphology was obtained compared to the coarser morphology seen in fibers spun from THF (or THF/DMF) solvents. The higher volatility of $CHCl_3$ (vapor pressure 21.2 kPa) compared to THF (vapor pressure 17.9 kPa) does not allow enough time for coarsening of the phases (Wei, M., et al. 2006b). However, the dielectric properties as well as surface tension of these solvents are also very different and may have also influenced the phase morphology. Vapor pressure of the solvent also plays an important role in generating surface features such as porosity; Wendroff and colleagues (Bognitzki et al. 2000), electrospinning PLLA solutions, found CH_2Cl_2 (vapor pressure 46.6 kPa) to yield surface porosity, but $CHCl_3$ (vapor pressure 21.2 kPa) did not (see Chapter 9). Using solvent mixtures where the fraction of the higher-boiling solvent is varied to obtain fiber mats of different porosity and morphology has been reported (Kidoaki et al. 2006; Larsen et al. 2004b).

4.2 ENVIRONMENT

The majority of reported data pertain to electrospinning carried out in air, but it is advantageous to control the gaseous environment in the spinning

[6] Surface area to volume ratio $(A/V) = (2\pi rh)/(\pi r^2 h) = 2/r$. It is therefore inversely proportional to fiber diameter.

chamber. The rate of drying of the unstable jet controls the final diameter of the nanofiber. Larsen et al. (2004b), used a gas-jacketed capillary tip to surrounding the jet with nitrogen saturated with the spinning solvent to retard the rate of drying. Electrospinning 10 w/w% PLLA in CH_2Cl_2 in an electric field of 0.8 kV/cm, they were able to control the electrospinning process itself as well as the morphology of the fiber by this technique. At the higher applied voltages, stable Taylor's cones were obtained only when a solvent-saturated coaxial gas stream was used to control the drying rate of the jet. Also, the rate of electrospinning could be controlled by merely changing the flow rate of the saturated gas about the capillary tip. Similarly, the rate of drying can also be accelerated using the same approach (Yao et al. 2005). Using an external heat source, such as an industrial heat gun or even a 500 W lamp, can also help dry the fiber rapidly (Subramanian et al. 2005). For instance, in electrospinning aqueous solutions of hyalauronic acid (HA), a particularly viscous polymer solution (concentration 0.01–2.0 w/v%), a jacket of heated air (25–57°C) was used to decrease solution viscosity and increase the rate of drying of the fiber (Um et al. 2004).

A more important reason to control the environment is to minimize potential leakage of the surface charge on the jet into the surrounding air. The type of gas employed may either encourage or discourage the electrospinning process; for instance, with a capillary tip that is positively charged, using an electron-rich gaseous environment such as Xe will impede or arrest the process. A recent U.S. patent[7] has proposed using environments of highly electronegative gases (such as CO_2 or freons), delivered coaxially to the jet to discourage the loss of charges from its surface. This conservation of surface charge improves nanofiber quality and, under some operating conditions, may even determine electrospinnability. An air or gas stream delivered at a high pressure coaxially around the tip might be used to provide an additional drag force contributing to jet extension. When the drag force is dominant over the electrostatic force in driving jet extension, the process will be an assisted electrostatic spinning process and is referred to as electroblowing (Um et al. 2004; Wang, X., et al. 2005). However, to practically contribute to fiber extension, the velocity of the air or gas stream needs to match the jet velocity (Burger et al. 2006), which is difficult to achieve, even in laboratory experiments.

[7]David S. Ensor and Anthony L. Andrady. (Filed August 4, 2004). "Electrospinning in a controlled gaseous environment." U.S. Patent 7,297,305 (issued November 20, 2007).

4.3 COLLECTOR

4.3.1 Collector Geometry

The simplest and the most used collector reported in laboratory electrospinning is a stationary metal plate or a foil placed at a fixed distance from the tip. The conical spray pattern impinging on it results in a symmetric circular patch of nanofiber on the surface of the metal. As the collector is grounded, the residual charges on the as-spun fibers are rapidly removed, allowing the fibers to consolidate into a mat of high areal density. A moving collector surface allows some control in the areal density. Using a grounded moving belt or a rotating metal drum as the collector in electrospinning dates back to the early patents on the technology[8] and has been used by numerous researchers. Different collector geometries reported in the electrospinning literature were recently comprehensively reviewed by Teo and Ramakrishna (2006). Figure 4.9 illustrates the common collector geometries described in the recent literature.

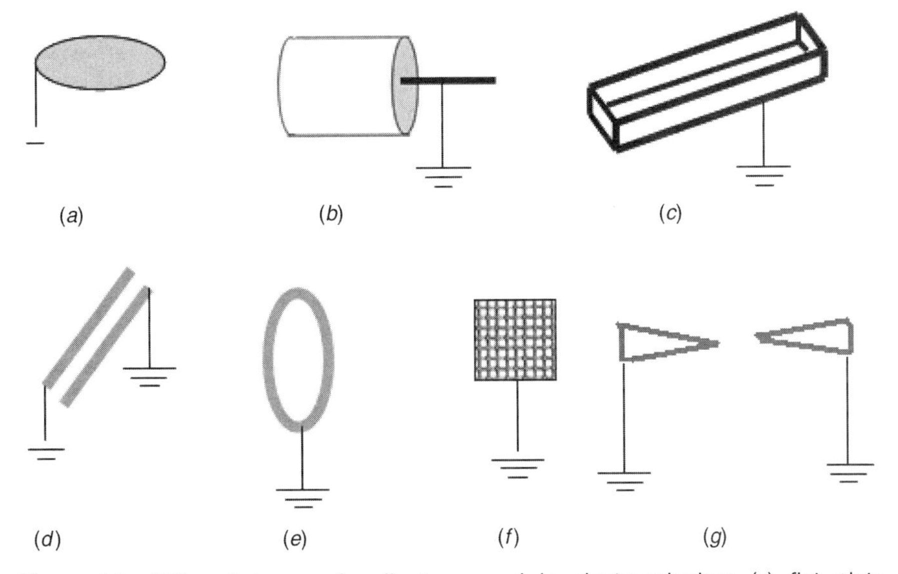

Figure 4.9 Different types of collectors used in electrospinning: (*a*) flat plate collector (Kidoaki et al. 2005); (*b*) rotating drum (Wannatong et al. 2004), mandrel (Mo and Weber 2004), rotating disc (Zussman et al. 2003); (*c*) rectangular, triangular, or wire cylinder frame (Katta et al. 2004); (*d*) electrode pair arrangements (Li, D., et al. 2003b); (*e*) single or multiple ring electrodes (Dalton et al. 2005); (*f*) mesh electrode (U.S. patent 6,110,590, Aug. 2000); (*g*) pair of cone-shaped collectors (Bunyan et al. 2006).

[8] J. E. Owens et al. (January 1970) U.S. Patent 3,490,115.

Using a rotating drum surface (or a moving belt) not only allows an even deposition of fibers, leading to a uniform nanofiber mat of controlled porosity, but it also allows for some degree of velocity-dependent alignment of nanofibers (Kim, K. W., et al. 2004). Noticeable fiber alignment, however, is not obtained at drum velocities below a threshold value under a given set of electrospinning parameters (Matsuda et al. 2005). Rotating mandrels are of interest because of these allow tubular constructs with potential use as vascular grafts to be spun (Boland et al. 2004a; Buttafoco et al. 2006; Inoguchi et al. 2006; Mo and Weber 2004; Stitzel et al. 2006). Tubular fiber mats with a range of diameters were fabricated by the technique which also yielded good control over wall thickness. In electrospinning poly(L-lactide-*co*-ε-caprolactone) [P(LL-CL)], the wall thickness was found to be a linear function of the duration of spinning (Inoguchi et al. 2006). Multilayered wall structures where each layer of nanofibers has a different orientation have been fabricated using mandrel collectors (Vaz et al. 2005).

This level of hierarchical organization via layer-by-layer processing shows great promise in developing structurally rich, detailed vascular grafts with the desired mechanical integrity. Placing knife-edges or pins maintained at a high negative voltage, either below (Teo and Ramakrishna 2005) or within (Sundaray et al. 2006) the hollow rotating mandrel further improved the alignment of nanofibers in the tubular mat. Open frame collectors (rectangular, pyramidal, or cylindrical) have been used for years to obtain aligned nanofibers. Available techniques for aligning nanofibers will be dealt with in detail in Chapter 9. The use of a pair of grounded ring electrodes placed equidistant from the capillary tip was reported by Dalton et al. (2005) to yield aligned nanofibers between them in space. Although frames of different geometry including the ring collectors do yield highly aligned, free-standing nanofiber mats it is not always easy to collect these for characterization or post-processing without disturbing their alignment. Recent work has demonstrated the utility of secondary or guiding electrodes in producing uniform highly-aligned nanofibers. These may be simple copper wire electrodes wrapped around a rotating drum (Bhattarai et al. 2005), steel blades placed below the rotating mandrel collector (Teo and Ramakrishna 2005, 2006; Teo et al. 2005), or using ring-shaped electrodes about the capillary tip (Jaeger et al. 1998; Kim, G.-H. and Kim 2006).

A first attempt at organizing collected nanofibers into yarn using SEM stubs as collectors was reported by Dalton et al. (2005). A pair of grounded stubs placed 6 cm apart were used as collectors to obtain a quantity of oriented poly(ε-caprolactone) (PCL) fibers between them. One of the grounded stubs was rotated at 2500 rpm to obtain samples of nanofibrous yarn 5 cm in length with an average diameter of 4.7 μm.

4.3.2 Collector Material

Contact of charged nanofibers with the grounded collector surface results in the removal of electrical charge on the fibers. However, only the charge on the first layer of nanofibers directly in contact with the metal surface is removed particularly rapidly. The discharge in subsequent layers will be slower because polymer nanofibers are good electrical insulators. As the process is influenced by the dielectric properties of the collector, the choice of collector material is an important factor in determining the characteristics of the nanofiber mat, especially their packing density. Kessick et al. (2004) showed that in electrospinning carboxymethyl cellulose (CMC), the dielectric properties of the collector markedly influenced the mat volume density. G. H. Kim and Kim (2006) electrospinning PCL from an 80/20 CH_2Cl_2/DMF solvent mixture, found that by using a supplementary electrode carrying an AC voltage, it was possible to minimize the effect of this variable of collector surfaces. This was due to charge reduction on fibers on passing through the supplementary field (Kim, G.-H. and Kim 2006).

In some experimental designs a nonconducting or conducting ungrounded auxiliary collector surface is interposed between the tip and the grounded plate (Mitchell and Sanders 2006). A significant and rapid build-up of electric charge in the mat may result from nanofibers collected on such a material. With the accumulation of several layers of nanofibers, the bulk of the mat may experience interlayer electrostatic repulsion, affecting the packing density or solidity of the mat. Mitchell and Sanders (2006) recognized the dielectric properties of the auxiliary collector to be a key factor that determines nanofiber mat quality.

Particularly interesting is the collection of nanofibers in a grounded shallow container carrying a liquid, usually water (Khil et al. 2005; Smit et al. 2005). The nanofibers contact the liquid on their passage towards a submerged grounded plate. The morphology of nanofibers collected in liquid is very different from that collected on grounded or auxiliary collector surfaces. When PLLA and PLGA solutions were electrospun into deionized water, dilute NaCl solution, or methanol, the average d (nm) of nanofibers, as well as their surface characteristics, varied significantly from those collected on grounded surfaces (Kim, H. S., et al. 2005). A liquid collector device might also be used when electrospinning from solvents that are not readily volatile; in place of solvent evaporation the fiber is generated by precipitation in solution as with conventional wet spinning. In electrospinning Kevlar® [poly(p-phenylene terephthalamide)], the nanofibers were spun from sulfuric acid solutions into a grounded water bath to precipitate the polymer (Srinivasan and Reneker 1995).

4.4 APPLIED POTENTIAL

4.4.1 Applied Voltage *V*

Applied voltage *V* (kV) provides the surface charge on the electrospinning jet. Instability and stretching of the jet therefore increases with the applied voltage (Buchko et al. 1999; Fridrikh et al. 2003; Shin et al. 2001a, 2001b) leading generally to smaller fiber diameters, *d* (nm). The consequent faster rates of solvent evaporation can lead to complete drying of the nanofibers. The expected decrease in *d* (nm) as *V* is increased has been reported for electrospinning of numerous polymers including PAN/DMF (Jalili et al. 2005), PS/THF (Megelski et al. 2002), PVA/water (Lee, J. S., et al. 2004), chitosan/dilute acetic acid (Spasova et al. 2004), silk-like protein/formic acid (Buchko et al. 1999), and DNA/ethanol solution (Takahashi et al. 2005). The effect of increasing *V* (kV) in reducing the average *d* (nm) of the nanofibers is illustrated in Fig. 4.10 for PAN nanofibers containing filler particles (of titania or carbon nanotubes) electrospun from DMF. For polyacrylonitrile (PAN) in DMF solution (9 wt%) the decrease in *d* (nm) with increasing *V* was reported to be linear (Kedem et al. 2005) with the regression coefficients $a = 107$ (nm) and $b = 2.1$ (nm/kV) with $r^2 = 0.99$. It is interesting to note, however, that this electric-field-attenuated change in fiber diameter is generally much smaller than that obtained by controlling the concentration of the spinning solution (Kalayci et al. 2004).

Several other reports on electrospinning of bisphenol-A polysulfone/DMAC (Yuan et al. 2004), hydroxypropyl, cellulose/ethanol or propanol

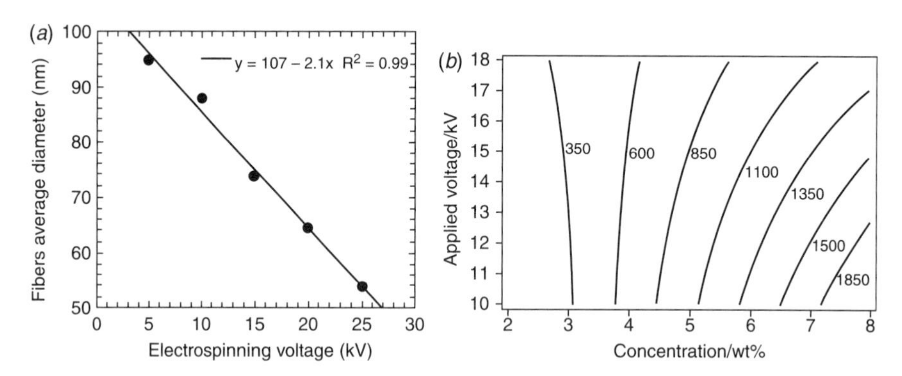

Figure 4.10 (*a*) The effect of applied voltage *V* (kV) on average fiber diameter *d* (nm) in electrospinning of PAN/carbon nanotube/TiO_2 composite nanofibers from DMF solution. Reprinted with permission from Kedem et al. (2005). Copyright 2005. American Chemical Society. (*b*) Contour plot of average fiber diameter *d* (nm) as a function of concentration and applied voltage in electrospinning PLA from chloroform/acetone (2/1 v/v) mixture. The numbers are predicted values of *d* (nm). Reprinted with permission from Gu and Ren (2005). Copyright 2005. John Wiley & Sons.

(Shukla et al. 2005), PDLA/chloroform : acetone (2:1 vol) (Gu and Ren 2005) and PVA/water (Lee, S. C., et al. 2002) solutions, however, suggested no significant effect of the applied voltage V (kV) on average fiber diameter d (nm). Also, values of d (nm) are even reported to increase with V (kV) under some processing conditions (Demir et al. 2002; Kidoaki et al. 2006; Tan, S.-H., et al. 2005). Baker et al. (2006), for example, reported a linear increase in d (nm) from 0.31 to 1.72 μm when V (kV) was varied from 5 to 25 kV in electrospinning 12.5–22.5% w/v solutions of PS ($r^2 = 0.94$).

This discrepancy in experimental observations suggests that the effects of applied voltage V (kV) on d (nm) need to be considered together with other parameters, particularly the feed rate and the gap distance L (Sukigara et al. 2003). Mass transfer increases with applied voltage as charged jets of liquid will travel to the grounded collector from the tip at a faster rate (Baumgarten 1971; Dersch et al. 2003; Khil et al. 2003; Shin et al. 2001a; Theron et al. 2004). This increase can be quite significant (Demir et al. 2002; Morota et al. 2004), and in the PEO/water system the mass flow rate (made evident by the current flow) increases almost linearly with the applied voltage (Deitzel et al. 2001). Demir et al. (2002), however, obtained a power law relationship between mass transfer and the applied voltage. Gap distance determines the time of travel (and the rate of drying) for the jet and therefore influences d (nm) as well. The apparent discrepancy in the reported effects of V (kV) on d (nm) noted above is likely due to different feed rates, gap distances and concentrations used in the different studies. Voltage-induced instability of the Taylor's cone may also result in changes in d (nm). At a constant gap and feed rate, very high values of V (kV) may result in the Taylor's cone receeding into the needle and the spinning occurs from within the needle (Deitzel et al. 2001; Jalili et al. 2005). The nanofibers formed were uneven and beaded.

4.4.2 Polarity of the Tip

The polarity of the capillary tip can have a significant impact on fiber quality and dimensions affecting both the average nanofiber diameter d (nm) and the areal density of the electrospun mats on the collector. In electrospinning nylon-6 from 85% formic acid (with added co-solvent at various volume percentages), a marked difference in average d (nm) between those spun from tips of positive and negative polarity was reported (Mit-uppatham et al. 2004a; Supaphol et al. 2005a). Table 4.3 summarizes the data. Average diameters from negatively charged capillary tips were found to be significantly larger at all concentrations studied (Mit-uppatham et al. 2004a; Kalayci et al. 2005). With PAN solutions in DMF, for example, nanofibers electrospun at −40 kV (as opposed to +40 kV) showed 52% larger values of average d (nm) (Kalayci et al. 2005). The decreased areal density

TABLE 4.3 The effect of polarity on the average diameter of PAN nanofibers electrospun from DMF solutions

		Average Nanofiber Diameter (nm)			
Solvent	Polarity	10 wt%	20 wt%	30 wt%	40 wt%
m-Cresol	+	110 ± 7	166 ± 11	170 ± 10	201 ± 17
	−	181 ± 17	286 ± 26	331 ± 32	334 ± 28
Acetic acid	+	94 ± 8	106 ± 10	120 ± 12	236 ± 49
	−	232 ± 25	232 ± 75	248 ± 70	266 ± 66
Ethanol	+	91 ± 8	115 ± 15	—	—
	−	200 ± 22	221 ± 30	—	—

Source: Based on data from Supaphol et al. 2005a.

(Fig. 4.11) of fiber mats spun from a capillary tip of negative polarity was attributed to increased numbers of charges per unit jet segment (Supaphol et al. 2005a). The mobility of negative and positive species in viscous polymer solutions being different, a difference in charging efficiency under different polarities is not unexpected.

As already referred to elsewhere, electrospinning can also be carried out using an AC voltage on the capillary tip. Although very limited data are available, the mats obtained appear to have a higher degree of fiber alignment. As seen in Fig. 4.12, on PEO nanofibers mats electrospun from water, the average fiber diameters for nanofibers electrospun using an AC potential can be relatively larger compared to when a DC voltage was used for the purpose (Kessick et al. 2004).

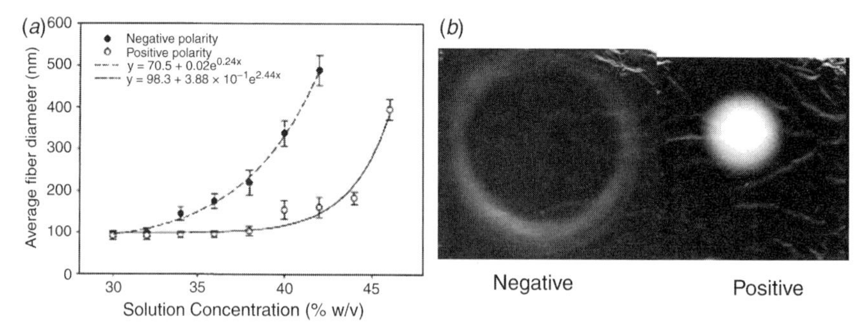

Figure 4.11 (*a*) The average nanofiber diameters obtained by electrospinning nylon-6 solutions in 85% formic acid solutions from a capillary tip maintained at a positive or a negative polarity. (*b*) A 30 s collection of the nanofiber sample in each case. A 40 wt% solution of polymer in 85% formic acid was electrospun using a gap distance of 10 cm and an applied voltage of +21 and −21 kV. Reprinted with permission from Supaphol et al. (2005a). Copyright 2005. John Wiley & Sons.

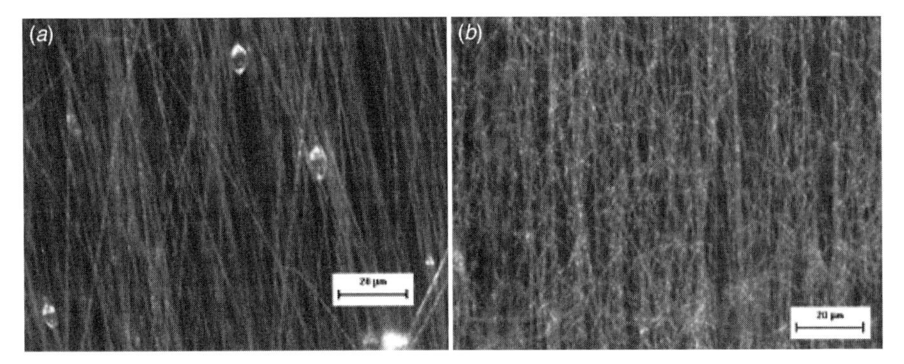

Figure 4.12 PEO nanofiber mats electrospun under identical conditions except (*a*) +7500 V AC was applied and (*b*) 7000 V DC was applied, showing differences in fiber alignment. Reproduced with permission from Kessick et al. (2004). Copyright 2004. Elsevier.

4.5 FEED RATE

This is the rate at which the polymer solution is pumped into the tip to replenish the Taylor's cone. Ideally, the feed rate must match the rate of removal of solution from the tip. Continuous nanofibers of uniform diameter are obtained under such conditions.[9] At lower feed rates electrospinning may only be intermittent with the Taylor's cone being depleted (with the cone even receding into the needle in some cases), but at higher feed rates larger fiber diameters and beads often result. Increasing the feed rate under conditions where applied potential is not a limiting factor results in the average fiber diameter d (nm) increasing with the feed rate (Kidoaki et al. 2006). The importance of the feed rate in determining nanofiber morphology has been widely reported (Buttafoco et al. 2006; Jeun et al. 2005; Zhang, C. X., et al. 2005b).

Experimental measurements indicate the volume charge density q_v on the jet to decrease exponentially with feed rate (Fridrikh et al. 2003; Theron et al. 2004). It is likely that higher feed rates lower the rate of replenishment of charges on the surface of the droplet. Theron et al. (2004), however, suggest charge replenishments to be governed by the drift velocity of ions and therefore to be independent of feed rate. The lower values of q_v are therefore likely to be a result of high rates of withdrawal of charges as well as polymer solution from the droplet surface at the higher feed rates.

[9] In electrospraying, the smallest average droplet sizes and narrowest droplet size distributions are generally obtained under the same conditions.

TABLE 4.4 The effect of capillary tip orifice size on electrospinning of (lactide-*co*-ε-caprolactone) copolymer

Needle	Diameter (mm)	Observation (Mo et al. 2004)	Fiber Diameter, *d* (nm) (Katti et al. 2004)
16G	1.19	—	240
18G	1.2	Beads and clogging	
20G	0.84	—	145
21G	0.8	Occasional clogging	
22G	0.7	Clogging rarely	
	0.58	—	135
27G	0.4	No clogging	

Source: Mo et al. 2004. Estimated from Fig. 4.2 (Katti et al. 2004).

4.6 CAPILLARY TIP

Metal needles[10] as well as those fabricated from nonconducting materials such as glass or plastic have been used in electrospinning.[11] Using a sharp, pointed electrode generally provides more efficient charging of the solution (Table 4.4) (Berkland et al. 2004). Practical considerations in selecting tip diameters are the throughput rates desired and possible interference from clogging due to solvent evaporation (Lee, K. H., et al. 2003a, 2003b). Mo et al. (2004) electrospun poly(L-lactide-*co*-caprolactone) P(LL-CL) from a particularly volatile solvent, acetone, and found even larger diameter needles (1.2 mm, 0.8 mm) to frequently clog and/or yield beads at a solution delivery rate of 2 mL/h. Their research, and that of Katti et al. (2004), indicates that, in general, smaller diameter capillaries yield fibers of smaller diameter. Yet, pumping a viscous liquid through a needle of small internal diameter may not always be practical (Zhao 2004). The optimal ratio of the length of the needle to its diameter has been suggested (Larrondo and St. John Manley 1981b) to be 4.5 and was successfully used by Mitchell and Sanders in their recent melt electrospinning study (Mitchell and Sanders 2006).

An interesting modification to the regular capillary tip reported by J. L. Li (2005), is the insertion of a nonconducting fiber into the lumen of the capillary (the fiber does not touch its walls). The modification resulted in two

[10]The needle or tube that delivers the polymer solution in electrospinning is also referred to as the "spinneret", "tip" or "nozzle" in the electrospinning literature. We prefer to use the term "capillary tip" or "tip" in this work.

[11]A capillary tip is not essential for electrospinning. Yarin and Zussman (2004) reported upward-directed needleless electrospinning using a two-layer solution where the upper layer is a polymer solution and the lower one a ferromagnetic suspension. A magnetic field applied over the surface creates spikes on the ferromagnetic liquid that perturb the upper layer, resulting in electrospinning of the polymer solution (see also Section 3.4).

changes: it eliminated backflow in the liquid meniscus and allowed the electric field to be used solely to accelerate the jet; it also resulted in enhancing the shear stress on the liquid surface flowing along the fiber. The use of this tip was reported to result in reducing the applied potential needed to obtain electrospinning.

Although a great majority of capillary tips used in reported studies are static, several innovations have explored the utility of movable tips. For instance, a capillary tip might be moved across the length of the drum collector to obtain an even deposition of nanofiber (Kidoaki et al. 2005). Innovative rotating capillary tips have also been reported. A simple design is an electrified polymer solution or melt contained in a cylindrical container with capillary tips extending radially from its periphery. The chamber is mounted on a rotating head within a cylindrical grounded metallic collector. On spinning the container about its long axis, the solution is forced into the needles by a centrifugal source and results in an electro-assisted dry-spinning process. The advantage of the setup is that it allows a degree of nanofiber alignment (depending on the rpm used), and by moving the collector (cylindrical mesh or metal foil) up and down vertically, a large area of homogeneous nanofiber mat can be obtained.[12]

A variation of this design uses a porous cylinder carrying the electrified polymer solution, which is spun rapidly about its long axis (Fig. 4.13)

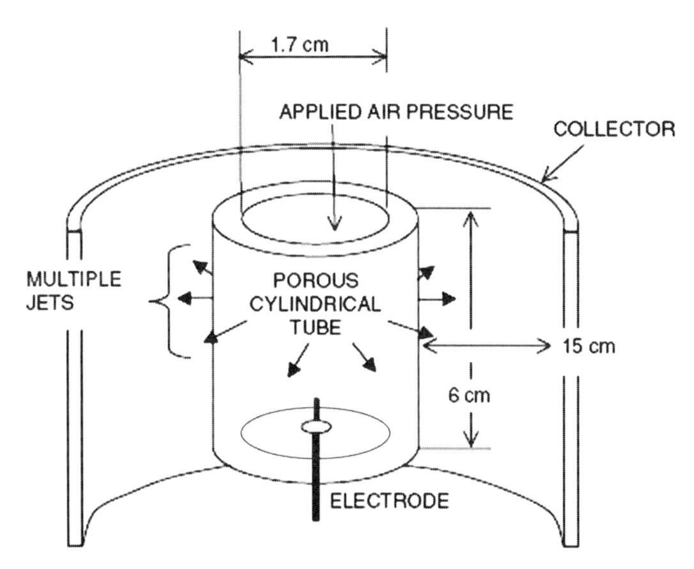

Figure 4.13 The design of a porous cylindrical needleless apparatus for electrospinning of polymers (Dosunmu et al. 2006).

[12]Anthony L. Andrady, David S. Ensor, and Randall J. Newsome. (Filed August 4, 2004). "Electrospinning of fibers using a rotatable spray head." US Patent 7,134,857 (issued November 14, 2006).

(Dosunmu et al. 2006). The solution is pushed through the porous walls by applied air pressure and the droplets that permeate through the wall are electrospun into nanofibers. In yet another variation of this approach recently commercialized in Europe,[13] electrospinning is obtained without using capillary tips; this technology (commercialized as the Nanospider™) was discussed in Chapter 3.

4.7 GAP DISTANCE

The gap distance L (the distance from the terminus of the capillary tip to the surface of the collector) defines the strength of the electric field as well as the time available for evaporation of the solvent before the nanofibers reach the collector surface. Increasing the gap distance L (cm), leaving other parameters constant, generally reduces fiber diameters (Baker et al. 2006), but depending on other parameters can at times increase fiber diameters (Kidoaki et al. 2006; Lee et al. 2004) or even halt the electrospinning process altogether. In electrospinning PS from chloroform (17.5 wt%), increasing L from 5 cm to 25 cm at $V = 15$ kV resulted in a linear decrease in average nanofiber diameter from 1 to 0.66 μm (Baker et al. 2006). Decreasing L (cm) (but keeping it large enough to prevent corona discharge) will affect both fiber surface morphology as well as the average diameter (Yao et al. 2005). A gap that is too small can lead to "wet" fibers that fuse on the collector (Hsu and Shivkumar 2004a; Jalili et al. 2005) (see Fig. 4.8). Depending on the feed rate selected, higher electric fields yield thinner nanofibers, but with too short a gap distance available for development of whipping instability, developing an unstable Taylor's cone (and concomitant defect formation) is more likely. Y. Hong et al. (2006) studied the average pore size of PVA nanofiber mats electrospun from aqueous solution (collected over 2 s intervals). As L (cm) decreased, the nanofiber mat quality progressively deteriorated from a nanoporous membrane to a network of fused nanofibers.

4.8 RELATIVE IMPORTANCE OF VARIABLES

The electrospinning process and resulting nanofiber mat morphology is clearly influenced by a large number of variables as discussed above (and by a few others such as polymer tacticity (Lyons et al. 2004), branching (McKee et al. 2004b, 2005) and fillers (Naebe et al. 2007) not included in this discussion). It is tempting to attempt a ranking of these process variables on the basis of their relative importance in controlling specific outcomes such

[13]The technology was invented and patented by The Technical University of Liberec, Czech Republic, and was recently commercialized under the trade name Nanospider.

as very small nanofiber diameters or a porous nanofiber surface. Although a few general rules such as "increasing polymer concentration will increase the average nanofiber diameter" universally hold, even such generalizations must be made with caution because of the complexity of the process. Any trends may not only be specific to a given polymer–solvent combination, but they may also be influenced by the specific set of process variables in operation. For example, as pointed out earlier, increasing the applied voltage V may have no effect, increase fiber diameter, or decrease fiber diameter, depending on the process regime in which electrospinning is carried out. Such relative assessments of selected electrospinning parameters for specific polymer/solvent systems are discussed in the literature (Gu and Ren 2005; Jalili et al. 2005, 2006; Kang et al. 2002; Katti et al. 2004; Sukigara et al. 2003; Tan, S.-H., et al. 2005; Theron et al. 2004). For example, in PLLA dissolved in a DMF/pyridine mixed solvent, the key parameters that control fiber morphology were reported to be polymer concentration, electrical conductivity, and the average molecular weight of the polymer (Tan, S.-H., et al. 2005). These observations for the most part apply only within the context of specific values of the variables used in the respective studies.

4.9 EXAMPLES OF REPORTED DATA

With so many variables involved in electrospinning, a complete description of a given experiment must necessarily include at least the key process and materials variables. Tables 4.5 to 4.8 present some examples of polymers reported to have been successfully electrospun.

TABLE 4.5 Example 1: Electrospinning of collagen type 1 nanofibers ($d = 100 \pm 40$ nm)

Polymer	Calf Skin Collagen
Solvent	HFP
Concentration	0.083 g/mL
Additive	—
Collector	303 steel drum
Rotation (rpm)	4500 rpm
Molecular weight	$\overline{M}_w \sim 700{,}000$ (g/mol)*
Applied voltage	+25 kV
Gap distance	125 mm
Feed rate	5 mL/h
Needle tip diameter	18 gauge (1.2 mm)
Environment	Air

*Not reported in the publication.
Source: Matthews et al. 2002.

TABLE 4.6 Example 2: Electrospinning of PEO nanofibers (d = 200–800 nm)

Polymer	PEO
Solvent	Water
Concentration	10 wt%
Additive	—
Collector	Aluminum sheet
Rotation (rpm)	—
Molecular weight	$\overline{M}_w = 300{,}000$ (g/mol)
Applied voltage	+7 kV
Gap distance	150 mm
Feed rate	0.025 mL/h
Needle tip diameter	0.51 mm
Environment	Air

Source: Megelski et al. 2002.

TABLE 4.7 Example 3: Electrospinning of PLGA nanofibers (d = 200–800 nm)

Polymer	PLGA (75 : 25)
Solvent	DMF
Concentration	45 wt%
Additive	Heparin (wt 20%)
Collector	Aluminum
Rotation (rpm)	—
Molecular weight	$\overline{M}_w = 126{,}000$ (g/mol)
Applied voltage	−12 kV
Gap distance	150 mm
Feed rate	0.26 mL/h
Needle tip diameter	0.51 mm
Environment	Air

Source: Casper et al. 2005.

TABLE 4.8 Example 4: Electrospinning of PLLA nanofibers (d = 800–2400 nm)

Polymer	PLLA
Solvent	CH_2Cl_2
Concentration	9–5 wt%
Additive	Pyridinium formate (0.2%)
Collector	Aluminum
Rotation (rpm)	—
Molecular weight	$\overline{M}_w = 670{,}000$ (g/mol)
Applied voltage	+40 kV
Gap distance	140 mm
Feed rate	1.3 mL/h
Needle tip diameter	0.51 mm
Environment	Air

Source: Jun et al. 2003.

5

CHARACTERIZATION OF NANOFIBERS AND MATS

Characterization of nanofiber mats is undertaken primarily to correlate test metrics with the useful properties of the material and to ensure consistent high quality of the product during manufacture. With single fiber measurements the intent is usually to generate fundamental data that allow a better understanding of the relationship between the structure and the properties of nanofibers. In principle there are a large number of different characterization techniques that can be used with nanofibers. With organic polymer nanofibers, the entire gamut of available polymer characterization techniques can be applied.

A basic limitation, however, is the lack of proven methodologies; conventional fiber testing methodologies are best suited for characterizing fibers with the average diameter d (nm) in at least the tens of micron range. Challenges associated with testing nanofibers were recently summarized (Tan and Lim 2006) and include: (a) manipulating extremely small fibers, (b) finding a suitable mode of observation, (c) sourcing for accurate and sensitive force transducers, (d) sourcing for accurate actuators with high resolution and (e) preparing samples of single-strand nanofiber. Some of these limitations are less serious than expected. For instance, nanomanipulators with resolutions in the 100–200 nm range are becoming available and sample preparation techniques that allow single fibers to be isolated are being reported in the literature. Load cells that are capable of measuring forces in the nN range are commercially available.[1] Clearly, the techniques are still in their infancy but will undoubtedly become increasingly sophisticated as the demand for characterization builds up in the near future.

A practical approach to characterization of nanofiber mats needs to be based on the end use intended for the material. Where nanofiber mats are intended as air filters for personal protection, for instance, tests that assess air permeability, particle penetration, porosity distribution, and residual electrical charge on fibers may be particularly meaningful. If the same material is intended for sensor applications, then the electrical conductivity, optical characteristics, equilibrium swelling by organic vapors, and chemical reactivity might be the more relevant measurements. Where the intended application is likely to subject the mats to handling or to be otherwise stressed during use, their mechanical properties such as tensile strength and ease of deformation may be the more pertinent measurements.

Although electrospinning technology has been around for quite some time, it is the improvement of the process itself rather than the development of characterization methodologies that has been the primary focus of research. It is only after the recent resurgence in interest in nanofibers that techniques to best characterize these are beginning to emerge and structure–property correlations are being seriously pursued. This effort is ongoing, and characterization techniques for nanoscale materials are continually being improved. What is reviewed here is therefore more of a starting point on the methodologies rather than a mature compendium of techniques. Table 5.1 summarizes the range of different characterization techniques for nanofibers reported in the recent literature.

In the following sections, selected characterization tests based on several common anticipated uses for nanofiber materials are presented. The assessment of interstitial porosity[2] properties, analysis of fiber dimension by microscopy, and mechanical integrity of fiber mats are the more important metrics in several application areas. Some of these will eventually be adopted as standard test methods in the field — both the American Society for Testing Materials (ASTM) and the International Standards Organization (ISO) have initiated efforts to standardize the nomenclature and test methods relating to nanomaterials, including nanofibers. Here, the reader will be afforded a flavor of the rather limited research literature on single-fiber measurements.

[1]Nanoindentors and a nano-universal testing machine platform with a load resolution of 50 nN and a displacement resolution better than 0.1 nm are available from MTS Instruments (Oak Ridge, TN).

[2]The terms “pore” and “porosity” are used in this chapter to mean the interstitial porosity associated with a fiber mat, where the pores are defined by a set of nanofibers lying at the periphery of an interstitial volume. This is contrasted with “fiber porosity,” which refers to porous features that occur on the surface or within individual nanofibers (dealt with in detail in Chapter 9).

TABLE 5.1 Common characterization techniques nanofiber materials reported in the recent literature

Polymer	SEM	TEM	XRD	Tensile Properties	Raman Spectroscopy	FTIR	Thermal Methods	Other Methods	Reference
Silk	x		x	x	x	x			Ayutsede et al. (2005)
Silk	x		x			x		AFM	Wang, M., et al. (2004a)
PANI/Silica	x		x				x		Choi et al. (2005)
PVA	x		x	x			x	Swelling	Ding et al. (2002a, 2002b)
PVA/Filler	x		x			x	x	Swelling	Gong et al. (2003)
PVA/NiO	x		x			x			Guan et al. (2003a, 2003b)
PVA/ZnAc	x		x			x	x		Yang, X. H., et al. (2004)
PVA/Silica	x		x			x	x		Shao et al. (2003)
PS/Clay	x	x					x	AFM	Ji et al. (2006b)
PLGA/PEG-C	x			x			x	NMR	Jiang et al. (2004b)
PLLA/Clay	x		x					Porosity	Lee, Y. H., et al. (2005)
PAN	x				x		x		Wang, Y., et al. (2003)
PAN	x		x			x	x		Gu et al. (2005b)
MEH-PPV	x		x		x	x		Photoluminescence	Madhugiri et al. (2003)
PVDF	x		x	x			x		Zhao, Z. Z., et al. (2005)
PEO/Collagen	x	x		x				NMR	Huang et al. (2001b)
P(LLA-CL)	x	x						NMR/porosity	Kwon et al. (2005a)
P(G-LA)	x		x	x				SAXS	Zong, X. H., et al. (2002)

(*Continued*)

TABLE 5.1 *Continued*

Polymer	SEM	TEM	XRD	Tensile Properties	Raman Spectroscopy	FTIR	Thermal Methods	Other Methods	Reference
TMC-CL	x		x				x		Jia et al. (2006)
PCL/Gelatin	x	x					x	AFM/XPS	Zhang, Y. Z., et al. (2006b)
PU	x			x				Porosity	Kidoaki et al. (2006)
PC	x	x						AFM/Optical	Megelski (2002)

SEM, scanning electron microscopy; TEM, transmission electron microscopy; XRD, x-ray diffraction; FTIR, Fourier transform infrared; PANI, polyaniline; PVA, poly(vinyl alcohol); PS, polystyrene; PLGA, poly(lactide-*co*-glycolide); PEG-C, poly(ethylene glycol)-chitosan; PLLA, poly(L, lactide); PAN, poly(acrylonitrile); MEH-PPV, poly[2-methoxy-5-(2′-ethyl-hexyloxy)-1,4-phenylenevinylene]; PVDF, poly(vinylidene difluoride); PEO, poly(ethylene oxide); P(LLA-CL), poly(L-lactide-*co*-ε-caprolactone); P(G-LA), poly(glycolide-*co*-lactide); TMC-CL, poly(trimethylene carbonate-*co*-caprolactone); PCL, poly(ε-caprolactone); PU, polyurethane; PC, poly(carbonate); AFM, atomic force microscopy; NMR, nuclear magnetic resonance; SAXS, small angle x-ray scattering; XPS, x-ray photoelectron spectroscopy.

5.1 MAT POROSITY AND PORE SIZE DISTRIBUTION

The interstitial porosity of a nanofiber mat is the fractional void space contained within it and is therefore a morphological property that is unrelated to the material characteristics of the nanofibers.[3] It is the pore size distribution and three-dimensional geometry that determine the key properties such as filtration or scaffold performance of nanofiber mats. Although no generally agreed set of definitions exists, porous materials can be classified in terms of their pore sizes as follows [see also ASTM F2450-04 "Standard Guide for Assessing Microstructure of Polymeric Scaffolds for Use in Tissue Engineered Medical Products" or Tomlins et al. (2004)]:

Micropores $\sim$0.5 nm to 2 nm
Mesopores $\sim$2 nm to 50 nm
Macropores $\sim$50 nm to 200 nm
Capillaries $\sim$200 nm to 800 nm
Macrocapillaries $>$800 nm

Fractional void volume, or total interstitial porosity Π, is defined in terms of the apparent density of the fiber mat, ρ_{mat} (g/cm^3), and bulk density of the polymer, ρ_{pol} (g/cm^3), of which it is made:

$$\Pi = 1 - (\rho_{mat}/\rho_{pol}) \tag{5.1}$$

$$\rho_{mat} = \rho_{pol}\pi r^2 L/V_{mat}, \tag{5.2}$$

where r is the mean fiber radius and L/V_{mat} is the length of nanofiber per unit volume of the mat. The void volume includes through-pores that would allow the passage of a fluid from one surface of the mat to the other across its bulk, as well as the blind pores or closed cells entrapped within the individual fibers. Porosity of nanofiber mats is almost exclusively interstitial and is due to numerous through pores. The total surface area of nanofibers per unit volume, s, is then given by

$$L/V_{mat} = -(\rho_{mat}/\rho_{pol})/\pi r^2 = (1 - \Pi)/\pi r^2 \tag{5.3}$$

$$s = 2\pi r L/V_{mat} = 2(1 - \Pi)/r. \tag{5.4}$$

[3]This does not, however, mean that the nature of the polymer has no effect on the porosity of the electrospun mats under a given set of processing conditions. The type of polymer affects fiber diameters as well as the yield in electrospinning, resulting in mats of different volume density invariably impacting the porosity of the mat (e.g., Ma, Z. W., et al. 2005a). Pore size distribution is known to change with concentration of the electrospinning solution (Ryu et al. 2003).

Direct measurement of porosity using equation (5.1) requires the experimental determination of the bulk mat volume and the volume of solid polymer in the mat. Although direct measurement has been used with some success (He, W., et al. 2005a; Ma, Z. W., et al. 2005d) to estimate Π, the values obtained invariably depend upon the accuracy of the measurement of mat thickness (usually made using a micrometer). Significant errors in Π are likely to result, as the uneven and convex fiber surface geometries are not fully taken into account in thickness measurements made with a micrometer (Kidoaki et al. 2006), leading generally to overestimation of the mat volume. Direct measurement of the apparent volume of nanofiber mats by liquid displacement is an alternative means of assessing interstitial porosity (Karageorgiou and Kaplan 2005). In this case, the mat is placed in a volume V_1 of a wetting liquid that does not dissolve or swell the polymer nanofibers. Hexadecane was successfully used as the liquid (Jin and Hsieh 2005a; Li and Hsieh 2005a) with a poly(vinyl alcohol) (PVA) nanofiber mat using samples sizes of about 11 mm $\times$ 18 mm weighing approximately 4.6 mg (Jin and Hsieh 2005a). The system was subjected to several pressure/vacuum cycles to ensure intrusion of liquid into all pores. The volume of the liquid-impregnated mat, V_2, as well as the volume of liquid left over, V_3, were estimated gravimetrically. Porosity was then estimated as

$$\Pi = (V_1 - V_3)/(V_2 - V_3). \qquad (5.5)$$

These simple measurements yield only an average estimate of Π, but cannot be employed to assess the distribution of pore volumes in a fiber mat. Intrusion porosimetry yields this additional information.

In the key application area of tissue engineering, close control over the pore-size distribution can be critical to success in viable scaffold development. Cellular ingress to the interior of nanofibrous scaffolds as well the efficacy of cell–fiber interactions leading to adhesion and proliferation of cells are well known to be related to pore size distribution (Karageorgiou and Kaplan 2005; Murugan and Ramakrishna 2006). In scaffolds used in bone regeneration, better osteogenesis and related vascularization occurs at average scaffold pore diameters $>300\ \mu m$ (see Chapter 7). In morphogenesis (BMP-induced osteogenesis), the pore size of scaffolds was shown to be an important variable (Kuboki et al. 2001). The close control over the average pore size of nanofiber mats, however, remains somewhat elusive at this time, partly because of the incomplete understanding of the interplay between process and material variables (Gu et al. 2005a), as discussed in Chapter 4. Good models for the pore-size distribution of nanofiber mats are therefore lacking in the literature. Also, the available techniques have not

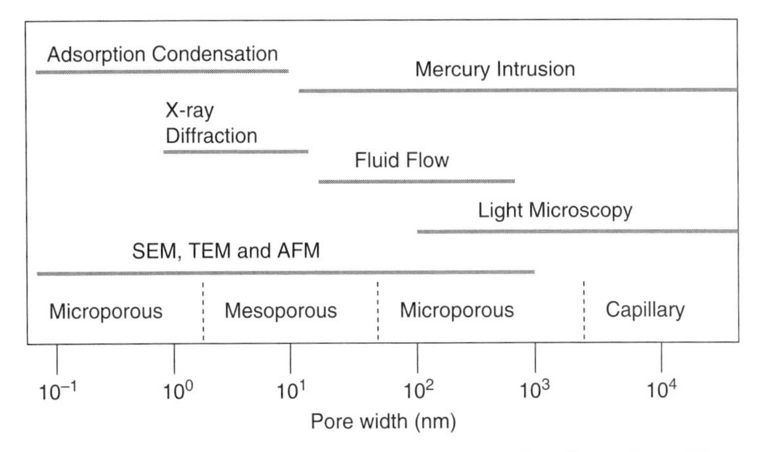

Figure 5.1 Techniques for characterization of porosity. (Based on Meyer et al. (1994) but updated.)

been fully and consistently used in pore-size characterization of mats used in scaffolding studies, making it difficult to compare results between studies.

Various general techniques available for characterization of porous materials are illustrated in Fig. 5.1. (Some of the techniques are more appropriate for stiffer materials and may not apply to nanofibers of all organic polymers.)

5.1.1 Mercury Intrusion Porosimetry

The porosity of a material is conveniently measured using intrusion porosimetry. In this technique, a nonwetting liquid (usually mercury) that does not dissolve or swell the nanofibers comprising the mat is forced into the mat to assess the pore volume. Forcing mercury into the pores of the mat requires the application of a pressure p to the mercury column in contact with the sample. As p is gradually raised, increasingly smaller pores in the mat are intruded by the mercury. The relationship between the applied pressure p and the pore radius r accessed is given by the Washburn equation (Karageorgiou and Kaplan 2005):

$$r = -2\gamma \cos\theta/p, \tag{5.6}$$

where θ is the contact angle (141.3° for mercury) and γ is the surface tension (484 mN/m for mercury) of the liquid. The technique measures only the fraction of porosity accessible by the mercury (Π_{mercury}) and therefore excludes closed pores, small mesopores ($r < 3$–4 nm) that are inaccessible by mercury, as well as very large pores (hundreds of micrometers in size) readily flooded by mercury even before application of pressure to initiate the measurement. Both the through pores as well as blind pores are generally accessible by this technique (Ioannidis and Chatzis 1993). Mercury intrusion measurements typically use high applied pressures (of up to thousands of psi) and may therefore distort the inherent pore volume distribution of mats made of soft polymer materials, potentially even collapsing the softer nanofibers into different cross-sections.

Typical values of Π in nanofiber mats as measured by mercury intrusion are quite high (Kim, S. H., et al. 2003; Kwon et al. 2005b; Min et al. 2004b; Yang, F., et al. 2004). For instance, that of poly(lactide-*co*-glycolide) (PLGA) copolymer (85 : 15/LA : GA) electrospun from (THF : DMF/1 : 1) solution was reported to be 91.63%, corresponding to 9.69 mL/g of pore volume (Li, W. -J., et al. 2002). The nanofiber diameters in this instance ranged from 500 to 800 nm, and the pore sizes varied from 2 to 465 μm. Poly(L, lactide) (PLLA) nanofiber mats electrospun from CH_2Cl_2/DMF mixed solvent yielded a porosity of 78% with an average pore diameter of 21.5 μm (Yang, F., et al. 2004). Mats of a biodegradable copolymer, poly(*p*-dioxanone-*co*-L-lactide)-*block*-poly(ethylene glycol), were recently characterized by a mercury intrusion technique (Bhattarai et al. 2003), and a porosity of 85% with a median pore size of 8 μm, and pore dimensions as large as 100 μm were reported. Figure 5.2 shows a pore-size distribution typically obtained by this technique.

Pore dimensions vary with the volume density of nanofibers in the mat and therefore also with the yield and collection time (Hong, Y., et al. 2006), but

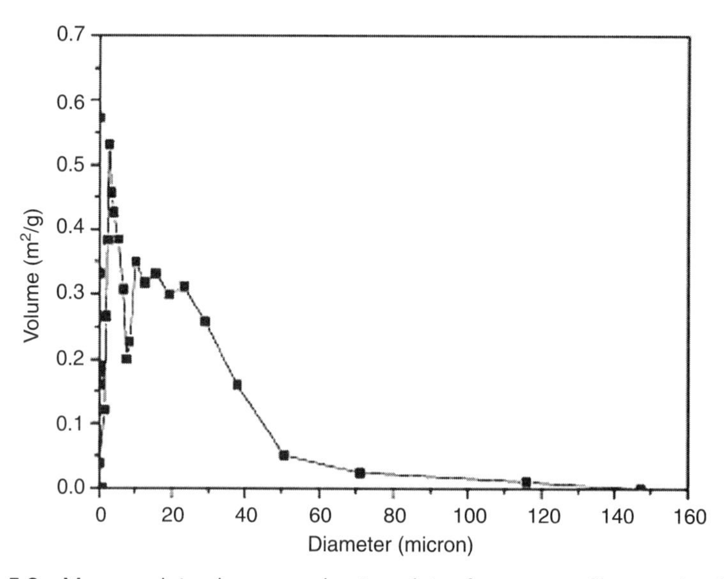

Figure 5.2 Mercury intrusion porosimeter data for a nanofiber mat of PLLA electrospun from CH_2Cl_2/DMF (70/30) (mercury pressure 1.23 psi and equilibration time 10 s). Reprinted with permission from F. Yang et al. (2004). Copyright 2004. Springer Science and Business Media.

TABLE 5.2 Porosimetric and modulus data illustrating the effect of changing solvent composition on electrospun segmented polyurethane nanofiber mat characteristics

Percent DMF	Fiber Radius (nm)	Porosity (%)	$(S/S_o)^a$	Young's Modulus (MPa)
0	3.11 ± 0.24	72 (1.0)	1	0.59
5	2.91 ± 0.29	57 (0.8)	1.06	0.64
10	2.35 ± 0.42	43.5	1.41	11.21
30	1.81 ± 0.41	44.6	2.52	2.43

[a] S denotes total surface area of the sample, and S_o value of S for samples electrospun from THF (0% DMF).
Source: Kidoaki et al. 2006.

are not directly related to the average fiber dimensions. Work by Kidoaki et al. (2006) has demonstrated the influence of the composition of the solvent mixture (DMF/THF) on the porosity and surface area of segmented polyurethane (SPU) nanofiber mats. Increasing the volume fraction of DMF in the mixture enhanced fiber bonding and decreased the average pore size as well as porosity Π. As expected, the mechanical integrity of the mat reflected by its Young's modulus also increased with the DMF content in spinning solvent. Table 5.2, based on data by Kidoaki et al. (2006), demonstrates the utility of porosimetry in characterizing nanofiber mats. As the likelihood of fiber bonding also affects porosity, it is also influenced by the gap distance L used in electrospinning. In mats of PVA electrospun from aqueous solutions, the average pore size increased sharply as L was changed from 2.5 to 6.0 cm, reflecting the change in mat morphology from a near macroporous membrane to a loose fiber mat (Hong, Y., et al. 2006). If other conditions are held constant, the time of collection and L are particularly good control variables for changing mat porosity.

5.1.2 Liquid Extrusion Porosimetry

Flow-through porosimetry avoids the use of very high applied pressures encountered in mercury intrusion measurements and is therefore well suited for characterizing relatively soft polymers. Liquid extrusion technique where the pressure employed is lower by as much as several orders of magnitude leaves the nanofiber geometry unperturbed during measurement (as long as no swelling of the fibers occurs).

In this technique a nanofiber mat is supported on a porous membrane and a layer of a wetting liquid (the contact angle between the liquid and the polymer is zero) is placed on its top surface. Gas pressure is applied over the liquid column and is gradually increased on the face of the mat until the gas is able to push the liquid through the largest of the pores in the mat, overcoming capillary forces. This allows a volume of liquid to be forced through the mat. The porous membrane supporting the mat is also saturated with the same liquid and its pore size is smaller than the smallest pore in the mat. Therefore, the liquid displaced from pores in the mat in turn displace the same volume of liquid from the liquid-filled pores of the membrane it is in contact with. This displaced volume of liquid and the applied pressure are accurately recorded. As the gas pressure is increased, progressively smaller pores are cleared of the liquid held in them. The differential pressure p is related to the pore diameter (at the most constricted diameter in the case of an irregular pore) and the volume of liquid extruded is indicative of the pore volume. The dependence of displaced volume on the pressure yields an estimate of the surface area. The relationship between p and the average pore radius r is given by

$$r = 2\gamma \cos\theta / p, \tag{5.7}$$

where θ is the contact angle and γ is the surface tension of the liquid used.

The volume of liquid displaced, V, can be related to the change in interfacial area at the liquid/polymer interface (Jena and Gupta 1999):

$$p\,\mathrm{d}V = \gamma \cos\theta\,\mathrm{d}S, \tag{5.8}$$

where the surface area S is obtained by integration of the function.

Any blind pores (e.g., surface porous features on individual nanofibers, Fig. 5.3) that do not allow flow-through of liquid cannot be assessed by the technique. The fraction of blind pores can therefore be indirectly estimated by comparing liquid extrusion data to the intrusion porosimetric data (which quantifies both open and blind pores).

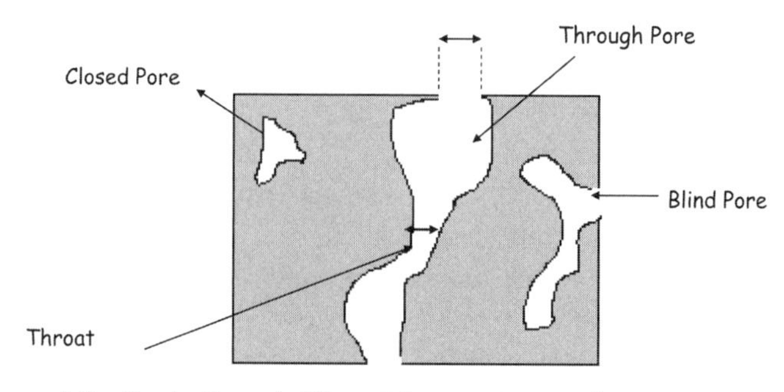

Figure 5.3 Illustration of different types of pores in a porous matrix.

The pore volume distribution function F_v of a mat, determined by liquid extrusion (or by mercury intrusion), is expressed as follows (Jena and Gupta 1999):

$$F_v = (\mathrm{d}V/\mathrm{d}\log D), \tag{5.9}$$

where V is the cumulative pore volume and D ($D = 2r$) is the pore diameter. Integrating this function over a specific range of values of D yields the pore volume that falls within that range.

The pore size distributions obtained by the mercury intrusion and liquid extrusion techniques are expected to be different. Mercury intrusion allows access to the pores from both sides of the mat and the entire pore volume is likely to be sampled. For pores that have large surface openings, the liquid extrusion generally tends to underestimate the pore volumes relative to those measured by intrusion porosimetry. Liquid extrusion measurements yield the pressure needed to push the liquid past the most constricted part (or the "throat") of the pore. The pore volume of the channel is estimated based on the throat diameter. This also introduces directionality to the liquid extrusion measurement. For a sample where porosity includes converging or diverging channels, the pore volumes (and pore dimensions) obtained from the liquid extrusion method will depend on the direction of the gas flow into the membrane. However, this is not expected to be a serious source of error in routinely characterizing nanofiber mats.

5.1.3 Capillary Flow Porometry

The technique is similar to the liquid extrusion technique in that the nanofiber mat is saturated with a wetting liquid and gas pressure is applied to one surface of the mat. The surface free energy of the liquid with the fiber mat needs to be less than that of the mat with the gas. As with liquid extrusion, the liquid column occupying through-channels will be displaced by the gas. In flow porometry, the gas displaces the liquid (and continues to flow through the emptied channel as well) and the flow rate of gas as a function of the differential pressure is recorded.

At lower pressures, the larger pores are cleared, allowing some gas flow. The lowest pressure at which this occurs is the "bubble point." The bubble point, a classical measurement in textile and paper technologies, is the point at which pressure is just sufficient to initiate flow. The pressures involved are less likely to result in significant distortion of the pore structure, but the magnitude of this error for fibrous polymer mats is unknown. The pressure needed depends on the surface tension of the liquid used, the surface energy of the nanofibers, as well as pore characteristics (such as the diameter and surface rugosity). As the gas pressure is gradually increased,

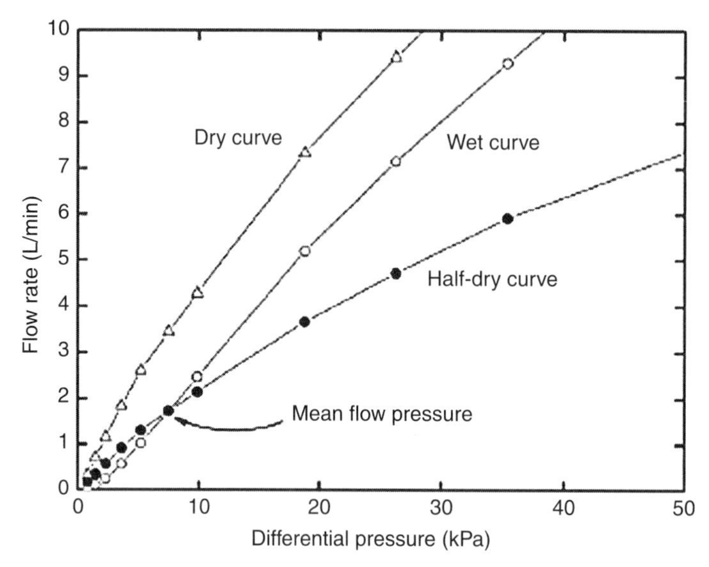

Figure 5.4 Flow rate vs applied gas pressure in capillary porometry for a porous polymer film sample. Reprinted with permission from Tomlins et al. (2004). Copyright 2004. National Physical Laboratory.

the flow rate also correspondingly increases, with the smaller pores in the mat being cleared of the liquid as well. The data yields a "wet" curve for the flow rate vs pressure. Finally, at a high enough pressure, a "dry" mat is obtained, where the gas flow essentially represents the permeability of the mat. The pore

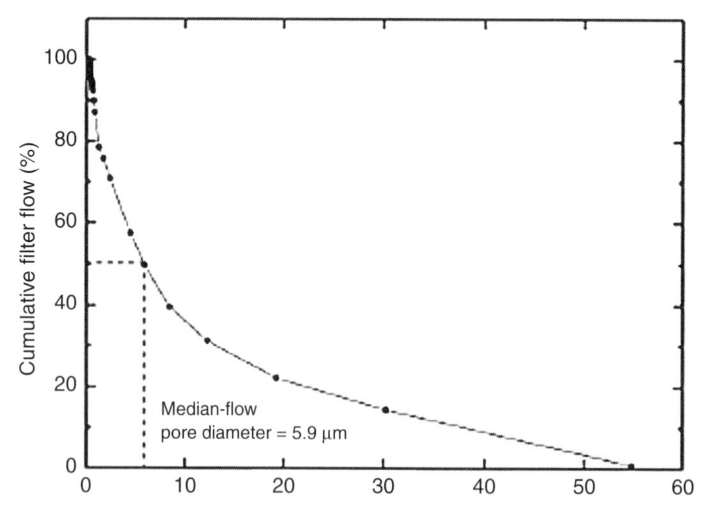

Figure 5.5 Cumulative distribution of filter flow pore sizes for the sample of scaffolding shown in Fig. 5.4. Reprinted with permission from Tomlins et al. (2004). Copyright 2004. National Physical Laboratory.

size distribution can be derived from the difference between the wet and dry curves.

The distribution of interstitial pore sizes obtained by this technique, however, refers exclusively to that of the throat of the pores (rather than to the average size of lumen). This might be the pertinent value in characterizing liquid filter media or studying the amenability of scaffolding to cell movements within it. By contrast, the full range of pore diameters encountered within a pore contributes to the mercury intrusion data.

The general features of wet and dry curves obtained from capillary flow porometry are shown in Figs. 5.4 and 5.5 for a porous poly(ε-caprolactone) film sample.

5.1.4 Brunauer, Emmett, and Teller (BET) Surface Area

The surface area of porous nanofiber mats determined using the Brunauer, Emmett, and Teller (BET) gas adsorption model has been reported for a limited number of electrospun polymers (Ding et al. 2003; Gong et al. 2004; Ryu et al. 2003; Zhang, Y. Z., et al. 2006c). Determinations that rely on gas adsorption involve several key assumptions:

1. Gas interaction with the polymer occurs with a constant heat of adsorption and exclusively due to van der Waals interactions between the gas and nanofiber surface (i.e., there is no significant sorption of the gas by the nanofibers).
2. Adsorbed molecules on the surface do not interact with each other.
3. Additional layers of gas molecules can be deposited on the surface of a complete or incomplete monolayer with the heat of adsorption being equal to heat of liquefaction of the gas.

The classical Langmuir model for sorption of a monolayer of gas molecules on a solid surface is described by the following simple expression:

$$V/V_{\mathrm{m}} = Cx/(1 + Cx), \tag{5.10}$$

where V is the volume of gas adsorbed, V_{m} is the volume corresponding to complete monolayer coverage of the surface by the gas, C is a constant, and x is the relative pressure (P/P_{o}) of the gas (where P_{o} is the saturation pressure of the gas). The BET model extends the Langmuir model to the case of adsorption of multiple layers of gas assuming there is no interaction between the layers, allowing the Langmuir isotherm to be applied to each layer. The following well-known BET isotherm describes the model

(c is the BET constant analogous to the Langmuir constant):

$$\frac{1}{V(x-1)} = \frac{1}{v_m c} + \frac{(c-1)x}{v_m c}, \tag{5.11}$$

where V is the volume of gas adsorbed at pressure P, v_m is the volume of gas at the standard temperature and pressure (STP) needed to form a monolayer, and $x = (P/P_o)$ is the relative pressure of the adsorbate. It can be shown that

$$c = \exp[(E_1 - E_L)/kT], \tag{5.12}$$

where E_1 is the heat of adsorption for the first layer of gas and E_L is that of the subsequent layers.

This form of the equation suggests a linear plot of $[1/V(x-1)]$ vs x; linearity is observed in practice when $0.05 < x < 0.35$. The value of v_m is calculated from the slope and intercept of the plot [$v_m = 1/(\text{slope} + \text{intercept})$]. The total surface area S_T of the solid is then given by

$$S_T = (v_m s N)/22.4, \tag{5.13}$$

where N is Avogadro's number and s is the cross-sectional area of the adsorbate molecule.

Experimentally, a sample of the nanofiber mat sealed in a chamber is maintained at a constant temperature well below its glass transition temperature and the chamber is evacuated. An adsorbate gas (usually N_2 or a N_2/He mixture, although other gases can be used) is admitted into the chamber in several increments until the entire surface of the porous mat is covered by a monolayer of gas. To allow sufficient gas molecules to be adsorbed by the weak van der Waals interactions, however, the sample has to be cooled, normally to the boiling point of the gas (for N_2 gas to 77.35 K).

The measured BET surface areas of nanofiber mats are several orders of magnitude larger than those for conventional textile fabrics. For PEO nanofiber mats, values in the range 10–20 m^2/g have been reported (Deitzel et al. 2001a). Zhang, Y. Z., et al. (2006c) prepared a highly porous mat of poly(ε-caprolactone) (PCL)/gelatin blend nanofibers by leaching out the gelatin component in a post-treatment step. The BET surface area of this nanofiber mat material was measured using nitrogen at 77°K over the range of $0.05 < x < 0.30$. The sample was evacuated over a period of 24 h prior to measurement. The BET surface area of the as-spun nanofiber mat of 6.56 m^2/g increased to a value of 15.84 m^2/g on leaching out the gelatin. The decrease in BET surface area with average diameter d (nm) of nanofibers for nylon-6 spun from 15–30% solution in formic acid was recently reported (Ryu et al. 2003), and the values typically varied from ~14 to 33 m^2/g over the average fiber diameter range of 600–100 nm. BET studies on a layer-by-layer nanofiber structure of TiO_2/poly(acrylic acid) (PAA) (Ding et al. 2004a) inorganic nanofibers of TiO_2/SiO_2 (50:50 molar percent) (Ding et al. 2003); PVA/$H_3PW_{12}O_{40}$ composites (Gong et al. 2003); and poly(acrylonitrile) (PAN)-based carbon nanofiber mats (Kim, C., and Yang 2003) have also been reported.

5.1.5 Other Approaches

Although not routinely used to characterize porosity, nuclear magnetic resonance (NMR) can also be used to study porous materials. The method relies on the fact that the difference in t_1 and t_2 relaxation times of protons in water (or other fluid) in porous materials will decrease with the average pore diameter.

TABLE 5.3 Summary of common techniques used to study the porosity of materials

Approach	Technique	Specific Techniques	Information Yielded
Image analysis	Microscopy	Optical Optical confocal Optical coherence tomography/ microscopy Scanning electron Scanning probe	Porosity, pore shape, pore size and pore size distribution
	Magnetic resonance imaging		
	Micro-X-ray tomography		For larger pore sizes
Intrusion methods	Porosimetry	Mercury intrusion porosimetry	Porosity, total pore surface area, pore diameter, pore size distribution
		Liquid intrusion porosimetry	
	Porometry	Capillary flow porometry	Median pore diameter, through-pore size distribution, permeability
Molecular probes	Diffusion of markers	Cyclic voltammetry	Permeability
		Molecular diffusion	

Source: Tomlins et al. 2004.

For instance, porous films of poly(3-hydroxybutyrate-3-hydroxyvalerate) (PHBV) studied by NMR relaxometry showed an increase in the t_1 and t_2-relaxation times with duration of immersion in water as a result of gradual intrusion of water into the porous structure (Marcos et al. 2006). Bhowmick et al. (2007) studied the same in PCL mats and found no significant difference in average porosity for mats with average fiber diameter d (nm) = 428, 1051, or 1646. An average porosity of 78 $\pm$ 3% was estimated using the technique. Recently the technique was successfully used to measure the void structure in textiles. NMR images of samples soaked in dilute $CuSO_4$ solution were analyzed and the void volume distribution determined from the fluid density autocorrelation function (Leisen and Beckham 2007).

Table 5.3 summarizes the general approaches to characterizing porosity of materials together with the information each is likely to provide. Those that do not apply to electrospun nanofibers are also included for completion.

5.2 NANOFIBER DIAMETERS AND PORE SIZES BY MICROSCOPY

Microscopic imaging is routinely used in the initial characterization of nanofiber mats. Although optical microscopy has sometimes been used to assess fiber diameter distributions in electrospun nanofiber mats, it is electron microscopy that is extensively used for the purpose. Optical microscopy typically allows a magnification of 1000×, and at best a resolution of only about 200 nm (Drummy et al. 2004). Using electrons (with a wavelength of only 6 pm as opposed to 450–600 nm for visible light) as the imaging radiation allows for a much higher resolution to be obtained. Scanning electron microscopy (SEM) and transmission electron microscopy (TEM) techniques are both particularly useful in understanding and quantifying fiber morphology. Depending on the particular instrument, resolutions of 1–20 nm are possible with SEM, and 0.1 nm for TEM, but both techniques essentially yield two-dimensional representations of nanofibers and pores. Also, only the surface of a pore can be observed in an image of the mat — pores not being necessarily cylindrical or even of regular geometry, surface dimensions often have no direct relationship to the average lumen diameter. SEM and optical methods therefore yield porosity information that can be very different from that obtained using other techniques.

Fundamentals of SEM (Hawkes and Spence 2008) or TEM (Fultz and Howe 2005; Williams and Carter 2004) techniques are well known and the reader is referred to recent texts for detailed information. In SEM, the nanofibers sample, coated with a thin layer of a conductive material such as gold sputtered over its surface, is placed in a beam of high-energy electrons

(0.1–50 keV) under vacuo. The beam generated by an electron gun is collimated by electromagnetic lenses into a controlled spot (1–5 nm) that is manipulated by a set of deflection coils and scans the surface of the nanofiber sample. An electron gun is essentially a cathode source — its composition defines brightness as well as the vacuum needed for the operation. (Simple tungsten filaments have a brightness of about 1 A/cm^2 steradian and can operate in 10^{-5} torr, but a field emission source has a brightness of 1000 A/cm^2 steradian, and requires a vacuum of 10^{-9} torr.) Interaction of the beam with the sample produces secondary electrons[4] that are detected (using a scintillator), amplified by a photomultiplier tube and displayed in the form of a high-resolution image of the sample surface.

Interactions of the primary electron beam with the sample produce some backscattered electrons as well (emitted at a near 180° to the original beam). These high-energy electrons can also be used to image the sample. As the backscattering obtained will be proportional to the atomic number of the scattering surface, these images yield compositional information as well. Particularly valuable in nanomaterials research is the coupling of SEM with an X-ray analyzer that detects and identifies the characteristic X-ray emissions from the sample in terms of its elemental composition. The technique is referred to as energy dispersive X-ray spectroscopy (EDS or EDXS).

It is also possible to image uncoated, as-spun polymer nanofibers, biological material, or even live cells using environmental SEM (ESEM) techniques. In place of the high vacuum, the electron beam in this case operates in an environment of water vapor (usually at a vapor pressures of 0.1–10 torr). Ionization of the water molecules prevents surface charge build-up on the sample, allowing nonconductive materials to be readily imaged without coating. Some loss in resolution due to scattering of incident electrons by water vapor, however, is to be expected. This compromise should be acceptable in most nanofiber imaging work. ESEM can also be used with X-ray spectroscopy.

With TEM, a part of the focused beam of high-energy primary electrons (typically 20 keV) is transmitted through the sample, and these are collected and processed to form the image. The interaction between electrons and nanofibers results in inelastic and elastic scattering of the transmitted electrons. The TEM images are constructed from the elastically scattered electrons and generally show structural features or defects at a high resolution and may also be used in the diffraction mode (even simultaneously) to obtain crystalline structure information. The inelastically scattered component is used in analytical microscopy such as in X-ray spectroscopy and electron energy loss spectroscopy.

[4]Interaction between the primary beam and the sample results in the generation of not only secondary electrons but also backscattered electrons, Auger electrons, X-rays, and photons (cathode luminescence).

Image analysis using suitable software is employed to quantify the pore sizes or fiber diameters from microscopic images. A key assumption implicit in using micrographs to quantify nanofiber characteristics is that the image is representative of the mat being studied (i.e. all parts of the mat have the same chance of being selected for imaging). Random selection of areas to be imaged (as opposed to deliberate selection of aesthetically pleasing ones) is important in this regard. A monochrome image from a charge-coupled device (CCD) with a radiometric resolution of 8 bits has a gray scale varying from 0 (black) to 255 (white).[5] In the image, the area covered by the nanofiber material can be distinguished by its relatively lower grayscale value relative to the open area (pore or interstice between fibers). Pixels that are near the edge of a nanofiber will naturally have intermediate grayscales. Quality image analysis requires a sharp transition in contrast in the image and therefore a well-focused image is essential. Also, in image analysis the "pore" areas and the "solid" areas are grayscale values. An appropriate point on the grayscale range that allows the program to distinguish between the two types of areas therefore needs to be provided by the operator or set by the program itself (this is called thresholding) (Russ 2002). A constant (global) thresholding factor might be applied to the entire image or local thresholding taking into account inhomogeneous grey-scale distributions, might be adopted (Gonzalez and Woods 2001). The software may then reduce the micrograph to a binary image to estimate the two types of pixel areas. The quantification is particularly sensitive to the manual choice of this threshold and its selection is very subjective. Nonsubjective statistical methods based on minimizing the statistical variance of the two types of pixels have been proposed (Gonzalez and Woods 2003; Ziabari et al. 2007). These assume the pore and solid pixels to be normally distributed. In a micrograph used to obtain a pore size distribution, what constitutes a pore must still be defined in terms of the minimum number of pixels. The magnification of the image used in the analysis therefore affects the accuracy of the measurement — images at lower magnifications generally give estimates of porosity closer to the theoretically expected value for a porous sample of poly(ε-caprolactone) (Grant et al. 2006).

An illustration of the types of data obtainable using automated analysis of fiber diameters is shown in Fig. 5.6. The micrograph is of an electrospun poly(ε-caprolactone) nanofiber mat imaged by SEM in the author's laboratory. Fovea Pro (Reindeer Graphics Inc., Asheville, NC), a 16-bit image analysis tool for Adobe Photoshop®, was used to generate the frequency distribution of fiber diameters shown in the figure. A sufficient number of such

[5] 8 bits yield $2^8 = 256$ digital values or shades of gray in a monochromatic image. A more sensitive imaging device will yield a wider tonal range for the image; for instance, a 16-bit image will yield $2^{16} = 65{,}536$ tonal values.

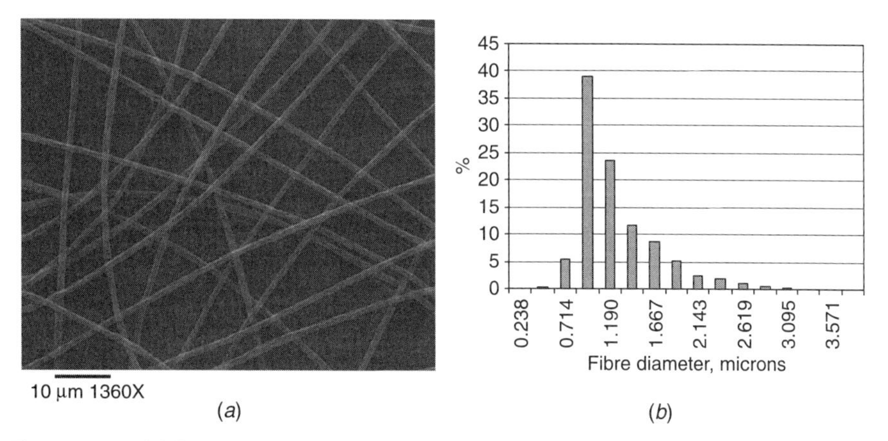

Figure 5.6 (*a*) SEM image of an electrospun nanofiber mat of poly(ε-caprolactone). (*b*) Frequency distribution of fiber diameter estimated from the image using image analysis software. (Courtesy of Research Triangle Institute, 2004.)

images need to be considered to obtain a reasonable estimate of the average fiber diameter of the mat. McManus et al. (2006), in their SEM characterization of bovine fibrinogen nanofibers electrospun from HFP solutions, averaged 60 measurements per micrograph in estimating the average fiber diameters (the measurements were calibrated using the micrograph scale). Others have used averages of 100 (Gu and Ren 2005; Kim, H. S., et al. 2006) or even 500 measurements (Patel et al. 2006) to arrive at a reliable average value.

A profilometric technique such as atomic force microscopy (AFM) or optical sectioning of a fluorescent dyed fibermat using laser confocal microscopy yields additional information on the height of the fibers. (These techniques, however, have their own limitations in yielding reproducible absolute dimensions.) As with most microscopic techniques, imaging is limited to a minute section of the mat sample and the findings often cannot be generally extrapolated as an average characteristic of the entire mat. Even in assessing nanofiber mat characteristics qualitatively, the need to study a sufficient number of random samples selected over the mat surface and use of appropriate statistical averages estimated to describe the material, cannot be overemphasized.

5.2.1 Atomic Force Microscopy Technique

Atomic force microscopy (AFM) is emerging as a popular technique for the characterization of nanomaterials. A sharp probe or tip affixed to the end of a cantilever is raster scanned over the surface of the sample to be imaged, in this

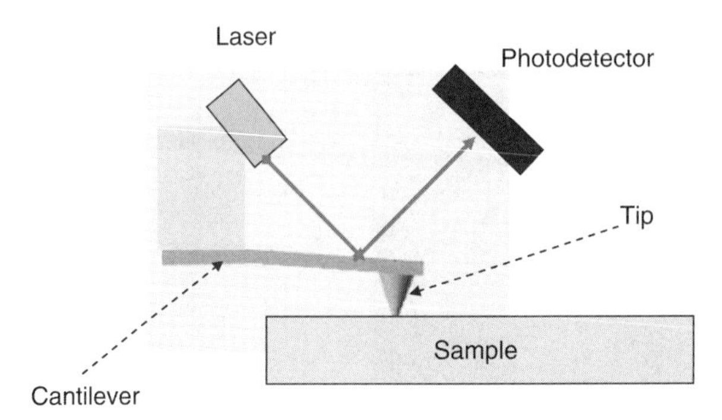

Figure 5.7 Schematic of a cantilever with tip in contact with the sample in an atomic force microscope. The electronics components are not shown for clarity (not drawn to scale).

case the nanofiber mat. The probe is pressed against the sample with a constant force during scanning. The movements of the probe are converted into a signal and used to generate a profile of the sample surface. A typical probe is made of monolithic silicon with a $\sim$15 μm tall tip of radius less than 10 nm, and the corresponding probe length and height might be about 500 μm and 50 μm, respectively. The cantilever and tip dimensions are so small that a low-power optical microscope is usually employed to position the tip accurately on the sample. Displacements of the tip, by as little as 1 nm, can be detected by a quality AFM instrument. The basic principal of AFM is illustrated in Fig. 5.7.

In profilometric use of AFM, variations in tip-to-surface distance during scanning of the mat surface are monitored by the deflection of a laser beam reflected off the surface of the cantilever onto a position-sensitive photodiode detector.[6] As the mat topography changes, the z-direction scanner moves vertically to maintain constant deflection of the tip. Images are generated from the data through a feedback loop between the photodiode arrangement and the piezoelectric scanners responsible for the translational movement of the tip. This allows even minute deflections to be detected by the optical system. Typically, imaging can be carried out in the contact mode or in the so-called "tapping mode," where the tip is made to vibrate at a low frequency and is therefore in intermittent contact with the sample surface. This latter mode allows imaging to be carried out without damage to the soft sample surface of polymers (or biological materials) due to dragging of the AFM tip over it. Even when using cantilevers with low spring constants ($k < 0.5$ N/m), the forces involved can be large enough to result in damage to polymers.

[6]Although this is a particularly accurate way of measuring cantilever deflection, light microscopy can be also be used to directly measure deflection, as demonstrated in the measurement of tenacity of PAN nanofibers (d (nm) $\sim$ 1200 nm), where a calibrated eyepiece was used to measure deflection (Warner et al. 1998).

Imaging, however, can also be achieved with the AFM tip not in contact with the sample surface at all; instead, the cantilever is oscillated at its bending frequency ($\sim$100 kHz) in close proximity ($\sim$10 nm) above the sample surface. The frequency and amplitude of the oscillating cantilever are monitored and the data used to generate an AFM image. A detailed discussion of the technique is beyond the scope of this work and the reader is recommended the following reviews of AFM techniques (Batteas et al. 2005; Birdi 2003).

Electrospun nanofibers of polyurethanes (Demir et al. 2002), poly(dicyclopentadiene) (Bellan et al. 2006), polyelectrolyte complexes (Ding et al. 2005a), polyacrylonitrile (PAN) (Ge et al. 2004; Hou et al. 2005), polystyrene (PS) (Ji et al. 2006b), poly(benzimadazole) (Kim and Reneker 1999b), poly(butylene terephthalate) (PBT) (Mathew et al. 2005), aramids (Srinivasan and Reneker 1995), DNA (Takahashi et al. 2005), and polylactide (PLA) (Yang, F., et al. 2004) have been imaged using AFM. Jaeger et al. (1996) imaged PEO nanofibers using silicon nitride, Si_3N_4, cantilevers ($k = 0.38$ N/m) to study the periodic morphological features on the surface of fibers. Others have studied the occurrence of beads in electrospun PEO nanofibers (Morozov et al. 1998) and the morphology of silk nanofibers (Tan and Lim 2006) using this technique. Using AFM to obtain high-resolution data on nanofibers requires the selection of a sharp tip of appropriate geometry. Typically a cantilever of low spring constant is employed ($k =$ 0.06 N/m was used in the PEO study) and multiple measurements are used in the analysis (over 100 images were used by Jaeger et al. (1996)) to obtain reproducible data. The sample response over a range of applied forces are studied to eliminate force-dependent artifacts (Morozov et al. 1998) in the data. Care must also be taken to ensure an appropriate angle of contact between the tip and the surface in order to avoid lateral ploughing of the tip into the soft sample surface (unless friction-mode data are desired).

AFM imaging of nanofibers is generally a difficult undertaking, and the fiber width obtained in the micrographs tends to be greatly exaggerated because of the convolution of the AFM tip-shape with the nanofiber geometry (Kim and Reneker 1999b). Srinivasan and Reneker (1995) recommend measuring the diameters perpendicular to the plane of the substrate. AFM measurements of fiber diameters on poly(benzimidazole) (PBI) nanofibers showed that fiber diameters when measured as the apparent width in the AFM image to be twice that measured in the vertical direction. In general, the smaller the real dimension, the larger will be the exaggeration in the AFM measurement (Morozov et al. 1998). Assuming the nanofibers to be dry and cylindrical, the difference is attributable to the fuzzy edges typical of AFM micrographs (Fig. 5.8).

Even when using a sharp AFM tip for the measurement, resolution can be poor, as nanofibers tend to be easily displaced by the movement of the probe

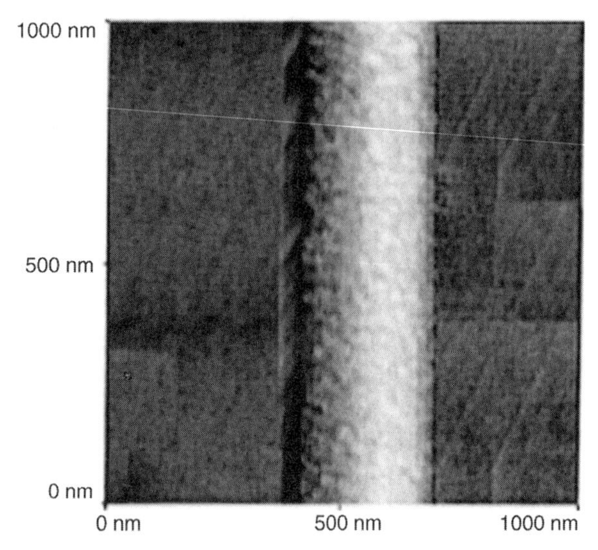

Figure 5.8 An AFM image of an electrospun polyamide nanofiber. Reprinted with permission from Q. F. Wei et al. (2006). Copyright 2006. John Wiley & Sons.

over them, leading to fuzzy AFM images of the smaller-diameter fibers. The difficulty can be overcome by placing nanofibers on a substrate that can hold them in place during measurement by adhesion or through non-bonded interactions. Adhering the nanofibers to a base using a layer of epoxy has been successfully demonstrated (Tan et al. 2005a), but the sorption of adhesive components by the nanofibers is a concern. Recently, Han and Andrady (2005) imaged electrospun polymer nanofibers spun onto surfaces

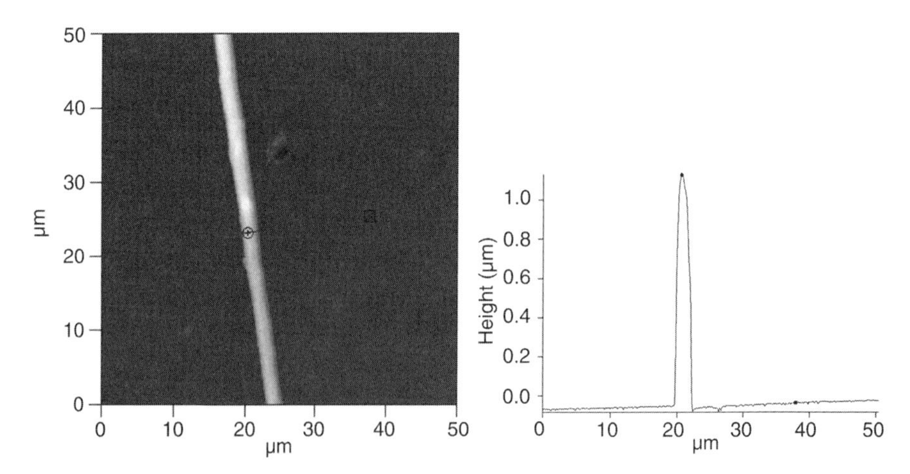

Figure 5.9 AFM image of an acrylic nanofiber deposited on a PGMA layer. (Courtesy of RTI International, 2005.)

coated with a layer of poly(glycidyl methacrylate) (PGMA) to avoid this problem. Nonbonded interactions between the PMMA nanofibers and PGMA surface minimized any displacement of nanofibers by the tip during imaging and yielded high-quality AFM images (Fig. 5.9).

5.3 MECHANICAL PROPERTIES OF MATS

The techniques available to measure the mechanical properties of films and textile materials are well known and in most instances can be applied with some modification to nanofiber mats. However, some caution in interpreting the data is warranted. The most common such technique is tensile property measurement using the same general experimental technique used with film or woven textile samples. Randomly oriented nanofiber mats are cut into the rectangular (Ding et al. 2004c; Li, M. Y., et al. 2005; Zong, X. H., et al. 2003a) or typical dumbbell-shaped test samples (Boland et al. 2001; Luu et al. 2003; Pedicini and Farris 2003) and tested using a universal testing machine to obtain the tensile properties. Wang and colleagues, for instance, successfully measured the mechanical properties of $\sim$100 μm thick poly(vinyl alcohol) (PVA) nanofiber mats electrospun from 8–10 wt% aqueous solutions, using a gauge length of 1 cm and a crosshead speed of 2 mm/min (Wang, X. F., et al. 2005). This is feasible with the thicker nanofiber mat samples.

Handling fragile fiber mats during sample preparation, however, is a concern, as the mechanical integrity and the geometric arrangement of the fibers can be compromised in the process. The alternative simple procedure described by Ramakrishna and colleagues (Huang et al. 2004) for handling gelatin nanofiber mats has considerable merit in this regard. A paper template (as shown in Fig. 5.10) with parallel strips of double-sided adhesive tape attached to it as shown is gently placed on the fiber mat (with the adhesive surface against the mat). A single-sided tape is then placed to secure the adhered part of the web to serve as end tabs, and the assembly cut into test strips (gauge length 30 mm and test piece width 10 mm).

The test strips are mounted in the grips of a universal testing machine fitted with a load cell in the appropriate force range and the test conducted as with a film sample. The extension ratio of the mat is readily measured using a pair of marks on the sample to define a gauge length, but the accurate assessment of stress in the sample is complicated by the porosity of the mat. Mats are inherently porous even in the case of ideal constructs based on uniform, unbeaded, and evenly deposited nanofibers. As discussed earlier, $\rho_{pol}\pi r^2 L \neq V_{mat}$, and most of the volume of a test piece is air. The volume fraction of polymer in the

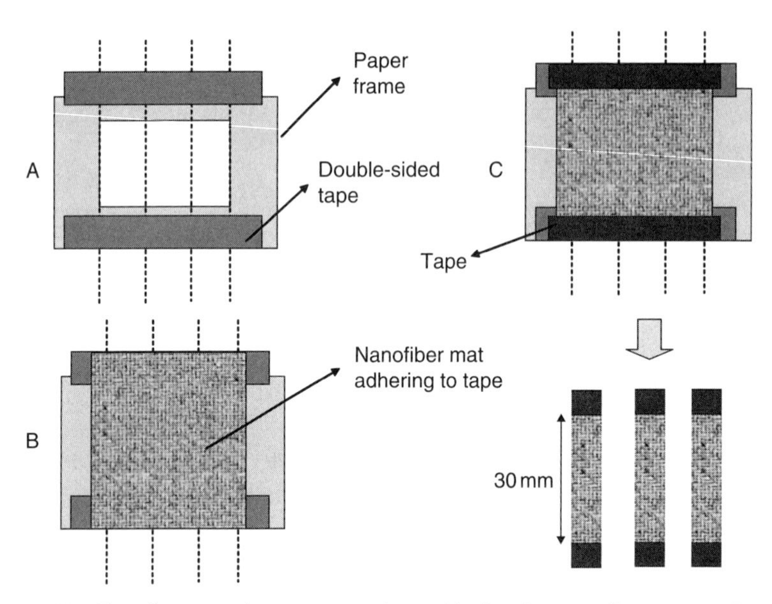

Figure 5.10 Tensile test piece preparation with fragile nanofiber mats. Redrawn from Huang et al. (2004).

test piece has to be accurately established and when comparing mechanical property data for different mats generated even in a single experiment.

Data can be normalized to a constant volume density of fibers in the sample. (Schreuder-Gibson and Gibson 2006 reported their data normalized to 30% volume fraction.) A simple correction based on areal density has been used (Ayutsede et al. 2005).

Stress expressed in units of (g/tex)[7]

$$\text{Stress (g/tex)} = \frac{\text{Force (g)}}{\text{Width (mm)} \cdot \text{Areal density (g/sq} \cdot \text{m)}}, \tag{5.14}$$

where the areal density is the measured mass of the test piece divided by its area.

Instead of measuring the thickness, an "equivalent thickness" might be estimated from the known mass, m, the density ρ of polymer, and the area A of the die used to cut the tensile test sample (McKee et al. 2005).

$$\text{Equivalent thickness} = m/A.\rho \tag{5.15}$$

[7]The tex is a unit of linear density commonly used in the textile industry, and is the weight in grams of a fiber 1 km in length (or mg per meter of fiber).

TABLE 5.4 Tensile properties of thermoplastic polyurethane nanofiber mats

Solvent	Modulus (MPa)	Tensile Strength (MPa)	Elongation (%)
THF	3	9.5	360
DMF	12	54.5	160

Source: Schreuder-Gibson and Gibson 2006.

A related issue is heterogeneity of the mat — the variation in volume density of the nanofibers at different points in the test piece. Points where the nanofiber deposition is particularly sparse or dense (or beaded) will naturally influence tensile properties in a manner essentially similar to that due to an air bubble or an inclusion in a film or laminate sample (Inai et al. 2005a). Visual examination of the test pieces to avoid obvious defects is therefore a necessity. Establishing evenness criteria based perhaps on optical measurement in selecting test pieces for mechanical testing might be desirable. In their recent book (Ramakrishna et al. 2005), reviewed the available information on tensile properties of nanofiber mats.

Factors unrelated to the chemical nature of the polymer used in electrospinning affect the tensile properties of the fiber mats. Nanofibers of the same polymer electrospun from different solvents often display very different mechanical properties (see Table 5.4). Pollethene nanofibers electrospun from THF and DMF, for instance, show very different tensile properties; just as films cast from these two solvents also show similar differences (Schreuder-Gibson and Gibson 2006). This is due to the morphology of nanofibers (and films) being affected by the different rates of evaporation of solvent and the consequently different kinetics of development of the relevant phase morphologies. The volume density of fibers in the mat, the mix of nanofibers (or the polydispersity of fiber diameters), degree of fusion of the individual nanofibers, existence of imperfections and branching of fibers can all be expected to affect the tensile properties of a mat electrospun under different process conditions even from the same polymer/solvent system.

5.3.1 Mat-Related Variables

The tensile deformation of a fiber mat is far more complicated than that of a polymer film; correcting for the void content is often inadequate to completely account for the stress-deformation data. Unlike with polymer films, subjecting nanofiber mats to tensile deformation will result in nonuniform stresses being developed resulting in rearrangement of nanofibers within the mat to accommodate the strain. Ensuing strain-dependent changes in the concentration of load-bearing nanofibers (essentially the slack segments of the fiber becoming taut due to imposed strain) and eventual breakage of a few of the fibers likely precedes segments prior to the overall catastrophic failure of the test piece. Strain rate is therefore an important parameter in tensile testing; faster test rates compared to the rate of rotation/re-alignment of nanofiber segments consequent to strain generally lead to higher moduli (Inai et al. 2005a). No comprehensive models of the process, however, are available.

When an isotropic mat of nanofibers is uniaxially strained, load is not equally born by all nanofibers. The initial load will be borne by the fraction of fibers parallel in alignment to the axis of the strain. Those oriented at a small angle ϕ to the axis will rotate and twist in an attempt to realign along the axis of strain. Depending on the strain, an increasing fraction of these will become load-bearing. Fibers that lie perpendicular to the axis are not load-bearing, and will buckle under the strain. SEM images of extended fiber mats support this simple qualitative picture (see Fig. 5.11, where

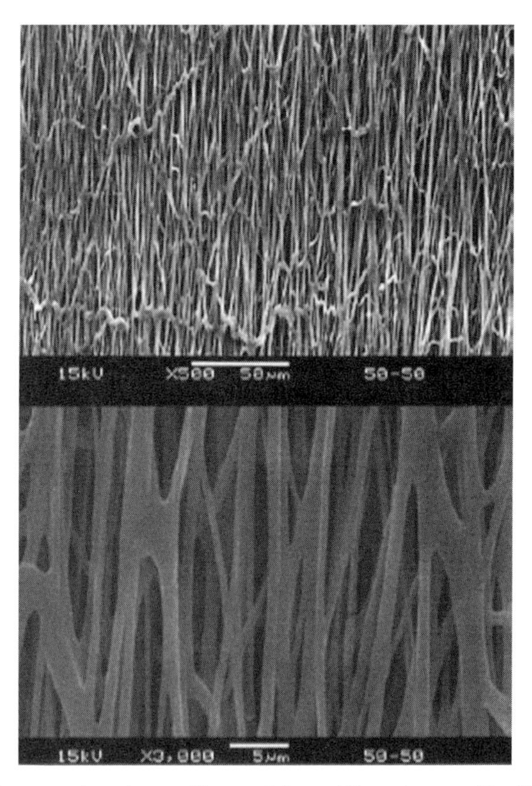

Figure 5.11 Micrographs of nanofibers at two different magnifications showing the high degree of fiber alignment resulting from uniaxial strain of a nanofibers membrane (at point C indicated in Fig. 5.12). Reprinted with permission from Inai et al. (2005a). Copyright 2005. John Wiley & Sons.

different alignments of fibers are apparent. Note that it also shows some fusion of fibers making the situation even more complex).

This effect due to fiber alignment is illustrated by the marked change in tensile properties of the mat with the speed of rotation of the cylindrical collector (changing the alignment/orientation of nanofibers; see Chapter 9). K. H. Lee et al. (2003a) found the tensile modulus of PCL nanofiber mats to increase by 40% (to 4.67 MPa), their tensile strength to increase by 50% (to 2.1 MPa), and the extensibility to remain unchanged when the velocity of collection surface was increased from 1.3 to 3.2 m/min. A further increase in the collector velocity to 4.5 m/min, however, decreased these properties. The decrease was in part attributed to the reduced dimension of pitch in the machine direction that accompanies the higher revolutions per minute (RPMs). Others (Fennessey et al. 2006; Pedicini and Farris 2003) have also reported the tensile properties of aligned nanofiber mats to be higher compared to comparable isotropic mats.

Figure 5.12 shows typical stress–strain curves for two nanofiber mats of copolymers of P(LLA-CL) electrospun from 10% acetone solution from a capilliary tip 0.2 mm in diameter, using a flow rate of 1.0–1.5 mL/h and an applied voltage of 15 kV (Inai et al. 2005a). The modulus, ultimate properties [tensile strength (MPa) and extensibility (%)] of the two mats are different because of the different copolymer composition of fibers. The one with the higher LLA content displayed better tensile properties. However, the two curves may not be directly comparable, as the mats may not have the same volume fractions of nanofibers and fiber morphologies.

The stress–strain curves for the fiber mats typically show an initial high modulus at low strains [up to about 4.5% strain for P(LLA-CL) mats], followed by a lower modulus until failure occurs. With a fibrous mat

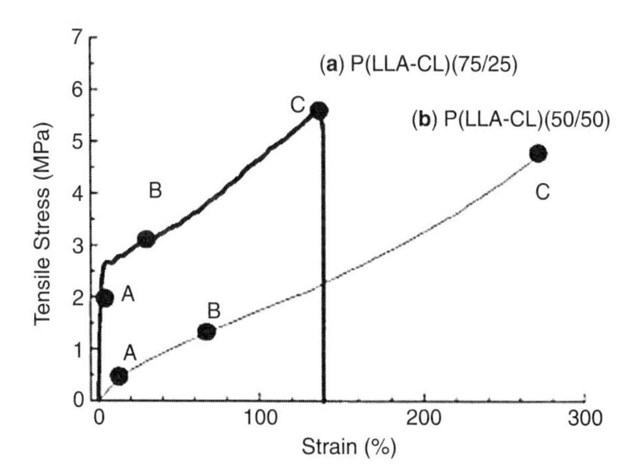

Figure 5.12 Stress–strain curves for P(LLA-CL) copolymer nanofiber mats. Reproduced with permission from Inai et al. (2005a). Copyright 2005. John Wiley & Sons.

morphology it is reasonable to qualitatively argue that at low strains the deformation will be resisted by the relatively few short fiber segments uniaxially extending between the grips of the test machine. The fraction of such chains will be higher in an oriented mat where the test piece is mounted with the strain axis parallel to the axis of alignment. These short fibers yield or rupture allowing the majority of the fibers in the mat sample to eventually take up the applied load, giving rise to the main section of the stress–strain curve. Fibers aligned at an angle relative to the direction of principal strain undergo rotation accommodating the strain, as already discussed. Increasing fractions of fibers align and stretch along the direction of uniaxial deformation as strain is increased (Huang, C. B., et al. 2006b; Inai et al. 2005a). Micrographs of the sample taken at points A, B, and C (see Fig. 5.11 for images at C) on the curve showed fiber-level detail in support of this notion. Fiber alignment increased from A to B to C on the stress–strain curve. The abrupt decrease in modulus in the curve was likely due to the catastrophic breakage of those nanofibers initially aligned axially. Micrographs of the sample at point C show a very high degree of orientation of fibers in the direction of applied strain (Inai et al. 2005a). Although not measured in this particular study, some fiber thinning by drawing and possible strain-induced crystallization from the imposed strain, might have also taken place in the sample.

In common with laminate samples studied by tensile testing, even small inhomogeneities in the material (scratches, air bubbles, micronicks, and particulate inclusions in the case of films) have a dramatic effect on the tensile behavior of the mat (Inai et al. 2005a). In the case of nanofiber mats, such inhomogeneities are well known to exist in the form of beads, microdrops of solution dropped on to the mat during spinning, uneven nanofibers, and "fused" nanofibers. Varying either of these in a sample will clearly affect the tensile properties of the mat. Electrospun gelatin nanofibers (from 5 to 12.5% in HFP) studied recently show that the presence of beads influence the tensile data (Huang et al. 2005). However, the mats studied are often not characterized well enough to delineate the effect of beads and other imperfections on tensile properties.

With bicomponent fiber mats (where the mat is made of a mixture of two types of nanofiber) the compositional characteristics are reflected in the tensile data. With poly(vinyl alcohol)/cellulose acetate (PVA/CA) mats with different fractional ratios of nanofibers of two types of fibers, the tensile properties showed a gradual change with mat composition (Ding et al. 2004c). The shape of the stress–strain curve remained essentially unaltered but showed higher moduli at the higher PVA content (Fig. 5.13). The fiber mats with different compositions had different average fiber diameters and possibly different porosities. Note that the initial high modulus region is also apparent in these stress–strain curves.

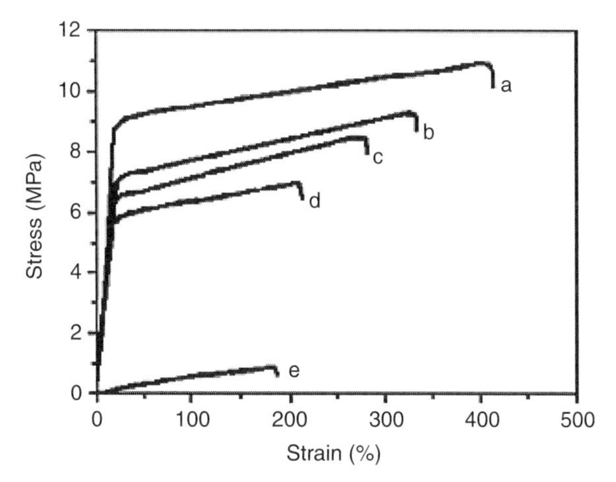

Figure 5.13 Tensile stress–strain curves for bicomponent nanofiber mats of PVA/CA nanofibers. Mat composition is indicated by the ratio of number of tips with the polymers PVA versus CA used to spin the mat: (*a*) 4/0, (*b*) 3/1, (*c*) 2/2, (*d*) 1/3, and (*e*) 0/4. Reprinted with permission from Ding et al. (2004c). Copyright 2004. Elsevier.

5.4 SINGLE-FIBER CHARACTERIZATION

Characterization of individual nanofibers is critical to understanding their physics because invariably it is the fiber characteristics that govern the properties of their mats. But, their minute dimensions makes this a daunting task. Nanofibers of particular interest, those with dimensions smaller than the wavelength of light, are especially difficult to isolate or test. It is not practical to extract fiber-level information by manipulation of test data generated from even the best-characterized fiber mats. Several innovative techniques for the direct assessment of mechanical properties of individual nanofibers are being developed and are discussed in this section.

New techniques developed for nanofibers (and nanotubes) are beginning to appear in the literature. While the following discussion concentrates on mechanical properties, other properties including electrical (Kitazawa et al. 2007), thermal (Motoo et al. 2005), and optical (Balzer et al. 2003) characterization of single fibers have been reported. Some of the conventional test methods used in textile research with mesofibers have also been successfully adapted for use with nanofibers. It is feasible for instance to carefully install a single electrospun nanofiber in a nanotensile test machine designed to measure their modulus and ultimate properties of fibers (Inai et al. 2005a; Tan et al. 2005b). The technique was described by Inai et al. (2005b) characterizing a highly-oriented poly(L-lactide) (PLLA) nanofiber electrospun from methylene chloride/methanol (80/20) solution. Nanofibers were directly collected on an open paper frame with a gauge length of 20 mm. The frame was

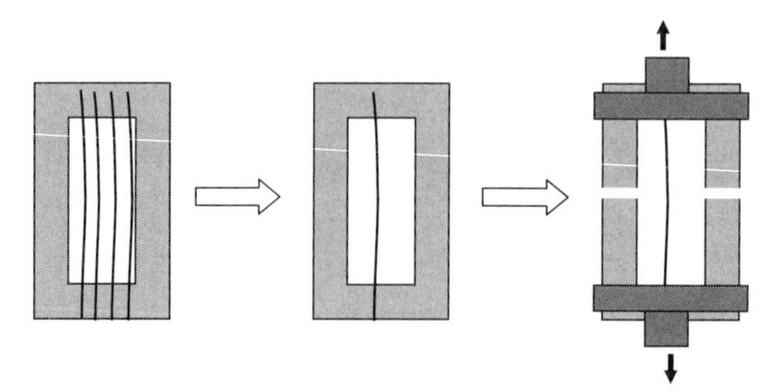

Figure 5.14 Illustration of technique used to measure the tensile properties of a single nanofiber using conventional test equipment.

attached to a rotating disk spun at a velocity of 63–630 m/min to facilitate orientation of the fibers. All but a single fiber was subsequently removed from the paper frame under an optical microscope and the single-fiber assembly installed in a tensile tester (Nano Bionics MTS) with a load cell in the 500 mN range and 50 nN resolution. The sides of the paper frame were cut (see Fig. 5.14), and the test was conducted at a strain rate of 25% per min. Electrospun nanofibers of poly(ε-caprolactone) (PCL) (Tan et al. 2005b, 2006) tested on a paper frame yielded[8] a tensile modulus of 120 $\pm$ 30 MPa and an ultimate extensibility of 200 $\pm$ 100% (Tan et al. 2005b). However, the procedure is beset with a myriad of practical difficulties and the variability in the data is usually quite high.

Inai et al. (2005b) used the technique to demonstrate the effects of chain orientation on nanofiber tensile properties (Table 5.5). Chain orientation in polymers can be established using conventional techniques such as X-ray diffraction or infrared (IR) dichroism, and its effect on tensile properties can be very significant. The tensile properties of PLLA (M_w = 300,000 g/mol) nanofibers collected on a high-velocity (630 m/min) vs a low-velocity (63 m/min) rotating drum collector showed an improvement in their tensile strength and modulus resulting due to better alignment and chain orientation. The effect is further illustrated in Fig. 5.15, showing typical stress–strain curves obtained for a pair of nanofibers collected on a drum collector moving at different velocities.

Cantilever techniques that use the AFM for these types of measurements avoid some of the problems associated with handling the nanofibers that are inherent in conventional tensile test methods. The approach is particularly

[8]The data, when compared to that for gravity-spun or melt-spun PCL microfibers, suggest the electrospun nanofibers to have a higher degree of chain and crystallite orientation. The change in mechanical properties with fiber diameter was similar to that obtained in "cold drawing" of conventional fibers (Tan et al. 2005b).

TABLE 5.5 Change in tensile properties with alignment of single nanofibers of PLLA

Take-Up Velocity (m/min)	d (nm)	Modulus (GPa)	Tensile Strength (MPa)	Extensibility (%)
63	890 ± 190	1.0 ± 1.6	89 ± 40	1.54 ± 0.12
630	610 ± 50	2.9 ± 0.4	183 ± 25	0.45 ± 0.11
600[a]	34,000	3.9	192	2.2

[a]Data for melt-spun PLLA fiber. d = average fiber diameter (nm).
Source: Inai et al. 2005b.

sensitive and has been used previously in individual molecule force spectroscopy measurements (Hugel et al. 2002).

5.4.1 Using the AFM for Single-Nanofiber Measurement

Although AFM is used mainly as a profilometric imaging method and a probe technique to study the surface chemistry of materials, it can also serve as an invaluable piece of equipment in measuring mechanical properties of fibers and even single macromolecules. Some of the sample-handling issues inherently associated with conventional tensile techniques are

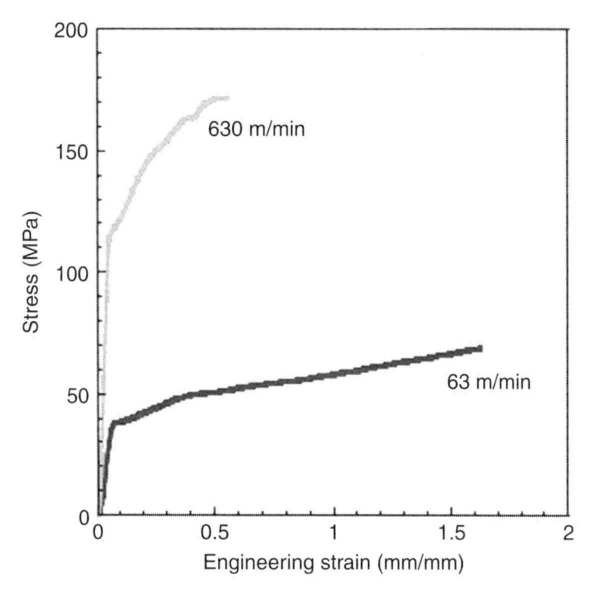

Figure 5.15 Tensile stress–strain curves for single nanofibers electrospun under the same conditions but collected on a drum rotating at low and high speeds (strain rate 25%, 20 mm gauge length). Reprinted with permission from Inai et al. (2005a). Copyright 2005. Institute of Physics.

avoided when using AFM for this purpose. For example, tensile properties of single-walled and multiwalled carbon nanotubes have been elucidated using AFM measurements (Demczky et al. 2002). It suggests the utility in applying the same technique to the relatively larger polymer nanofiber characterization process.

5.4.1.1 *Nanoindentation* In the nanoindentation technique (Fig. 5.16), with the AFM operating in the force mode the tip is brought into contact with the fiber surface and a force is applied to indent the surface of the nanofiber. The indentation is force-controlled, with a maximum load of P_{max} applied to the sample. For the technique to yield meaningful data, the tip diameter needs to be very much smaller than that of the nanofiber and all data must be

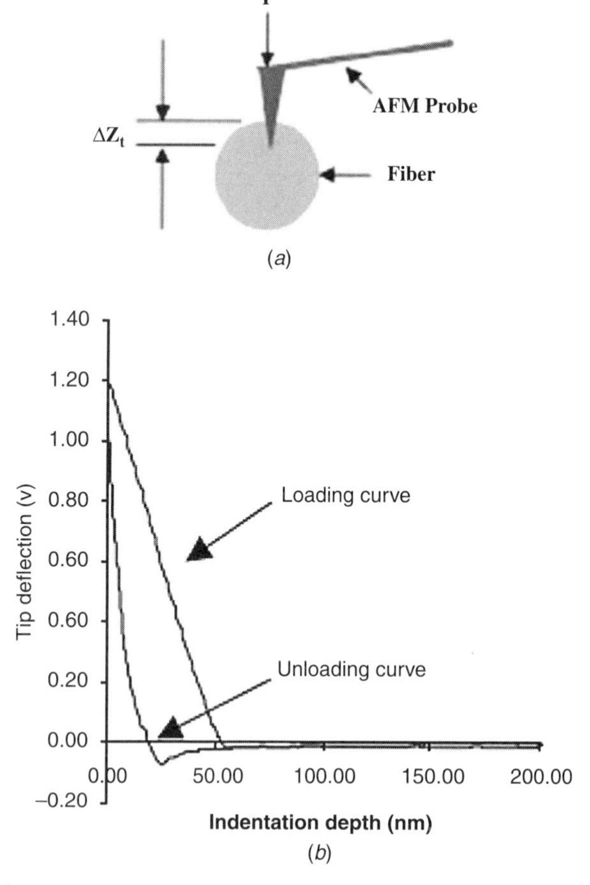

Figure 5.16 (*a*) Tip and fiber interaction during indentation measurement. (*b*) Indentation deflection curve. Reprinted with permission from M. Wang et al. (2004a). Copyright 2004. Elsevier.

generated at the same strain rate. The indentation depth ΔZ (measured as the deflection of the cantilever) varies linearly with the applied load P. The slope of the curve at P_{max} is then directly proportional to the effective Young's modulus E^* of the nanofiber material[9]:

$$\text{Slope} = \{dP/dZ\}_{max} = 2E^*(A/\pi)^{1/2}, \quad (5.16)$$

where A is the contact area of the tip. However, uncertainties related to tip geometry and the cantilever force constant generally preclude the use of equation (5.16) to directly calculate E^* (Wang, M., et al. 2004). Instead, a material of known E^*_R is used as a reference sample and E^* for the nanofibers estimated from the ratio

$$\frac{\{dP/dZ\}_{max}}{\{dP/dZ\}_{R_{max}}} = \frac{rE^*}{r_R E^*_R}, \quad (5.17)$$

where r and r_R are the contact radii from the measurements carried out on the sample and the reference material and $\{dP/dZ\}_{R_{max}}$ is the slope obtained with the reference material.

The utility of this technique is illustrated by the work on single nanofibers of degummed native silk fibroin carried out by M. Wang et al. (2004). The study yielded a value of $E^* = 13.66 \pm 0.85$ GPa (1 Hz strain rate) for the silk nanofibers. The value agreed well with that obtained from independent triboindentor (Berkovitch tip) measurements ($E^* = 14.38 \pm 1.83$ GPa). The effect of water and methanol extraction on the modulus of the nanofibers was also investigated by the group using this technique. Others have used the technique with poly(D,L-lactide-*co*-glycolide) (PDLGA) nanofibers (Xin et al. 2007) and PLLA nanofibers (Tan et al. 2005).

5.4.1.2 Bending Test Using an AFM tip to apply a deflecting force to a nanofiber lying across a 10–100 μm wide depression (5–10 μm deep) machined into a hard substrate such as silicon can be used to measure the modulus E^*. Assuming the ends of the nanofiber to be firmly anchored to the flat surface and the fiber is not sagging into the depression, the force F exerted by the tip at the middle of the fiber results in a tip deflection ΔZ that can be converted into the value of fiber displacement δ. It can be

[9]The effective modulus E^* is related to the elastic modulus of the material E_1 and that of the tip E_2: $1/E^* = (1-v_1^2)/E_1 + (1-v_2^2)/E_2$, where v_1 and v_2 are the Poisson ratios of the sample material and the tip, respectively (Mack et al. 2005).

shown that the force F is related to deformation δ as follows (Zhou et al. 2005):

$$\delta = F(L^3/\alpha E^* I), \tag{5.18}$$

where I is the moment of inertia (second moment of the area), L is the suspended length of the fiber, E^* is the Young's modulus, and α is a constant. This technique has been used on PEO nanofibers (100–300 nm; Bellan et al. 2005) and poly(dicyclopentadiene) (Bellan et al. 2006) to obtain values of E^* of 0.76 GPa and 11 ± 5 GPa, respectively. A three-point bending test (Fig. 5.17) carried out on TiO_2-filled poly(vinylpyrollidone) (PVP) nanofibers draped over a mesoscale pore, using essentially the same AFM method, was also recently reported (Lee, S. H., et al. 2005). The AFM-based approach was used with polyacrylonitrile (PAN) nanofibers (Gu et al. 2005c), inorganic titania nanofibers (Lee, S. H., et al. 2005), and composite PVA nanofibers (Shin, M. K., et al. 2006) as well.

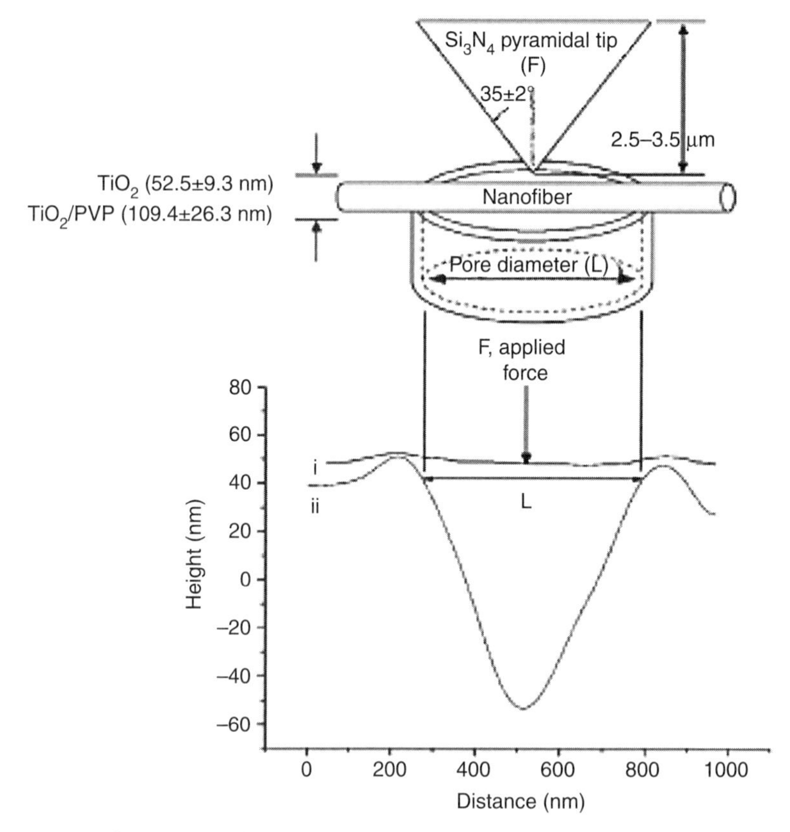

Figure 5.17 Schematic of the sample arrangement for the three-point bending test and actual AFM data on fiber (i) and pore (ii). Redrawn from S. H. Lee et al. (2005).

5.4.1.3 Uniaxial Extension Generation of a complete stress–strain curve for a single nanofiber using AFM is particularly challenging (Tan et al. 2005a). Tan et al. reported a technique where aligned nanofibers of PEO ($\sim$700 nm) stretched across a pair of glass cover slips affixed to a glass slide mounted on an inverted microscope assembly was used. A piezoresistive AFM cantilever (spring constant 8 N/m) was attached to a point on the nanofiber via a short length of glass fiber using a drop of cyanoacrylate adhesive. The arrangement (shown in Fig. 5.18(*a*)) allowed the nanofiber to be extended uniaxially by slow translation of the sample stage of the microscope and the deflection of the cantilever was recorded accurately. The deflection data, converted to changes in the electrical resistance of the piezoresistive cantilever, were used to quantify the magnitude of the uniaxial load (force resolution 0.2 μN). The extension data were estimated from an

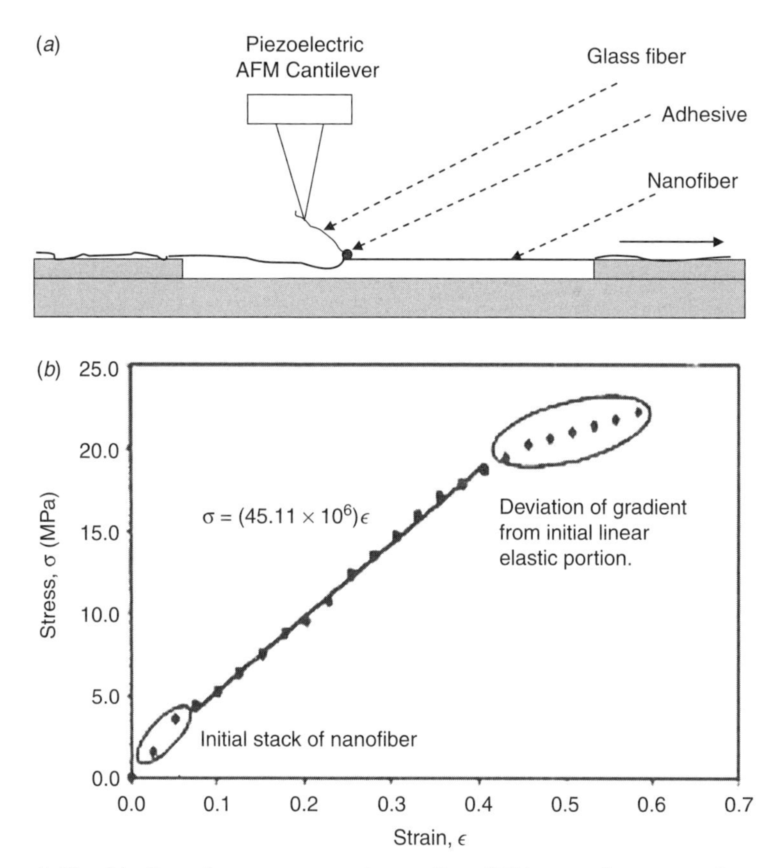

Figure 5.18 (*a*) Sample arrangement on the AFM sample stage for tensile measurement. (*b*) Stress–strain curve for a single nanofiber ($d = 700$ nm) of PEO. Reprinted with permission from Tan et al. (2005a). Copyright 2005. American Institute of Physics.

image of the fiber recorded by a CCD camera. By stepwise translation (in 0.01 mm increments) of the stage, a load–extension curve was generated at different strains and the results plotted as in Fig. 5.18.

The stress–strain curve showed good linearity over a considerable strain, but did not show a typical yield point characteristic of polymer films and macrofibers. An elastic modulus of 45 MPa was determined for the PEO nanofibers (Tan et al. 2005a). A similar approach based on using AFM for assessing the single-fiber tensile properties of PAN was reported by Buer et al. (2001).

5.5 NANOFIBER CRYSTALLINITY

5.5.1 Differential Scanning Calorimetry (DSC) Technique

Differential scanning calorimeters are designed to hold a sample of the nanofiber sealed in a metal sample pan (usually made of aluminum) and a reference metal pan carrying no sample, at exactly the same temperature. The higher thermal capacity of the sample pan will demand more thermal energy relative to reference pan to attain a given temperature. As the temperature of the insulated enclosure holding the pans is slowly increased, the difference in power demand by sample and reference pans is accurately monitored as a function of temperature and yields thermal data on the nanofiber sample in the sample pan. Events such as first- and second-order transitions, chemical reactions, or solvent evaporation can generally be readily observed as distinctive patterns in thermal behavior. This approach to calorimetry is often referred to as power-compensation type DSC. Alternatively, the sample and reference pans might be placed in an insulated chamber in good thermal contact with each other via a metal plate (silver or constantan alloy). Thermal imbalance induced by the presence of a nanofiber sample in the sample pan causes heat to flow between the sample and reference pans, and changes in enthalpy can then be determined by accurately monitoring the difference in temperature ΔT between the pans. With nanofiber samples the significant thermal event anticipated is the melting of the crystalline fraction, an endothermic process. Loss of residual solvent from the nanofiber web may also take place and will constitute an endothermic event.

In a typical experiment where the temperature is raised from T_1 to T_2, a plot of heat flow (mW) vs temperature is obtained. Generally, the change in enthalpy ΔH is related to the area A under the curve:

$$\Delta H = kA, \tag{5.19}$$

where k is an instrument constant. As the temperature is increased to the crystalline melting temperature of the nanofiber sample (T_m) the latent heat associated with the thermodynamic transition results in an endothermic peak in the DSC curve. The location of the peak on the temperature axis is determined by the crystallite chemistry. The area under the peak on the DSC curve ($J \cdot {}^\circ K/s \cdot g$) divided by the heating rate (°K/s) yields the joules per unit mass (J/g) associated with melting. From the known mass g of the nanofiber mat used and the known latent heat of the crystalline fraction, an estimate of crystallinity can be reliably determined.

As seen from Table 5.1, the DSC technique is widely used to study electrospun polymer nanofibers and the data have been compared with that for a cast film of the polymer. In conformity with findings from X-ray diffraction studies, the nanofibers show a high degree of chain orientation compared to that in a cast film of the same polymer. High degrees of molecular orientation during electrospinning have been reported for electrospun nanofibers of PLLA (Zong, X. H., et al. 2002), poly(benzimidazole) electrospun from dimethylacetamide (DMAc) solutions (Kim and Reneker 1999b), PEO (Deitzel et al. 2001a; Larrondo and St. John Manley 1981a), and PAN (Buer et al. 2001). This appears to be a general phenomenon with most polymer nanofibers, especially at the smaller fiber diameters. Orientation in nanofibers can also be detected by optical birefringence measurements. DSC studies yielded indirect evidence of chain orientation in the case of electrospun PAN nanofibers. Cast films of PAN generally undergo a cyclization reaction on heating in nitrogen, yielding a sharp exothermic peak in the DSC thermogram at 293°C. Nanofibers similarly heated show the same transition,

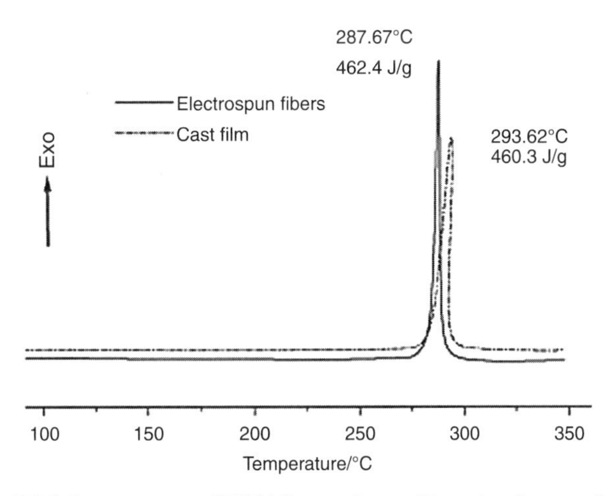

Figure 5.19 DSC thermogram of PAN film and nanofibers in nitrogen (heating rate 10 K/min). Reprinted with permission from Gu et al. (2005b). Copyright 2005. Elsevier.

but at the much lower temperature of 288°C (Fig. 5.19; Gu et al. 2005b), suggesting that cyclization is facilitated by molecular orientation of the fibers in the latter samples (Mathur et al. 1992). These highly oriented domains in nanofibers are metastable and may slowly revert to crystallites; the process can be accelerated by annealing. As pointed out by Gu et al. (2005b), however, this same shift could also result from a large fraction of the acrylonitrile repeat units being arranged at the nanofiber surface. DSC analysis of polyester nanofibers of poly(ethylene terephthalate) (PET) and poly(ethylenenapthalate) (PEN) also showed the crystallization temperatures for the nanofibers to be lower than that for the bulk polymer materials (Kim, J.-S., and Lee 2000).

5.5.2 X-Ray Diffraction Methods

In X-ray diffraction (XRD) studies, a beam of X-rays is aimed at the sample and the ensuing diffraction pattern is observed. Because of their high energy, X-rays are hard to focus, and slit collimators are usually employed for the purpose. The monochromatic X-ray beam is multiply scattered from the various lattice planes in the sample at specific angles. Constructive interference from the scattered X-rays by an ordered lattice (such as in a crystal) results in well-formed sharp diffraction patterns. The diffraction pattern produced is essentially a fingerprint of the atomic level periodicity in the polymer sample. It is a particularly useful nondestructive technique that yields information on partial crystallinity (crystallite sizes and orientation), the chain orientation and phase composition of the materials studied. In nanofiber work, XRD methods are useful in studying the crystalline nature of the fibers, change in crystalline morphology during annealing, as well as chain orientation in as-spun or oriented fibers. Electrospun fiber samples collected on a glass slide can be directly studied by XRD without further sample preparation.

According to Bragg's law, the difference in path length between X-ray beams scattered from two adjacent lattice planes of a crystal will be a multiple of the X-ray wavelength, λ:

$$n\lambda = 2d \sin\theta, \tag{5.20}$$

where n is an integer, d is the distance between the two crystal planes, and 2θ is the scattering angle or the angle between the incoming and outgoing beams (lying in the same plane). The larger the diffraction angle 2θ, the smaller will be the length scale probed by the technique. Wide-angle X-ray diffraction (WAXD) is generally used to determine crystal structure on the atomic length scale. (The smallest feature measurable using the technique is $\lambda/2$.)

The complementary technique of small-angle X-ray diffraction (SAXS), where the angle is typically varied from 0.1° to 10°, is used to explore the microstructure on somewhat larger features of colloidal dimensions. The XRD techniques provide scattering patterns as opposed to direct information on the structural features. As the scattering patterns obtained are not necessarily unique to a given morphological feature of interest, caution must be exercised in interpreting XRD data. In semi-crystalline polymers, XRD methods might be used to estimate the percent crystallinity, with the percentage determined by integration of the one-dimensional XRD peaks. Plots of intensity versus 2θ for semi-crystalline fibers consist of an amorphous "halo" signal with peaks from crystalline regions of the polymer superimposed on it. Fractional peak area of the latter is used to estimate percent crystallinity.

X. H. Zong et al. (2002) used two-dimensional WAXD and SAXS to study nanofibers of poly(D,L lactide) (PDLA) electrospun from DMF and PLLA from CH_2Cl_2/DMF (1 : 5). The WAXD data indicated, as expected, little or no crystallinity in the PDLA nanofibers, but the presence of a high level of order in the as-spun mats. Electrospinning yielded metastable amorphous nanofibers of PLLA (a semicrystalline polymer), consistent with the notion that electrospinning being a rapid process generally retards development of crystallinity. The timescale of the whipping instability process is too rapid in most instances to allow crystallite growth during the spinning process (Bognitzki 2001a), but the rapid extension of the jet results in high levels of orientation, as made evident by birefringence measurements (Fong and Reneker 1999). Birefringence increased nearly linearly with fiber diameter for $d = 1$–6 μm, but the degree of orientation obtained was not particularly impressive (Kalayci et al. 2004). The reduced crystallinity in electrospun nanofibers was also demonstrated for electrospun PLLA and PDLA (Zong, X. H., et al. 2002) as well as copolymers of lactide and glycolide such as PLGA (10 : 90) (Zong, X. H., et al. 2003a, 2003b). Low levels of crystallinity were also observed with other polymer nanofibers: in poly(*m*-phenylene isophthalamide) (PMPI) (Liu, W., et al. 2000), crystallizable PEO (Deitzel et al. 2001a), poly(glycolide) (PGA), nylon-6 (Dersch et al. 2003; Liu, Y., et al. 2007), polyesters (PET, PEN and their blends) (Kim, J.-S. and Lee 2000), and in PAN (Jalili et al. 2006). Higher levels of orientation generally results in facile crystallization on annealing. Orientational effects are influenced by solvent properties. This was suggested for PLLA nanofibers electrospun from CH_2Cl_2/MeOH (80/20 wt%) or CH_2Cl_2/pyridine (60/40 wt%). With solutions of PLLA in dichloromethane (7.5 wt%), increasing the solution conductivity (using cosolvents) yielded fibers with higher levels of molecular orientation as measured by thermal and XRD techniques (Inai et al. 2005a). The cold crystallization temperature T_c decreased and the

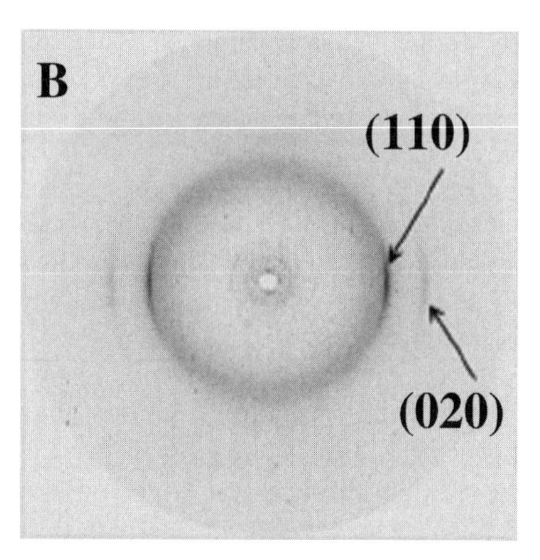

Figure 5.20 WAXD pattern obtained for nanofibers of poly(glycolide-*co*-lactide) (PLGA) (GA/LA: 90/10) annealed at 60°C under a uniaxial strain of 250%. Reproduced with permission from Zong et al. (2003). Copyright 2003. Elsevier.

intensity of the crystallization exothermic peak also decreased with increased solution conductivity.[10]

An increase in chain orientation might also be expected as the applied voltage V (kV) is increased, resulting in an increasingly rapid extension rate of the jet. For instance, the percent crystallinity of ethyl-cyanoethyl cellulose nanofibers electrospun from THF solution increased with the applied voltage, at least up to 50 kV, when all other variables were held constant (Zhao et al. 2004). Other process variables such as the take-up speed of the fiber mat on the rotating collectors and materials variables such as the the solvent, polymer characteristics (such as the average molecular weight) (Lee, J. S., et al. 2004) and the concentration (Inai et al. 2005a; Zong, X. H., et al. 2003a) also affected either the as-spun crystallinity of the nanofibers, or their propensity to crystallize on subsequent annealing. Figure 5.20 illustrates the WAXD pattern developed in nanofibers of PLGA (GA/LA: 90/10) on annealing at 60°C.

Annealing at a higher temperature can dramatically increase the amount of crystallinity in polymers, including nanofibers (Inai et al. 2005a). On annealing PDLA noncrystalline PEO nanofibers electrospun from CH_2Cl_2/DMF (1 : 5), at a temperature of 55°C for 24 h, two characteristic crystalline

[10]However, significant crystallinity on melt electrospinning of the polyesters (at 270°C and 290°C) has been reported for poly(ethylene terephthalate) and poly(ethylene naphthalate) and their blends. Dersch et al. (2003) and others (Veluru et al. 2007) also found significant crystallinity in electrospun nanofibers as well.

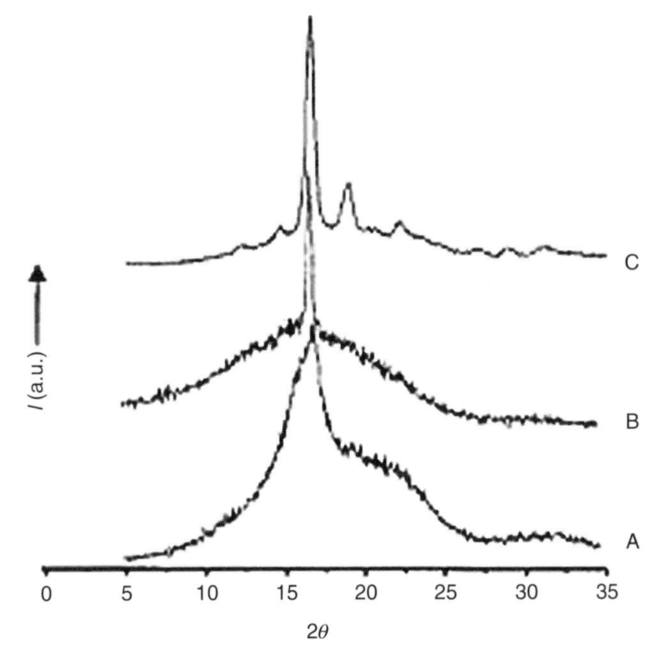

Figure 5.21 WAXD patterns for P[D-L]-*b*-PEG copolymer: (*a*) electrospun nanofiber mats; (*b*) the same mats annealed at 50°C for 2 h; and (*c*) the copolymer powder. Reproduced with permission from Bhattarai et al. (2003). Copyright 2003. John Wiley & Sons.

peaks at $2\theta = 16.4^\circ$ and $2\theta = 18.7^\circ$, which can be indexed as 110 and 131 reflections (pseudo-orthorhombic unit cell[11] in the α-form of the crystals), were observed (Zong, X. H., et al. 2002). Annealed samples studied by SAXS suggested scattering features typical of a microfibrillar structure. The SAXS data also supported the absence of the shish-kebab morphology (Buchko et al. 1999) observed in similar systems. Also, electrospun, essentially noncrystalline poly(glycolide-*co*-lactide) [P(LA-GA)] (GA/LA: 90/10) nanofibers yielded a degree of crystallinity of 23, 30, and 40% when annealing was carried out at 70, 80, and 90°C, respectively (Zong, X. H., et al. 2003b). Similar observations are reported for PVDF electrospun from dimethylacetamide (Choi et al. 2004). Herman's chain orientation function[12] for the nanofibers, however, decreased from −0.22 at 30°C to −0.08 at 90°C due to thermal motion of the chains disrupting chain orientation or

[11]The unit cell is the smallest volume of a crystal that retains its basic crystalline structure and is somewhat analogous to molecules of a chemical compound.

[12]Herman's orientation function *f* is a measure of chain orientation usually relative to the nanofiber axis; its value varies from zero for random orientation to unity for perfect alignment. It is calculated as follows: $f = 3(\cos^2\theta - 1)/2$ where θ is the angle between a polymer chain segment and the nanofiber axis.

possibly due to conversion of such oriented domains into crystallites. Similar development of crystallinity has been reported for nanofiber mats of a biodegradable copolymer, poly(*p*-dioxanone-*co*-L-lactide)-*block*-poly(ethylene glycol) P[D-L]-*b*-PEG. The comparison of WAXD data shown in Fig. 5.21 for electrospun nanofiber mats of the copolymer before and after annealing shows the characteristic sharp crystalline peaks developed (Bhattarai et al. 2003). Dersch et al. (2003), working with electrospun polyamide-6 and PLA nanofibers, however, found no significant difference in their percent crystallinity compared to the less rapidly quenched and even melt-extruded fibers.

Increasing the fractional crystallinity by adding nucleating agents did not work particularly well with nanofibers of polyamide-6. Although crystallites embedded in the matrix of nanofibers did show relatively high degrees of local orientation, the distribution was found to be inhomogeneous. Also, the crystallinity of the electrospun polyamide-6 was predominantly of the γ-form (as opposed to the common α-form typically found in as-received polymer) (Dersch et al. 2003). This finding is in agreement with those based on Raman spectroscopic studies on polyamide-6 nanofibers as reported by Stephens et al. (2004). The formation of the metastable γ-form crystallites is indicative of the high level of deformation undergone by the polymer during electrospinning. On annealing, polyamide-6 electrospun nanofibers increased in crystallinity (and also underwent reversion of the crystallites from the γ-form to the common α-form).

6

COMPOSITE NANOFIBERS

It is a general practice in polymer technology to compound inorganic (and sometimes even organic) fillers into a polymer matrix to either reduce the cost of a formulation or to improve its mechanical properties.[1] Fillers used in the latter case are reinforcing fillers and must be of small enough average particle size and of adequate surface compatibility with the matrix to effectively play this crucial role. Ideally, the particle size should be smaller than the interchain distances in the polymer matrix to avoid the introduction of points of local stress into the material. For instance, in elastomers only filler particles smaller than about a micrometer result in significant levels of reinforcement, with better composite properties obtained at even smaller particle sizes. Larger particles of filler (>10 microns) typically reduce the mechanical properties of composites. Qualitatively, the mechanism of reinforcement in composites is via the transfer of stresses propagating through the polymer to the higher-modulus filler particles. High specific surface area and larger aspect ratio of the filler as well as good compatibility between the filler and polymer will determine the efficiency of load transfer and invariably the degree of reinforcement. Reinforcing fillers are often surface treated to alter their chemistry (e.g., silica might be treated with

[1] Fillers like precipitated calcium carbonate might be used in a plastic composition to reduce the overall cost of the formulation or to improve its optical properties. With fillers such as carbon black (used in tire formulations) or glass fibers (used in fiber-reinforced unsaturated polyester, GRP), the primary role of the filler is to improve the mechanical properties of the composite.

Science and Technology of Polymer Nanofibers. By Anthony L. Andrady

<1% by weight of aminosilane) to allow better wetting or closer interaction of filler with the polymer. Carbon nanotubes and carbon fibers have been extensively researched to improve their function as potential reinforcing fillers in polymer composites.

Composite materials owe their exceptional mechanical and other useful properties to the existence of an extensive interface fraction localized at the phase boundary between filler and bulk resin. The larger the fractional interface (i.e., the smaller the particle or fiber dimension), the more pronounced will be its influence on the properties of the composite. With a compatible filler material the interface is more complex than a simple two-dimensional contact region between the particle and polymer. The interface "layer" formed around the particle has a finite thickness, and within it the material properties are very different from those in the bulk (Pukánszky 2005). These properties depend on interactions that are specific to the polymer/filler system. Experimentally determined thicknesses of the polymer/inorganic filler interface typically range from 0.004 to 0.16 μm (Pukánszky 2005). Therefore, the use of nanoparticle fillers with high specific surface area (as opposed to conventional fillers) that maximize the fractional interface area is particularly desirable in the design of composites.

The properties of the interface can be studied using spectroscopic techniques such as X-ray photoelectron spectroscopy (XPS), Raman spectroscopy, positron anhilation spectroscopy (PAL), or Auger electron spectroscopy (AES). The fractional interface in poly(acrylonitrile) (PAN)-derived carbon fibers,[2] embedded with multiwalled CNTs (MWCNTs) and prepared using an electrospinning technique, was recently investigated (Chakrabarti et al. 2006) using PAL. Increasing the carbon nanotube (CNT) fraction in the carbon–carbon composite resulted in a corresponding increase in the distinct positron trapping sites in the form of vacancy-type defects at the interfaces. Numerous examples of reinforcement of thermoplastic (Kumar et al. 2002) and thermoset (Mamedov et al. 2002) polymers by CNTs are reported in the literature.

Nanocomposite conventional mesoscale fibers (textile fibers that carry nanoparticulate filler) are produced via conventional fiber-spinning techniques by incorporating well-dispersed nanoparticles into the spinning dope. For instance, an intercalated poly(ethylene terephthalate) (PET)/organo-montmorillonite (MMT) nanocomposite prepared by *in situ* polymerization of the polyester in the presence of MMT clay was successfully melt spun into microfibers (Guan, G.-H., et al. 2005). Melt-spun conventional fibers of

[2]Electrospinning petroleum-derived isotropic pitch solutions in tetrahydrofuran (THF) (or as melts) and carbonization of the resulting microfibers also yields carbon fibers (Park, S. H., et al. 2003, 2004a).

nylon-6/MMT composite fibers (Yoon et al. 2004) and PET-based composite fibers containing polyhedral oligomeric silsesquioxane (POSS) nanoparticles (Zeng et al. 2005c) were recently reported. The latter was made by either melt blending POSS with PET at 5 wt% loading level or by *in situ* polymerization with 2.5 wt% reactive POSS followed by spinning. Inclusion of nanoscale fillers even at this low level resulted in significantly improved mechanical properties of the composite fiber (Zeng et al. 2005b). Other nanocomposite microfibers have been reported in the literature including polypropylene/carbon fiber (Ran et al. 2004), cellulose/MMT (White 2004), fluoroethylene-propylene copolymer/MWCNT (Chen et al. 2006), PAN/silver (Yang, Q. B., et al. 2003) and polypropylene/silver (Yeo et al. 2003). Single-walled CNTs (SWCNTs) are a particularly attractive reinforcing filler because of their excellent mechanical characteristics. Incorporation of nanoscale material into conventional fibers is straightforward, as the fiber diameters are several orders of magnitude larger than the particle size.

Nanocomposite nanofibers (nanoscale fibers that include nanoparticulate fillers),[3] however, present a more complex situation as nanofiber diameters allow only a very limited size range of candidate nanoparticles to be accommodated within the composite fiber. However, as long as proper dispersion of filler is ensured, the experimental procedure is not complicated. A range of different nanomaterials has been successfully electrospun in polymer solutions to yield composite nanofibers. Most of the relevant literature pertain to experimental details on how these mats were generated but often with insufficient emphasis on interpreting nanofiber properties or morphology in terms of their interface structure. A simple example of composite nanofibers is afforded by the butyl rubber/carbon black (150–350 nm) composite nanofiber ($d \approx 1-12\,\mu m$) system. Blends of butyl rubber (a copolymer of polyisobutylene with isoprene) and a medium thermal carbon black (250–350 nm particle size) were electrospun from THF solutions (viscosity range of 0.05–0.82 Pa-s) (Viriyabanthorn et al. 2006). The elastomeric polymer solutions with no carbon black were not even electrospinnable into nanofibers, but blending 25–75 phr (parts per hundred rubber) of carbon black into the solution allowed composite nanofibers to be electrospun. Reinforcing fillers such as carbon black are known to function in effect similarly to chemical crosslinks, and physically linking the polymer chains into longer network structures. Normalized 300% tensile modulus of nanofibers mats (corrected for mat density) as a function of carbon black loading was shown to agree well with that predicted by the modified

[3]Assuming the nanoparticles to be fully within the fiber, the term "composite nanofiber" implies a "nanocomposite nanofiber," as microparticles cannot be accommodated within a nanofiber. It is therefore convenient to use the simpler term in referring to these.

Halpin–Tsai equation.[4] The different types of composite nanofibers reported in the research literature fall into three broad classes: those based on carbons, on silicate clays, and on metal particles.

A great majority of the composite nanofibers are of a conventional structure where the nanoparticles are contained within the matrix of the nanofiber. However, another class of composites (referred to here as exocomposites for convenience) consists of the nanoparticles partially embedded in and decorating the surface of the nanofibers. These might be prepared by post-treatment of the nanofiber or by electrospinning in a "dusty" environment, allowing nanoparticles to come into contact with the moist elongating jet. These "decorated" nanofibers may be of interest in applications such as the rapid delivery of poorly soluble bioactive materials via a water-soluble nanofiber exocomposite mat, in optical devices, or as chemical/biological sensors. These will be briefly discussed at the end of the chapter.

6.1 CARBON NANOTUBES IN NANOFIBERS

Ever since Sumio Iijima discovered them in 1991, CNTs, with their intriguing chemical structures, have become one of the most fascinating of nanomaterials. Based on their unusual properties, CNTs are expected to contribute to advances in several different application areas, especially in molecular electronics where they can serve as an almost one-dimensional quantum conductor (nanowire) (Hertel et al. 1998). Other applications of nanotubes, such as hydrogen storage, electrochemical supercapacitors, transistors, field-emitting devices, and nanoscale sensors (Lu, X. B., et al. 2008) have been proposed. Carbon nanotubes rank among the highest-modulus and strongest fibers known[5] (more than 100 times stronger than steel) and are therefore expected to be excellent reinforcing fillers in plastics (Cho and Daniel 2008; Shen et al. 2007). With a range of surface functional chemistries already explored, these can be custom designed for optimal compatibility with different polymer matrices. Their flexibility (which imparts toughness to the composite), light weight ($\rho \approx 1.33$–$1.40\,g/cm^3$), thermal conductivity, and electrical conductivity make them unique reinforcing materials. Table 6.1 illustrates the reinforcing effects of vapor-grown short carbon fibers in conventional polypropylene fibers. CNTs in volume are available as unoriented materials that can only be used as isotropic

[4]The Halpin–Tsai equation (Halpin 1969) expresses the mechanical properties of a composite in terms of those of the filler and the matrix. Modifications of the Halpin–Tsai model to better describe clay-filled nanofiber composites have recently been reported by Ramakrishna et al. (2006).

[5]The theoretical Young's modulus of SWCNTs is ~1 T Pa and 0.3–1.0 T Pa for MWCNTs. The maximum strength is 200 GPa, compared with 2 GPa for steel.

TABLE 6.1 Tensile properties and electrical resistivity of melt-spun conventional composite fibers of polypropylene/carbon fiber (spun at 220°C and a draw ratio of 1.0)

Carbon Fiber Volume Fraction (%)	Tensile Strength (MPa)	Young's Modulus (GPa)	DC Resistivity (Ω m)
0	17.2	1.1	10^{13}
5	49.3	3.2	6.2×10^{9}
10	54	4.2	6.8×10^{-1}
15	56.2	4.9	2.2×10^{-2}

Source: Gordeyev et al. 2001.

fillers. Embedding them in aligned nanofibers (with high level axial orientation) in a mat allows them to be oriented to some extent within a composite. Theoretically estimated and measured properties of CNTs are summarized below

Specific gravity[6] (g/cm^3)	0.8 (SWCNT)	1.8 (MWCNT)
Elastic modulus (T Pa)	~1 (SWCNT)	0.3–1.0 (MWCNT)
Strength (Gpa)	50–500 (SWCNT)	10–60 (MWCNT)
Resistivity ($\mu\Omega/cm$)	5–50	
Thermal conductivity[6] ($W\ m^{-1}K^{-1}$)	3000	
Thermal stability (°C)	> 700	
Specific surface area (m^2g^{-1})	10–20	

Micromechanical computations suggest at least an order of magnitude improvement in specific modulus to be achievable by reinforcement of nanofibers with CNTs (Ko et al. 2006). The rheological percolation threshold for CNTs in polymers is generally low, allowing small fractions of CNTs to effect large enhancements in mechanical properties (e.g., a threshold mass fraction of 0.12% was determined for SWCNT/poly(methyl-methacrylate) (PMMA) nanocomposites based on storage modulus G' measurements at 0.5 rad/s; Du et al. 2004). Using minimal amounts of CNTs in a polymer also renders it electrically conductive. Incorporating a mere 1 vol% of SWCNTs in polyimide enhances conductivity by as much as 10 decades (Park, C. R., et al. 2002), as the electrical percolation threshold of CNTs is also achieved at very low volume fractions. In nanofiber mats of poly(vinylidene fluoride) (PVDF) electrospun from DMF, the percolation threshold for electrical conductivity was reached at only 0.04 wt% of CNTs (Seoul et al. 2003). This is still higher than the 0.015 wt% reported for thin films of the

[6]Indicates a theoretical estimate. Data is based on Xie et al. (2005).

same polymer and may be due to the orientation of the nanotubes along the fiber axis during electrospinning (Dror et al. 2003). In polyetherimide (not nanofibers but compression molded test pieces) 1–3 phr of CNTs reduced the resistivity of the polymer by two orders of magnitude (Kumar et al. 2007).

Carbon nanotubes generally tend to exist as bundles or even networks of aggregates because of strong nonbonded interactions. To exploit their full potential as fillers, however, techniques that achieve near-complete dispersion of nanofibers and improve their compatibility with the polymers, need to be developed. Data on single-fiber measurements that illustrate the reinforcing effect of CNTs in fibers are sparse in the literature (Fornes et al. 2006; Moore et al. 2004). Recent studies on melt-spun conventional composite fibers of polypropylene illustrates reinforcement by CNTs (Moore et al. 2004).

Structurally, the simpler SWCNTs are crystalline sheets of carbon atoms rolled up and connected at the seam to form closed cylinders about 1.2–1.4 nm in diameter. As the carbon atoms are sp^2 hybridized, the tubular structure is essentially a rolled-up graphene and therefore three forms of nanotubes of different chirality are possible (each displaying a different "wrapping angle" a). The value of a also determines the characteristics of the SWCNT; $a = 0^\circ$ yields a zig-zag structure with semiconductor properties, and $a = 30^\circ$ yields the "armchair" structure typical of nanotubes with metallic character. Intermediate angles ($0^\circ < a < 30^\circ$) define yet a third type of nanotube. The more common, and the first-discovered type of CNT, however, is the multi-walled type (MWCNT) consisting of three or more single-walled tubes nested within each other yielding a thicker wall for the nanotube (~20 nm for ~30 nested tubes is typical). Benoit et al. (2002) reported an average diameter of ~14 nm with 18–20 walls as typical for the MWCNTs.

In electrospinning composite nanofibers, particular attention needs to be paid to achieving a good dispersion of nanotubes in the spinning solution, and achieving a high degree of axial orientation of individual CNTs to maximize mechanical properties of the nanofibers (Fig. 6.1). As the reported properties of nanofibers depend on these two factors, comparing reported experimental data on the efficiency of reinforcement from different studies even where the same weight fraction of the same type of CNTs were used, should be attempted with caution. The formidable processing challenge of fully dispersing CNTs into individual nanotubes to allow their high-volume industrial use has not been economically achieved as yet. Table 6.2 summarizes the reported information on composite nanofibers.

6.1.1 Dispersion of Nanotubes

Carbon nanotubes are insoluble in water or other solvents but can be suspended or dispersed in liquids. As-received samples of CNTs being agglomerated into

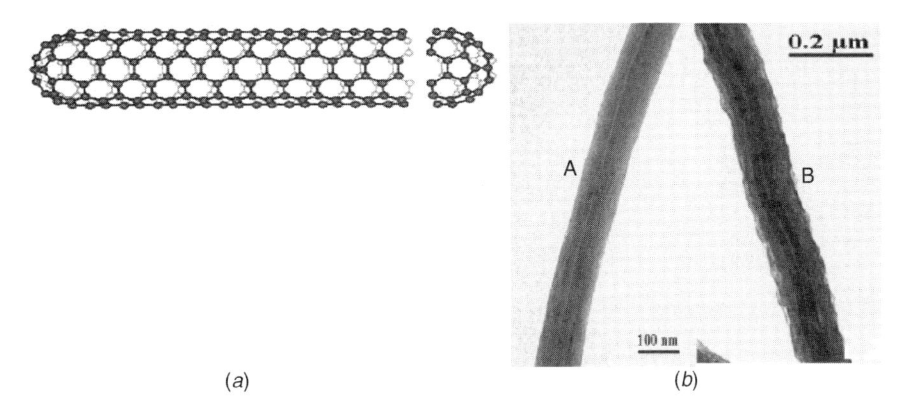

Figure 6.1 (*a*) Representation of a single-walled CNT showing the rolled-up graphene chemical structure. (*b*) Transmission electron microscopy (TEM) image of PAN/MWCNT composite nanofibers before (A) and after (B) calcination, showing CNTs oriented along the fiber axis. Reprinted with permission from Hou et al. (2005). Copyright 2005. American Chemical Society.

masses or "ropes" of nanotubes, and considerable amounts of energy have to be used to separate the material into individual tubes. Optimal reinforcement of the electrospun composite nanofibers depend on the extent of the dispersion of CNTs in the spinning solvent. Experimentally, this is achieved by either mechanical methods [either using high shear mixers (Xie et al. 2005; Chen, G.-X., et al. 2007) or more commonly by sonication (Sundaray et al. 2006)]. Sonication in the presence of a surfactant or a dispersing agent has also been used (Delozier et al. 2006). Sonication alone can achieve some degree of dispersion of the nanotubes within the composite nanofibers (Sung et al. 2004) depending on the frequency and duration (usually for 1–2 h or sometimes for longer periods) of treatment. With the MWCNT/PAN system (Kedem et al. 2005) and SWCNT/PAN system (Ko et al. 2006), sonicating the CNTs in dimethylformamide (DMF) for 3 h resulted in a good dispersion within the electrospun nanofibers. Sonication, however, can also potentially damage nanotubes, resulting in kinked, bent, or damaged CNTs in the composite nanofibers (Ayutsede et al. 2006; Dror et al. 2003). Good dispersion is not obtained in all systems; for instance, in the DMF/SWCNT systems, dry nanotubes did not disperse well by sonication alone (Du et al. 2003).

The highest modulus (~50 GPa) composite microfibers of polymer/SWCNTs are generally obtained using a combination of surfactant or an interfacial binding agent along with sonication to achieve particularly high levels of dispersion (Baughman 2002; Delozier et al. 2006). The same can be used with nanofibers as well; for instance, sodium dodecyl sulfate surfactant has been used for the purpose (Dror et al. 2003; Kim, H. S., et al. 2006). Typically, the CNT suspension in a suitable solvent is mixed with a

TABLE 6.2 Examples of composite polymer nanofibers filled with carbon nanotubes

Polymer	Filler	Dispersant	Solvent	Fiber *d* (nm)	Characterization	Reference
Silk (*B. mori*)	SWCNT	–	Formic Acid	35 ± 15–153 ± 99	SEM, TEM, FTIR	Ayutsede et al. (2006)
PAN	MWCNT		DMF	100–500	SEM, Raman, PAS	Chakrabarti et al. (2006)
PAN	MWCNT/TiO_2	S	DMF	20–140	TEM, HRSEM	Kedem et al. (2005)
PAN	MWCNT	S + F	DMF	100–300	TEM, FTIR, Raman, XRD, Tensile	Hou et al. (2005)
PAN	MWCNT	S + D		50–300	UV-Vis, AFM, TEM, Tensile, XRD	Ge et al. (2004)
PAN	MWCNT	S + D	DMF	500 ± 60	SEM, AFM	Kim, H. S., et al. (2006)
PAN PMMA	SWCNT	S	Nitromethane DMF	100–2000	TEM, Raman	Liu, J., et al. (2005); Liu, L., et al. (2007)
PVA	MWCNT	S + F (acid mix)	H_2O	100–200	SEM, TEM, Tensile	Jeong (2007)
	SWCNT	S	H_2O		SEM, TEM, Raman	Kannan (2007)

PC	MWCNT	S	$CHCl_3$	~350	SEM	Kim, G.-M., et al. (2005)
Nylon 6	MWNCT	S, D	DMF or H_2O	~500	SEM	Kim, H. S., et al. (2006)
	MWNCT	S + F (—COOH)	Formic Acid, DMF	–	SEM, FTIR	Jeong, J. S., et al. (2006)
	MWCNT	S + F (—COOH)	HFP	500–900	SEM, TEM, XRD, Tensile	Jose (2007)
PLA/PAN	SWNCT	–	DMF	–	TEM, AFM	Ko et al. (2003)
PBT	MWCNT	S	HFP	>2500	SEM, AFM, TGA	Mathew et al. (2005)
PEO	SWCNT	S + D	H_2O	–	SEM, TEM, XRD	Salalha et al. (2004)
PEO	MWCNT	S + D	H_2O	–	XRD, TEM	Dror et al. (2003)
PEO	MWCNT	S + D	H_2O	50–200	AFM, SEM, TEM	Liu (2007)

Dispersant: S – Sonication; D – Detergent; F – Functionalized.

polymeric (Delozier et al. 2006) or other amphiphilic compound and the mixture is sonicated to obtain a good dispersion. For example, a 0.35 w/w% suspension of MWCNTs in water, treated with 1 wt% of gum arabic and the mixture sonicated at 43 kHz for 1 h (Dror et al. 2003), achieved good dispersion. TEM images of PEO ($M_w \approx 900{,}000$ g/mol)/MWCNT composite nanofibers electrospun from these dispersions showed the embedment of nanotubes as individual entities within the fiber. Similar results showing aligned and separate nanotubes in the fiber matrix were reported for SWCNT/PEO composite nanofibers, using an amphiphilic copolymer poly(styrene-*co*-sodium maleate) as the dispersing agent. Some polyimides [those prepared from 2,7-diamino-9,9′-dioctylfluorene (AFDA) and either 3,3′,4,4′-oxydiphthalic anhydride or 3,3′,4,4′-biphenyltetracarboxylic dianhydride also act as good dispersants for CNTs (Delozier et al. 2006)].

However, unlike with SWCNTs, the inclusion of MWCNTs in the matrix reduced the degree of axial orientation of PEO chains and of crystallites in the nanofibers (Salalha et al. 2004). Again, the orientation of CNTs within nanofibers was found to depend on the extent of their dispersion in the spinning dope. In electrospun silk fibroin/SWCNT composite nanofibers (Ayutsede et al. 2006; Ko et al. 2004), however, about a 50% increase in the crystallinity of silk was reported at a level of only 2 wt% CNT content (Ayutsede et al. 2006). The presence of well-oriented nanotubes in the nanofiber matrix appears to encourage local ordering of polymer chains in the vicinity of the inclusions. Similar data were reported for coaxial electrospinning of silica nanoparticles in core material in poly(vinylpyrrolidone) (PVP)–polyacrylonitrile (PAN) core-shell nanofibers (Hong, Y. L., et al. 2005).

Modification of the surface chemistry of CNTs can also help achieve good dispersion. Surface oxidation of MWCNTs with 6M HNO_3 assisted by continuous sonication was successfully used to disperse nanotubes (without using surfactants) in the preparation of PAN ($M_w = 86{,}000$ g/mol)/MWCNT dispersions for electrospinning. This treatment introduces surface carboxylic acid groups on the nanotubes and helps their dispersion in DMF (Hou et al. 2005). Composite nanofibers ($d \approx 100$–200 nm) of PAN with oxidized MWCNTs, electrospun from DMF (0–20 wt%), and collected on a rotating drum showed no signs of delamination at the carbon/polymer interface (Ge et al. 2004). The smooth fiber surface typical of PAN, however, was roughened due to the inclusion of CNTs as observed in both TEM and atomic force microscopy (AFM) images. This is likely due to some nanotubes breaking the surface layer of the fibers (see Fig. 6.2). The same was reported for poly(butylene terephthalate) (PBT)/MWCN composites (Mathew et al. 2005) as well.

The inclusion of CNTs in electrospun nanofibers appear to increase the fraction of relatively thicker nanofibers in the collected mats (Mathew et al.

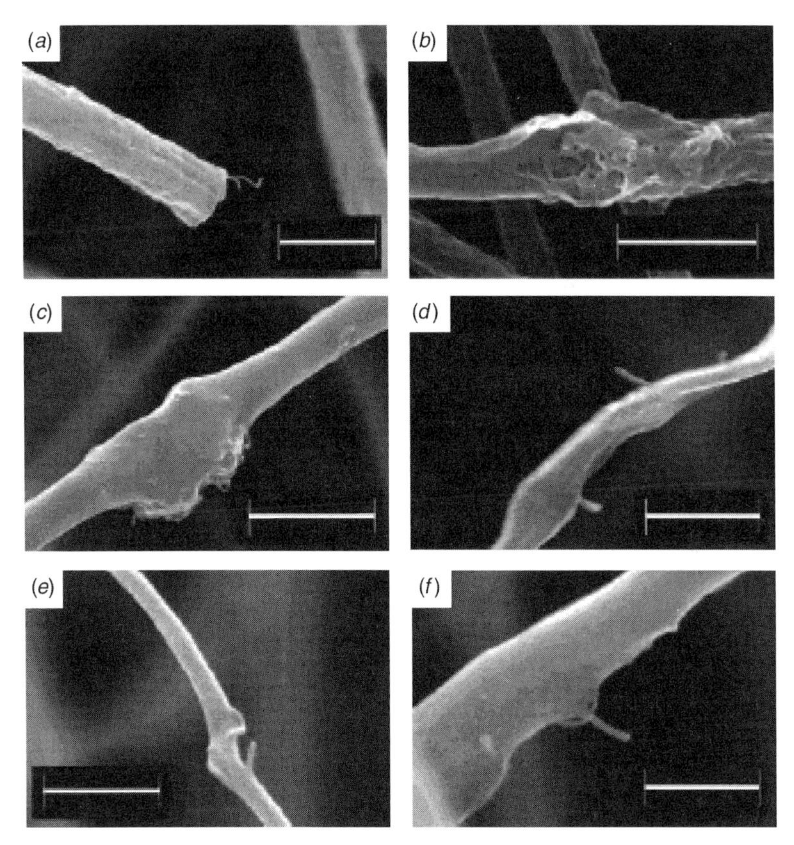

Figure 6.2 SEM images of electrospun nanofibers of PMMA/MWCNT showing several types of irregularities: (*a*, *b*) PMMA/MWCNT (99/1 wt%); (*c*–*f*) PMMA/MWNT (95/5 wt%). Scale bars: 1 μm for (*a*–*c*) and 500 nm for (*d*–*f*). Reprinted with permission from Sung (2004). Copyright 2004. American Chemical Society.

2005; Wang, Z.-G., et al. 2006). This effect is at least in part due to the presence of agglomerates of incompletely dispersed nanotubes that have to be accommodated within the nanofiber dimensions. [As expected, melt-spun conventional fibers such as PP/CNT also show a similar linear increase in denier with the fraction of CNTs in the composite (Erickson 2003).] However, Ra et al. (2005) reported the average nanofiber diameters to decrease with increasing MWCNT content. Under their processing conditions the increased electrical conductivity and consequent increase in surface charge due to the presence of CNTs in the jet may have led to smaller average nanofiber diameters. Some of the irregularities in composite nanofibers filled with MWCNTs are illustrated in Fig. 6.2. The increase in mechanical properties of the fiber mats at different loadings of MWCNTs is summarized in Table 6.3 (Ge et al. 2004; Hou et al. 2005). Improvement in

TABLE 6.3 Tensile properties of mats of composite nanofibers of PAN/MWCNT

PAN/ MWNT	Tensile Strength (MPA)	Modulus (GPa)	Extensibility (%)	Tensile Strength (MPA)	Modulus (GPa)	Extensibility (%)
	Ge et al. (2004)			Hou et al. (2005)		
100/0	265	4.5	17.8	4.57	1.8	10.7
97/3	312	6.4	12.8	6.57	2.5	8.6
95/5	366	9.8	9.9	8.00	3.1	2.5
90/10	370	10.9	8.2	4.86	3.7	1.3
80/20	285	14.5	4.4	3.71	4.4	0.9

Source: Ge et al. 2004; Hou et al. 2005.

mechanical properties reported is significant, with the modulus increasing more than threefold at 20 wt% CNT. Others have reported more modest changes in the composite mat modulus for the same loading range of MWCNTs (Hou et al. 2005; Ye et al. 2004). Similar data for single composite nanofibers have been reported (Fornes et al. 2006).

6.1.2 Orientation of Nanotubes

The best reinforcement of polymer nanofibers with CNTs requires their near-perfect orientation in the axial direction within the nanofiber. Conventional fiber spinning generally results in some degree of orientation of high-aspect-ratio fillers in the direction of the flow at the spinneret (Siochi et al. 2004). Given the rheology involved in electrospinning, individual nanotubes will be sucked into the jet and will undergo such orientation to even a greater degree (probabilistic models of the process have been proposed; Dror et al. 2003). Constraints imposed by the converging nanoscale jet help the process by reducing the number of available orientations for the nanotube in the flow field, encouraging their axial placement. The process is illustrated in Fig. 6.3. Figure 6.4 shows a TEM image of SWCNTs aligned along the fiber axis within composite PMMA nanofibers.

Ko et al. (2003) electrospun composite nanofibers of PAN with dispersed SWCNTs from DMF solutions and reported their orientation in the axial direction in the fiber. Using an AFM-based indentation technique (see Chapter 5), the modulus of the composite PAN fibers (as opposed to that of fiber mats) was measured. The modulus of the nanofibers increased linearly with the volume fraction of CNTs incorporated. Interestingly, the increases were also higher (by a factor of more than two) than that expected on the basis of the rule of mixtures calculated assuming a value of 1 TPa for

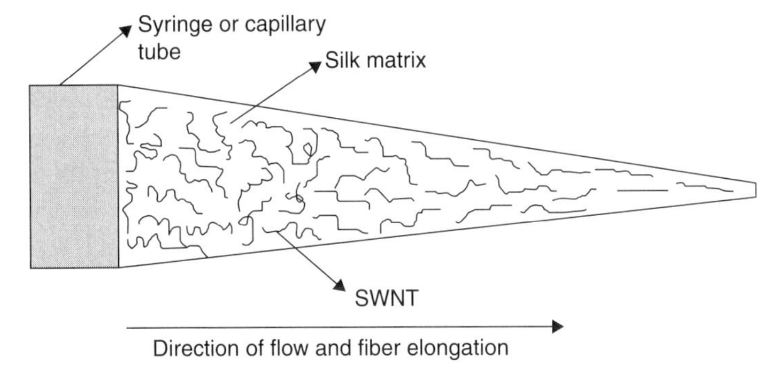

Figure 6.3 Schematic representation of CNT alignment during electrospinning. Reprinted with permission from Ayutsede et al. (2006). Copyright 2006. American Chemical Society.

the modulus of the SWCNTs. The observation was explained in terms of structural changes undergone by polymer chains in the presence of nanotubes. In electrospun PAN/MWCNT nanofibers X-ray studies showed the orientation of the carbon nanotubes within the nanofibers to be even higher than that of the PAN polymer crystal matrix (Ge et al. 2004). One of the strongest natural fibers (with a modulus of 4 GPa) is spider silk. The electrospun nanofibers of transgenic spider silk with only 1 wt% of well-dispersed SWCNTs showed a tenfold increase in their already high modulus, a fivefold increase in strength, and threefold increase in toughness (Ko et al. 2004). The exceptional level of reinforcement obtained in this case is indicative of

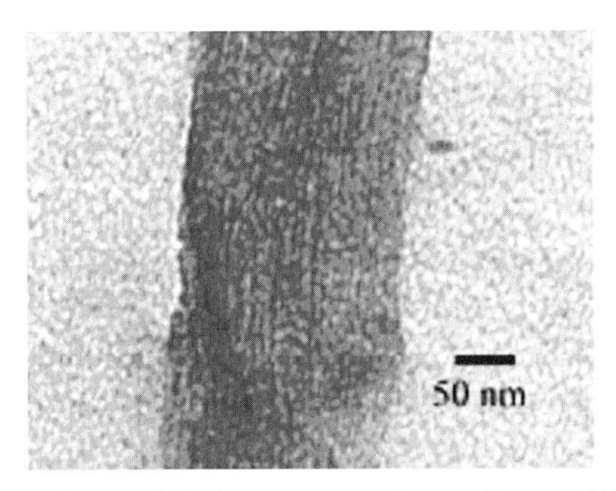

Figure 6.4 TEM image of electrospun composite nanofibers of PMMA/SWCNT showing highly aligned CNTs. Reprinted with permission from Sundaray et al. (2006). Copyright 2006. American Institute of Physics.

the excellent interface properties afforded by the CNTs in silk and is particularly impressive.

Both Raman spectroscopy (Benoit et al. 2002; Sundaray et al. 2006) and wide-angle X-ray diffraction (WAXD) (Chen, X., et al. 2006) are used to characterize composite nanofibers containing CNTs; polymorphs of carbon being Raman-active, Raman spectroscopy in particular provides a wealth of information on SWCNTs in composite materials. The Raman spectrum of a carbon film is dominated by the D-peak (1360 cm^{-1}) and a G-peak (1580 cm^{-1}) attributed to sp^2 carbon species. The presence of nanotubes generally results in a peak in the radial breathing mode (RBM) region (75–300 cm^{-1}) and tangential mode (1500–1700 cm^{-1}) of the spectrum (Ko et al. 2006). The frequency of peak in the RBM is inversely proportional to the nanotube diameter. Information on the orientation of CNTs within the matrix is reflected in the tangential mode.

The utility of the technique is illustrated by the Raman spectrum of PAN/MWCNT composite nanofibers shown in Fig. 6.5 (Hou et al. 2005). Strong Raman scattering (disorder-induced scatter) by MWCNTs in the matrix gives rise to D, G, and D′ signals. The peak at about 2300 cm^{-1} is due to the –CN group of the polymer. Raman spectroscopy also has been used to study CNTs in composite nanofibers of silk (Ayutsede et al. 2005, 2006), carbon fibers (Chung et al. 2005), and PAN (Hou et al. 2005).

As already suggested in Chapter 5, X-ray methods are also particularly useful in studying nanocomposites. The WAXD spectrum of the nanocomposites yields nonoverlapping peaks characteristic of the crystallinity inherent in the PAN polymer as well as that associated with CNTs (Hou et al. 2005).

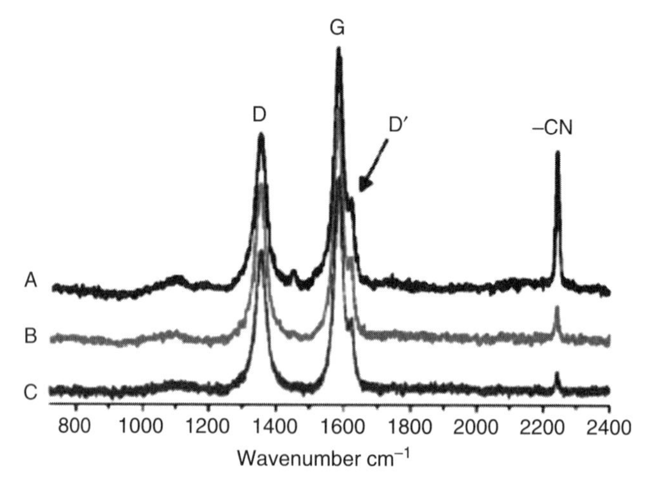

Figure 6.5 Raman spectrum of PAN/MWCNT composite nanofibers. A, B, and C refer to 5%, 10%, and 20% of the nanotubes in the fibers. Reprinted with permission from Hou et al. (2005). Copyright 2005. American Chemical Society.

Diffraction of the (002) crystal plane from MWCNT (at $2\theta \approx 27^\circ$) is quite distinct from the diffraction due to the (200) crystal plane from the PAN, which is seen at $2\theta \approx 15-20^\circ$. Figure 6.6 shows the WAXD results, comparing the spectra for PAN with those for PAN/MWCNT composite nanofibers with different volume fractions of CNTs dispersed (as indicated in the caption of the Fig. 6.6). Peaks due to CNT content generally tend to be sharper than those due to polymer (Hou et al. 2005). The peak corresponding to the (002) crystal plane of the MWCNT increases with CNT content and allows quantification of the nanotubes in composite.

WAXD is routinely used to demonstrate the linear increase in crystallinity of polymers with the volume fraction (0.5–2.0%) of nanotubes incorporated in the composite, for example, in silk/SWCN nanofibers (Ayutsede et al. 2006); in PAN/MWCNT (10–20 wt%) nanofibers (Ge et al. 2004); and PEO/SWCNT (~ 1 wt%) (Salalha et al. 2004). Poly(ethylene oxide) nanofibers electrospun from 3% solutions in ethanol/water (40/60 v/v) with 1% sodium dodecyl sulfate (dispersant) yielded the diffraction pattern shown in Fig. 6.7*a* (Dror et al. 2003; Salalha et al. 2004). The pattern for control nanofiber is consistent with PEO crystallites being aligned along the chain axis and therefore along the fiber axis as well. The image in Fig. 6.7*b* shows the same nanofibers with 0.35 wt% of MWCNTs. Clearly the presence of the CNTs very significantly reduced this orientation of crystallites. The origin of this detrimental effect of nanotubes on crystallinity has not been explained satisfactorily.

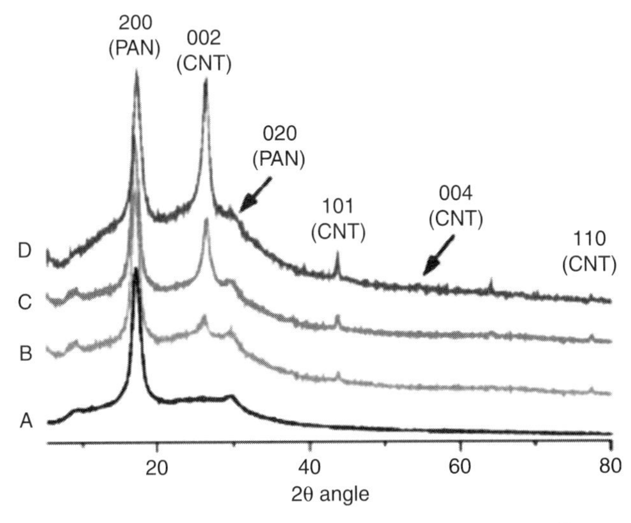

Figure 6.6 WAXD scattering in PAN/MWCNT composite nanofibers. A, B, and C refer to 5%, 10%, and 20% of the nanotubes, respectively. Reprinted with permission from Hou et al. (2005). Copyright 2005. American Chemical Society.

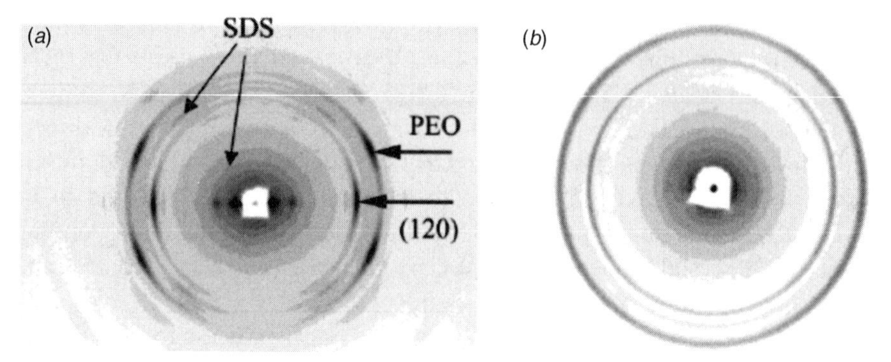

Figure 6.7 X-ray diffraction patterns for PEO nanofibers with the dispersant sodium dodecyl sulfate: (*a*) control nanofibers: (*b*) composite nanofibers with MWCNTs. Reprinted with permission from Dror et al. (2003). Copyright 2003. American Chemical Society.

6.1.3 Other Carbons

Carbon blacks have been used as reinforcing fillers in rubber products since the early days of the industry. The particle size of carbon and its surface chemistry were widely recognized to determine the extent of reinforcement obtained in their composites. A medium-grade thermal black (MT) for instance has an average particle size of about 350 nm and a surface area of $\sim 9\,m^2/g$. Carbon/rubber blends can be electrospun into fibers; however, the fibers tend to be in the micrometer range when large-particle blacks are used. Viriyabanthorn et al. (2006) electrospun butyl rubber/MT blends from THF solutions. Defect-free fibers were difficult to spin from butyl rubber, but the physical crosslinking afforded by carbon (25, 50, and 75 phr) allowed the polymer to be electrospun.[7] Even with electrospinnable polymer solutions of polyimides, the incorporation of carbon black into the dope resulted in narrower fiber diameters under the same electrospinning conditions (Lee, S. G., et al. 2002).

Exfoliated graphite carbon is a reinforcing filler material in polymer matrices. Mack et al. (2005) electrospun exfoliated graphite in PAN/DMF mixtures to obtain composite nanofibers containing up to 4 wt% graphite. Exfoliation was achieved by first reacting graphite with metallic potassium and reacting the resulting KC_8 product with ethanol. Young's modulus of the composite nanofibers $d \approx 300\,nm$ (measured by AFM nano-indentation technique) increased linearly with the weight fraction of graphite.

[7]Interaction of the carbon surface with rubber molecules results in an increase in solution viscosity η of butyl rubber solutions with the concentration c (phr) of added carbon black: $\eta = \eta_0(1 + 2.5c + 14.1c^2)$, where η_0 is the viscosity of polymer solution at $c = 0$ (Viriyabanthorn et al. 2006).

Petroleum-derived isotropic pitch precursors have also been electrospun either from THF solutions or from melt into microfibers and carbonized into carbon-fiber mats (Park, S. H., et al. 2004a, 2004b). Kessick and Tepper (2006) electrospun nanofibers of PEO ($M_w = 400{,}000$ g/mol in 8% aqueous solution), PVP ($M_w = 1{,}300{,}000$ g/mol in 8% aqueous solution), polyisobutylene (PIB) ($M_w = 500{,}000$ g/mol, 9 wt% in toluene), and poly(epichlorohydrin) (PECH) ($M_w = 700{,}000$ g/mol, 9 wt% in $CHCl_3$) containing ~15 wt% of carbon powder (Cabot Black Pearl 2000 grade). The change in each case in electrical resistance of these on contact with volatile organic chemicals (VOCs) was studied and their effectiveness as chemical sensors (in electronic nose applications) was explored.

6.2 METAL–NANOFIBER COMPOSITES

Unlike with CNTs or mineral fillers, metal nanoparticles are generally not intended to provide reinforcement in polymer nanofibers. Composite nanofibers with metal particles are of interest primarily because of their potential use in catalysis (Lewis 1993), sensors (Aussawasathien et al. 2005; Liu et al. 2004; Wang, Z.-G., et al. 2006), and in electrical applications (Sawicka et al. 2005; Song, M. Y., et al. 2005; Zhu et al. 2006b). However, the presence of metal particles can still have an impact on nanofiber morphology, particularly on the degree of crystallinity of the material. Carbonized PAN nanofibers that contain iron acetylacetanoate, for example, showed the evolution of the graphite crystal structure in the nanofiber at a relatively lower temperature (1300°C) compared to that for regular PAN fibers (~2000°C) (Park, S. H., et al. 2005). Nanocomposites might be prepared using either direct electrospinning of nanoparticles mixed in with the polymer solution, as with the CNTs discussed in the previous section, or also by liquid-phase (or gas-phase) post-reaction of the spun fibers to generate them in the nanofiber.

6.2.1 Direct Electrospinning

The simplest route to preparing metal nanoparticle/polymer composite nanofibers is to electrospin from a solution that contains preformed metal nanoparticles. The nanoparticles, however, need to be protected from agglomeration in solution during the spinning process by using a surfactant or a shell layer of capping molecules over their surface. Several examples in the literature illustrate the feasibility of this approach. Copper nanoparticles generated by reducing $CuCl_2$ with hydrazine in solution in the presence of poly(vinyl alcohol) (PVA) yielded a suspension of elemental copper nanoparticles that was then electrospun into PVA/Cu-nanoparticle composite nanofibers. The same technique was also used to prepare PVA/Cu-core/shell

(~100 nm/400 nm) nanocables by coaxial electrospinning (Li, Z. Y., et al. 2006b). The difficulty in dispersing nanoparticles is avoided to some extent by synthesizing the nanoparticles in the polymer solution itself prior to electrospinning. Similarly, PVP solutions in ethanol containing various concentrations of $AgNO_3$ were reacted with H_2S gas to obtain nanoparticles of silver sulfide *in situ*, and the mix was electrospun to obtain composite nanofibers of PVP/nanoparticle–Ag_2S. The method yielded crystalline nanoparticles (~15 nm) of β–Ag_2S phase as ascertained by X-ray diffraction methods (Lu, X. F., et al. 2005a). Nanocomposites with lead sulfide were also prepared using a similar procedure (Lu, X. F., et al. 2005a); the progress of this reaction can be monitored visually or with visible-radiation spectroscopy, as the nanofiber turns yellow as the reaction proceeds.

This approach has also been used in conjunction with coaxial spinning of core–sheath nanofibers (see also Chapter 9) to encapsulate nanoparticles of Pt/Fe in a poly(ε-caprolactone) (PCL) nanofiber core matrix (see Fig. 6.8). The Pt/Fe nanoparticles were synthesized by reduction of platinum acetylacetonate [$Pt(CH_3COCHCOCH_3)_2$] using 1,2-hexadecanediol with simultaneous decomposition of iron pentacarbonyl [$Fe(CO)_5$]. The ~4 nm nanoparticles, having a composition of ($Fe_{52}Pt_{48}$), were stabilized and suspended in hexane at a concentration of 5 mg/mL. This was used as the core material along with PCL ($M_n = 80{,}000$ g/mol) in 2,2,2-trifluoroethanol (TFE) (150 mg/mL) as the shell material (Song, T., et al. 2005). The solutions spun through a coaxial spinneret yielded a uniform encapsulation of the nanoparticles in the core of the nanofiber, as seen in the TEM image of Fig. 6.8.

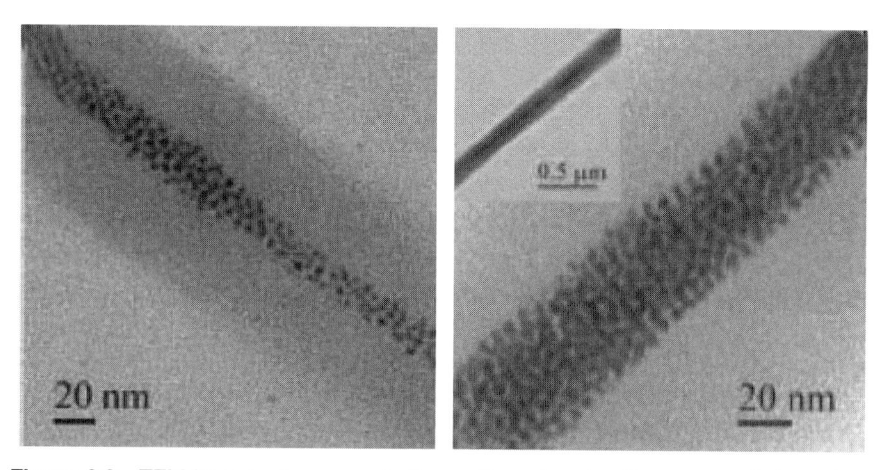

Figure 6.8 TEM image of a core–shell nanofiber showing the encapsulation of the $Fe_{52}Pt_{48}$ nanoparticles in the core region of the PCL nanofibers. Reprinted with permission from T. Song et al. (2005). Copyright 2005. Elsevier.

6.2.2 Reductive Post-Reaction

A second approach to nanoparticle synthesis is to reduce the metal salt incorporated into nanofibers in an *in situ* post-reaction within the nanofibers to yield discrete nanoparticles on its surface and in bulk. Heating fiber mats in a reductive atmosphere or calcination can be used to facilitate the reduction reaction. There is some tendency for metal nanoparticles to migrate on the surface of fibers at higher temperatures, leading to some inevitable sintering and aggregation.

The approach is illustrated by nanocomposite fibers of nanoparticle–Fe/carbon. Nanofibers of PAN were electrospun from DMF (6.7 wt%) solutions containing 3.3 wt% of dissolved ferric acetylacetanoate. Subsequent carbonization[8] of the nanofibers in an inert (Ar and H_2) atmosphere (Hou and Reneker 2004) at high temperatures yielded carbonized nanofibers with nanoparticles of elemental iron. These were in the size range of 10–20 nm for the most part and were embedded on the surface of the fibers. Essentially, the same approach was also used with polycarbonate (PC)–palladium acetate solutions, but on calcination of the electrospun polymer nanofibers yielded inorganic palladium oxide nanofibers (Viswanathamurthi et al. 2004a) rather than Pd–nanoparticle/PC. The oxygen in keto groups of the PC was speculated to have reacted with the metal acetate to yield the oxide. Palladium nanoparticles are of particular interest in industry because of their potential use in catalysis[9] (Briot and Primet 1991).

Reduction of the metal salt can also be carried out under milder conditions. For instance, poly(acrylonitrile-*co*-acrylic acid) P(AN-AA)–$PdCl_2$ with 0.63–8.3 wt% of the inorganic salt (based on polymer) was electrospun from 8 wt% DMF solutions (Demir et al. 2004). Nanofibers ($d \approx 165 \pm 35$ nm) were then reacted with aqueous hydrazine (0.5 vol%) in basic solution at ambient temperature to reduce the metal salt into Pd nanoparticles:

$$N_2H_4 + 2PdCl_2 \longrightarrow N_2 + 2Pd + 4HCl \qquad (6.1)$$

The technique yielded composite nanofibers with 50 mg Pd/g polymer with nanoparticle diameters in the range of 10–60 nm. The catalytic effectiveness of Pd in the composite nanofiber mat for hydrogenation reaction of dehydrolinalool (3,7-dimethyloct-6-ene-1-yne-3-ol) was assessed to be 4.5 times higher than that for a conventional Pd/alumina catalyst. Silver nanoparticles

[8]This is a multistep process involving low-temperature annealing in air (250°C/3 h); heating to 500°C in Ar; heating in an Ar : H_2 (3 : 1) atmosphere (550°C/4 h); and heating to 1100°C in Ar (carbonizing for 0.5 h) and cooling to 700°C.

[9]Half the annual palladium demand (2006) was in autocatalyst applications.

can also be similarly incorporated into PAN nanofibers via nanofibers electrospun from polymer-$AgNO_3$ solutions. Wang and colleagues reported nanofibers reduced with aqueous hydrazinium hydroxide to yield homogeneously dispersed Ag–nanoparticles (~10 nm) on the nanofiber (Wang, Y. Z., et al. 2005).

The silver salt can also be readily reduced photochemically by UV irradiation under a high-pressure Hg lamp (Li, Z. Y., et al. 2006a; Son et al. 2004b). The average size of the Ag–nanoparticles could be controlled (in the range 3.5 nm to 10 nm) by varying the molar ratio of acrylonitrile repeat units in PAN to that of the silver salt. Silver NPs deposited on the cellulosic nanofibers prepared using this method were shown to have antibacterial characteristics (Son et al. 2004b). Using a sol–gel precursor allows inorganic nanofibers (such as anatase nanofibers with Ag) to be produced using a very similar method (Lee and Sigmund 2006).

6.2.3 Gas-Phase Post-Reaction

It is advantageous to use a gaseous reactant to convert the metal salt into either metal or metal compound nanoparticles within nanofibers because it avoids the difficulties associated with the removal of residual reactants from the mat. The short diffusion distances within nanofiber dimensions ensures rapid reaction making this a very practical technique. The approach was successfully demonstrated by Lu and colleagues in the preparation of composite nanofibers of CdS nanorods in PVP. Cadmium acetate (100 wt% of polymer) was dissolved in the spinning solution and electrospun into nanofibers carrying the salt. The conversion of the acetate into the sulfide within the nanofiber was carried out by exposing the nanofibers to gaseous H_2S (Lu, X. F., et al. 2005b). The nanorods obtained were relatively large (50 nm diameter and with a length of 100–300 nm) and were dispersed randomly in nanofibers of 100–900 nm. It is not clear if this is a generic approach that works well with any polymer; specific interactions with PVP functionalities may have facilitated the process in this instance. FTIR evidence has suggested interaction between Cd^{2+} and the polymer carbonyl group, but no mechanism has been proposed (Lu, X. F., et al. 2005b).

A dip-coating method of depositing a metal oxide coating on preformed nanofiber mats from a sol–gel precursor system has also been described in the literature. This approach does not, however, always result in uniform coatings of the oxide, although 20–80 nm thick coatings (on ~100 nm fiber) have been reported (Drew et al. 2003a). Drew and colleagues reported a liquid-phase deposition process on PAN nanofibers coated with TiO_2 and SnO_2 layers. The fiber mats were immersed for 12–36 h in an aqueous mixture of equal volumes of aqueous 0.12 M hexafluorotitanate(IV) ammonium

[TiF_6-$(NH_4)_2$] and 0.2 M boric acid (H_3BO_3) (or equal volumes of aqueous 0.12 M hexafluorostannate(IV) ammonium [$SnF_6(NH_4)_2$] and 0.2 M H_3BO_3 solution in the case of SnO_2 deposition). Reaction of the titanate or stannate with water yields the oxide, while the boric acid consumes the HF formed during reaction (Drew et al. 2003a, 2005).

6.3 POLYMER–CLAY COMPOSITES

A majority of studies on composite nanofibers reinforced with clays has been on montmorillonite (MMT), a clay with plate-like particles having a chemical composition of hydrated sodium calcium aluminum magnesium silicate hydroxide $(Na,Ca)_{0.33}(Al,Mg)_2(Si_4O_{10})(OH)_2 \cdot nH_2O$ (value of n varies with the degree of hydration). The platelets of the clay have a high modulus (170 GPa), a high aspect ratio (1000 nm × 75–100 nm), a surface area of 750 m^2 per g, are hydrophilic, and occur in aggregated form or as tactoids. These have to be dispersed into individual platelets to exploit their high surface area in reinforcing polymers. As with CNTs, this is usually achieved by a combination of sonication and the use of chemical agents or surfactants. X-ray diffraction signals characteristic of MMT tactoids disappear on complete dispersion of the platelets in a liquid or polymer medium (Hong, J. H., et al. 2005).

Some grades of commercial MMT are surface modified with quaternary ammonium compounds to facilitate better dispersion and interaction with the polymer matrix. Organically modified Cloisite-30B (Southern Clay Products Inc.) is functionalized with a quaternary ammonium ion. It has a structure of NP—$(CH_2$–$CH_2OH)_2(CH_3)$–T, where NP refers to the clay nanoparticle, T represents the hydrogenated tallow (C-14 ≈ 65%, C-16 ≈ 30%, and C-18 ≈ 5%) and also contains Na^+ and Ca^{2+} ions associated with platelets. Dispersing 1–5 wt% of the clay into polyurethane-urea solutions in dimethylacetamide (DMAc)/THF (7 : 3 w/w) resulted in increased conductivity and therefore quality nanofibers (Hong, J. H., et al. 2005). Generally, the inclusion of clays in nanofibers tends to decrease the average fiber diameter (Zhou et al. 2006) possibly due to increased conductivity of the spinning solution due to the presence of MMT.

Ultrasonic-assisted blending of polymer with clay can yield well-dispersed solutions suitable for electrospinning. The clay is typically sonicated in the solvent (e.g., THF/DMF; 50 : 50) for about 3 h and mixed with a solution of the polymer (e.g., PS dissolved in the same solvent) and the mixture sonicated for a further 3-h period (Ji et al. 2006b). The resulting solution with a concentration of 5–20 wt% PS and 1–8 wt% of MMT clay (Closite B) was electrospun into nanofibers varying in average diameter from 150 nm to

4000 nm by adjusting the polymer concentration (Ji et al. 2006b). At a level of 4 wt% of clay, highly oriented, unagglomerated platelets were obtained in these nanocomposite fibers. The existence of clay not only increased the modulus of fibers as determined by shear modulation force microscopy (a technique based on AFM microscopy; Ge et al. 2000) but also affected the polymer morphology in the fiber as evidenced by the increase in T_g of the nanofiber. A similar approach to incorporating clay into nanofibers has been reported for PVA (Ristolainen et al. 2006), poly(L-lactide) (PLLA) (Lee, Y. H., et al. 2005), polyurethane-urea (Hong, J. H., et al. 2005), and polyamide-6 (Fong et al. 2002).

An alternative dispersion technique is to polymerize a monomer *in situ* onto the clay tactoids to obtain an intercalated polymer, leading to a particularly good dispersion of platelets. As the clay is hydrophilic it is convenient to use emulsion polymerization for the process. For instance, vinyl acetate monomer mixed with 10 wt% of the clay dispersed in water was polymerized using 0.1% of $K_2S_2O_8$ as the initiator, under an inert atmosphere in a reactor to obtain a composite polymer. The reaction was terminated by adding hydroquinone/NaCl solution and the poly(vinyl acetate) (PVAc) reaction product is hydrolyzed into the corresponding polyalcohol PVA (Ristolainen et al. 2006). The same general approach was successfully used with polyamide-66 (Ristolainen et al. 2006), poly(methyl-methacrylate-*co*-methacrylic acid) (Wang, M., et al. 2005), and PMMA ($d \approx 240$–540 nm) (Kim, G.-M., et al. 2005a).

As with CNTs, a high degree of platelet orientation in the fiber matrix is desirable as it results in high levels of reinforcement. As already pointed out, electrospinning results in some orientation due to the hydrodynamics of the process and constraints placed by the nanoscale dimensions of the fiber, but the nature of the solvent also plays a key role. SEM images of electrospun nylon-6 nanofibers illustrate the sensitivity of dispersion to solvent composition. The presence of even a small amount of DMF in the spinning solvent resulted in considerable agglomeration of the clay platelets within the nylon-6 nanofibers with 7.5 wt% clay (Fong et al. 2002) (Fig. 6.9). Arrows show nanofibers and their aggregates oriented along the fiber axis. Figure 6.10 shows a TEM image of MMT material.

Orientation of clay platelets in the axial direction has an impact on the mechanical properties of the nanofiber. In PMMA/MMT clay composite nanofibers, the mechanical deformation process was reported to be somewhat different from the brittle failure of comparable bulk nanocomposites. The deformation in the nanofibers occurs via shear flow that involves a nanoscale "necking" process (Kim, G.-M., et al. 2005a). However, in this instance the interpretation is complicated by the nanoporous morphology of the fiber electrospun from $CHCl_3$.

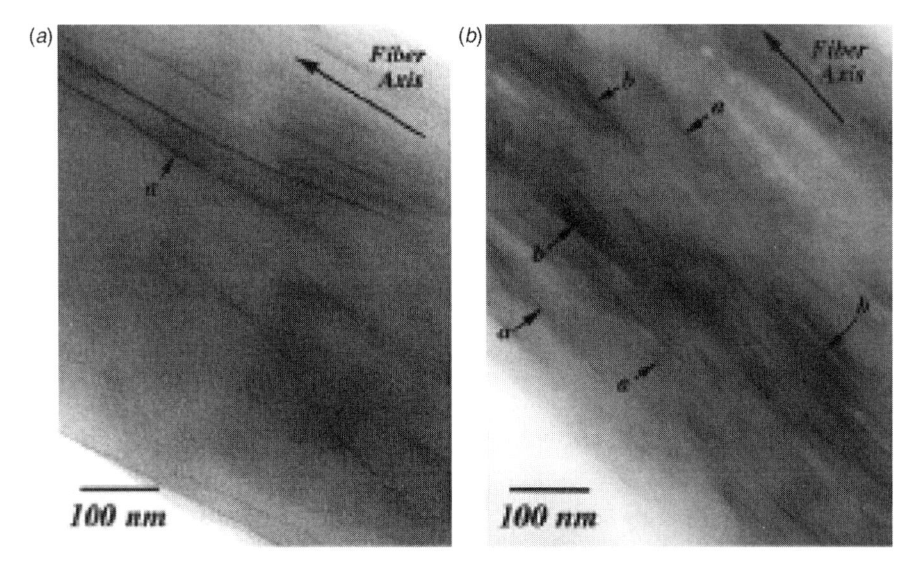

Figure 6.9 (*a*) Nanofibers of nylon-6 electrospun from 10 wt% solutions of 1,1,1,3,3,3-hexafluoro-2-propanol (HFP) with 7.5 wt% of Closite 30B. Arrows indicate clay in the fiber matrix; "a" indicates single sheets and "b" tactoids of clay. (*b*) Same, except the solvent used was HFP/DMF (95 : 5). Reproduced with permission from Fong et al. (2002). Copyright 2002. Elsevier.

The crystalline morphology of composite nanofibers is influenced by the orientation of macromolecules as well as by the presence of filler particles in the fiber. As with unfilled polymer nanofibers, the percent crystallinity of composite nanofibers also tends to be lower than that of the bulk material.

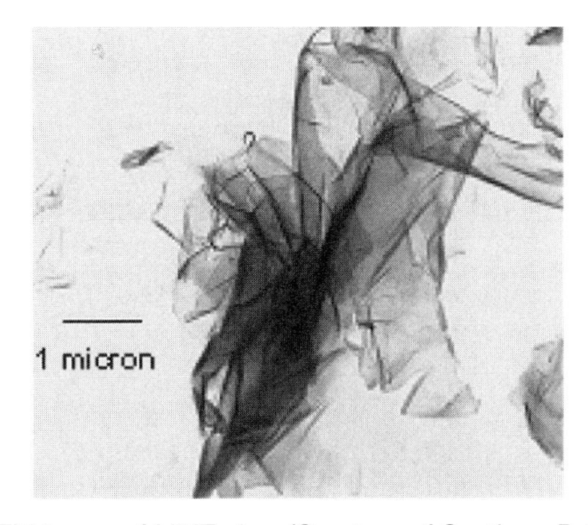

Figure 6.10 TEM image of MMT clay. (Courtesy of Southern Polymer Products.)

This is demonstrated for PVA/silica (Shao et al. 2003) and PVA/$H_4SiMo_{12}O_{40}$ (Gong et al. 2003) composite nanofibers. This is to be expected in any event because of rapid solidification of the jet during electrospinning as discussed previously. Differential scanning calorimetry (DSC) and x-ray diffraction (XRD) studies comparing cast films with electrospun polylactide (PLA) nanofibers show the latter samples to have both α-form (lamellar folded-chain crystallites) as well as the β-form (fibrillar form) of crystallinity. Elongational deformation is responsible for this latter form of crystallinity (Zhou et al. 2006). However, in PLA/MMT composite nanofibers the formation of the β-crystallites appears to be enhanced by the aligned clay yielding improved mechanical properties. Even at 3% clay the modulus of the fiber increased by 100%.

MMT is a reinforcing filler in polymers such as poly(urea urethane) (Ge et al. 2000) as evidenced by the very significant increase in mechanical properties of the composite nanofibers. Unfilled polymer nanofiber mats of polyurethane (PU) ($M_w = 150{,}000$ g/mol) were electrospun from 11 wt% solution in DMAc/THF (7 : 3 wt/wt) into nanofibers with $d \approx 150$ nm to 410 nm. The tensile properties of these mats are shown in Table 6.4, where the last digit in the nanofiber designation is the weight fraction of MMT in the polymer. Based on the WAXD patterns for the composite nanofibers, MMT appeared to be well dispersed, exfoliated, and oriented in the axial direction of the samples.

Zhou et al. (2006) studied the mechanical properties of electrospun mats of PLLA/MMT with 1–5 wt% of the nanoclay in composite fibers. The nanoclay and PLLA were intimately mixed in a high-speed mixer and the blend electrospun from 10–25 wt% chloroform solution to obtain the composite nanofiber mats. The yield stress (MPa) of the composite nanofiber mats improved more than threefold with the incorporation of 3 wt% of the clay, while the modulus doubled. This is consistent with the results in Table 6.4 for polyurethane nanofibers (Hong, J. H., et al. 2005).

TABLE 6.4 Summary tensile property data on PU/MMT composite nanofiber mats

Nanofiber	Young's Modulus[a]	Tensile Strength[a]	Extensibility (%)
PU	0.7	2.7	294
PU/O–MMT-1	1.5	4.6	254
PU/O–MMT-3	1.3	4.8	264
PU/O–MMT-5	1.7	5.1	243

[a]Kilograms force per square millimeter (kgf/mm^2).

Source: J. H. Hong et al. 2005.

6.4 DECORATED OR EXOCOMPOSITE NANOFIBERS

As opposed to conventional composite nanofibers where the particles (or the nanotubes) are buried within the fiber matrix, one where the filler materials reside primarily on the surface of the nanofibers, decorating it, can also be fabricated. These might be referred to as exocomposites to differentiate them from conventional composite nanofibers. In these, the filler cannot of course play a reinforcing role, but the construct may have uses in applications such as in biomedical devices or chemo-biosensors. Water-soluble polymer exocomposite nanofibers carrying nanoparticles of sparingly soluble pharmaceutical compounds can be used for rapid delivery of the drug in the stomach. The high surface area of the delivered nanoparticles would increase the bioavailability of the drug. A recent patent application addresses this application for nanofibers with drug-particle filler. The efficiency of catalysts might also be improved by this approach; a nanoparticulate structure maximizes the surface area available for catalytic reactions and the location of particles at the periphery of nanofiber ensures ready accessibility to reactants. Both post-treatment of fibers by physical methods and its modification using chemical approaches can be used to synthesize this class of composite nanofiber. In these, however, the prevention of surface aggregation of nanoparticles is critical to ensure optimal performance. The compatibility of nanoparticles with the polymer matrix is less of a concern than with conventional composites.

6.4.1 Nanofiber–Nanoparticle Composites

6.4.1.1 Dry Methods Post-treatment of nanofiber mats by a metal compound followed by reduction (already discussed in the previous section) at times can yield composites of this type in certain polymer/metal salt systems. Titania nanofibers (Li, D., et al. 2004b) as well as $MgTiO_3$ nanofibers ($d \approx$ 100–150 nm) (Aryal et al. 2006) can be surface treated by immersion in a solution of $HAuCl_4$ (with PVP) and UV-irradiation in the presence of a capping agent to obtain gold/titania composite nanofibers. In these cases, however, the gold particles for the most part migrate to and decorate the surface of the inorganic nanofibers. Other metals are also amenable to this technique, and composites carrying Ag, Pt, and Pd nanoparticles have been successfully prepared in the laboratory (Li, D., et al. 2004b). The decorated nanofibers in Fig. 6.11*a* was made by first electrospinning PAN (6.7 wt%) and ferric acetylacetonoate (3.7 wt%) in DMF in an electric field of 100 kV/m (applied voltage 30 kV). The nanofibers heated in a reducing atmosphere (up to 550°C) in a H_2/Ar mixture yielded the decorated morphology.

Sputter coating is a particularly convenient means of decorating the surface of fibers. A DC sputter coater using a high-purity silver target, for instance,

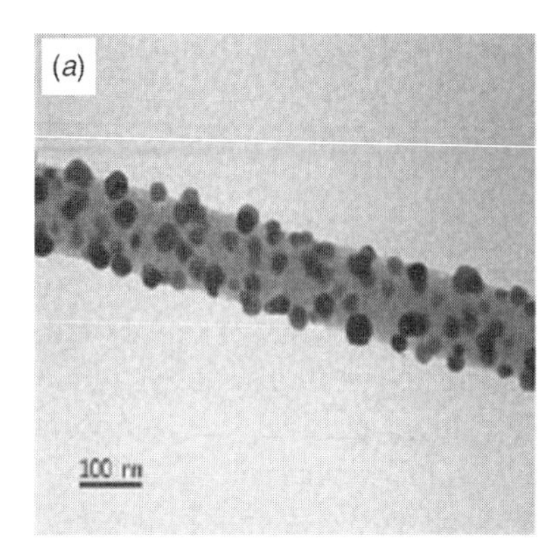

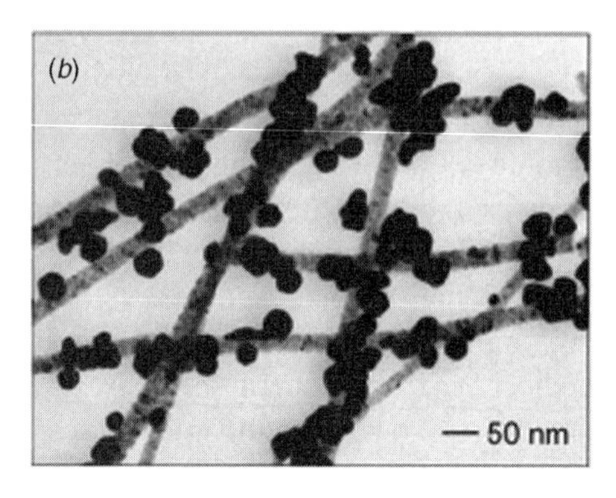

Figure 6.11 (*a*) Carbonized PAN nanofiber decorated by iron nanoparticles on its surface. Reproduced with permission from Hou and Reneker (2004). Copyright 2004. John Wiley and Sons, Inc. (*b*) TiO_2/gold composite. Reproduced with permission from Li, D., et al. (2004b). Copyright 2004. Elsevier.

allows rapid surface decoration of nanofibers. On nylon-6 nanofibers, a duration of treatment as short as 1 min decreases the surface electrical resistance of the fiber by at least two orders of magnitude (Wei, Q. F., et al. 2006). Physical vapor deposition (PVD) can also be used to obtain a surface coating of metals or metal oxide on the nanofibers. Carbon, copper, and aluminum have successfully been deposited by plasma-enhanced chemical vapor deposition (PECVD) on thermally stable poly(*m*-phenylene isophthalamide) (MPD-I) nanofibers (Liu, W. X., et al. 2002). Electroless coating of Ni on 2-acrylamido-2-methyl-1-propane sulfonic acid doped polyaniline (PANI) also has the same effect of lowering the resistance of nanofibers; in 2–10 μm fibers with a 100 nm coating, a decrease in resistance by three orders of magnitude was reported (Pinto et al. 2004).

Decorated nanofibers can be obtained by simultaneous electrospinning and electrospraying, where the nanofibers and nanoparticles are produced by a pair of capillary tips of different polarity oriented facing each other (Fig. 6.12 shows the arrangement of the tips). The oppositely charged materials are naturally attracted to each other, resulting in a decorated nanofiber. Figure 6.13 shows a decorated fiber of PS electrospun from 25 wt% DMF solution using a positively charged capillary tip.[10] The particles are PCL electrospun from dilute solution in CH_2Cl_2 from a negatively charged capillary tip, and the average particle size of PCL was controlled by adjusting the concentration of the PCL solution. The density of particles per unit

[10]The method is described in U.S. patent application # 20060264140 A1 (November 2006) entitled "Nanofiber mats and production methods thereof" (Andrady, Anthony L., Ensor, David S., Walker, Teri A., Prabu, Purva).

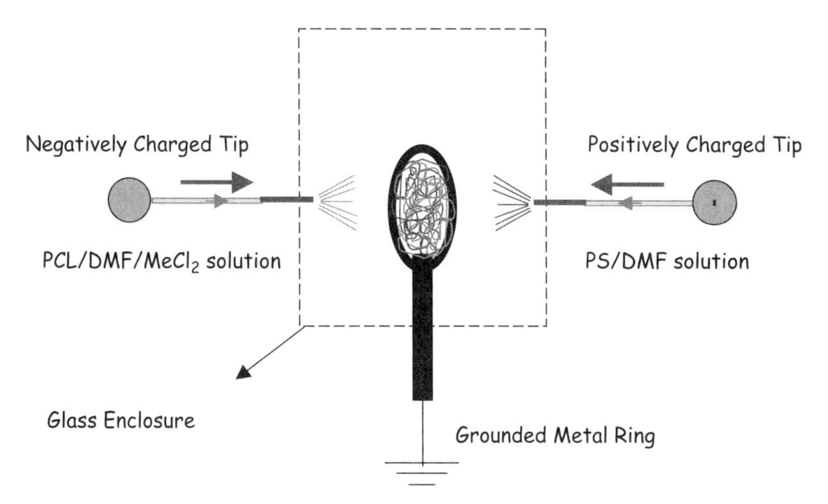

Figure 6.12 Arrangement of capillary tips for concurrent electrospinning and electrospraying to obtain nanofibers decorated with nanoparticles. (Courtesy of RTI International.)

mat area depends on the relative feed rates of polymer solutions to the pair of tips. The technique was also successfully used to decorate polymer nanofibers with insoluble nanoparticles such as silica; these were electrosprayed as suspensions in organic solvents. Where larger particle sizes are of interest, electrospinning into a nebulized cloud of the particle-forming solution might be employed.

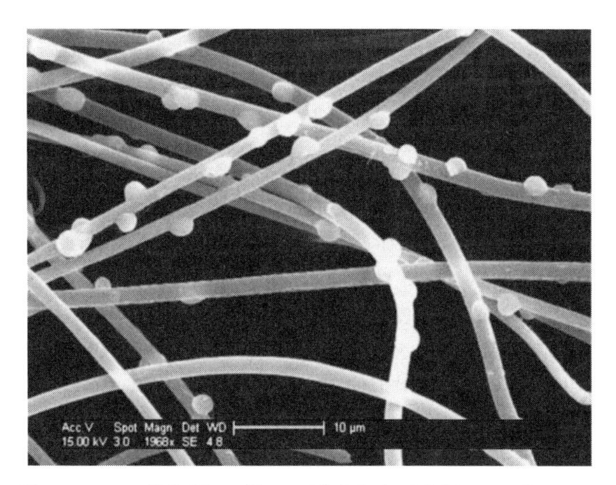

Figure 6.13 Electrospun PS fiber (25 wt% PS in DMF at a flow rate of 2.5 mL/h) decorated with PCL nanoparticles (1.0 wt% PCL in CH_2Cl_2 at a flow rate of 0.5 mL/h). (Courtesy of RTI International.)

A similar technique was used by Min et al. (2004d) to electrospin copolymers of lactide with glycolide (PLGA) nanofibers from hexafluoroisopropanol (HFP) solvent from one capillary tip and regenerated chitin in formic acid electrosprayed from a separate tip, onto the same grounded collector drum. In this case, however, both electrospinning solutions carried a positive charge.

6.4.1.2 Wet Methods Decoration of nanofibers by nanoparticles can also be achieved in liquid media. Poly(4-vinyl pyridine) (PVP) has an affinity towards metal ions due to the pyridyl moieties on the repeat units interacting with the ions. Nanofibers of either PVP or its blends with PMMA (PVP : PMMA 50 : 50 w/w) when immersed in aqueous 10 mM $NaAuCl_4$, for instance, turn bright yellow due to surface sorption of gold ions. The nanofiber mat with bound gold ions can be reduced in 50 mM $NaBH_4$ solution to obtain nanofibers ($d \approx 360$ nm) decorated with elemental Au nanoparticles (Dong, H., et al. 2006), The same technique was shown to work for Ag nanoparticles as well. More interesting is the observation that the same nanofibers had a strong affinity towards citrate-protected gold nanoparticles. Electrostatic attraction of negatively charged (because of citrate capping) gold nanoparticles to the positively charged amine functionalities on PVP nanofibers resulted in a richly decorated fiber surface. Swelling of the PVP nanofibers in an aqueous medium facilitated this process; attempting the same with the hydrophobic PMMA nanofibers, however, did not result in significant decoration of the fiber by nanoparticles (Dong, H., et al. 2006).

6.4.2 Nanofiber–Nanotube Composites

A post-treatment of nanofibers to fabricate MWCNT/polyamide-6 nanofiber exocomposites was recently reported in the literature. Polyamide nanofibers ($d \approx 500$ nm) were electrospun from 30 w/w% solution in 95% formic acid (Kim, H. S., et al. 2006). The MWCNTs were dispersed in water (with 0.3% Triton X-100 or sodium dodecyl sulfate) or in DMF by sonication for 7 h at 28 kHz and 600 W. The polyamide nanofiber mats were immersed in a (0.05 wt%) solution of the nanotubes in water for 60 s and then rinsed in water (60 s), yielding CNT-decorated nanofibers. The weight percent of the MWCNTs in the composite estimated by thermogravimetry was 1.5% for aqueous dispersion and ~4% for DMF dispersion. The mechanism of decoration might be related to the ability of the solvent to slightly swell the polymer surface, facilitating better embedment. The DMF dispersions resulted in better embedment of the nanotubes on the surface. At 1.5% loading the conductivity of the nanofibers was determined to be 3.5×10^{-2} S/cm. The high-resolution

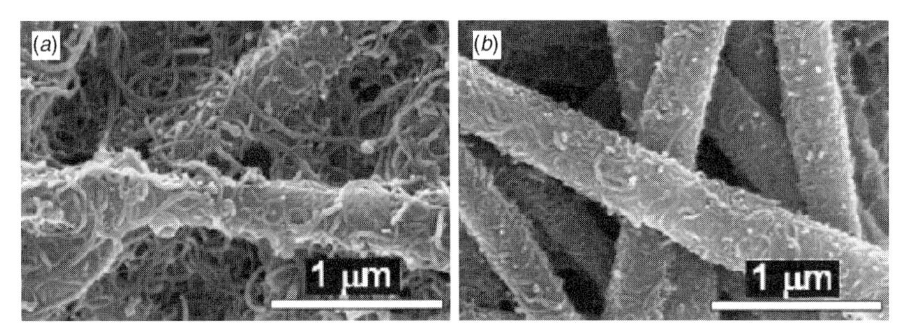

Figure 6.14 Nylon-6/MWCNT exocomposite nanofibers showing surface-embedded CNTs: (*a*) after exposure to CNT solution and (*b*) after sonication for 10 min in deionized water. Reproduced with permission from H. S. Kim et al. (2006). Copyright 2006. John Wiley & Sons.

SEM in Fig. 6.14 illustrates the surface characteristics of the exocomposites (Kim, H. S., et al. 2006).

Carbon nanotubes might also be grown (as opposed to being deposited) on the surface of inorganic nanofibers to obtain carbon–carbon composite nanofibers with an interesting unique morphology. Iron nanoparticles (10–20 nm) initially deposited on the surface of PAN-derived carbon nanofibers were successfully used to initiate the growth of CNTs on the nanofiber surface in a recent study (Hou and Reneker 2004). Nanotubes were grown from hexane vapor at 700°C and, as shown in Fig. 6.15, yielded a "brush-like" morphology with nanotubes a micrometer or more in length affixed to the nanofibers' surface.

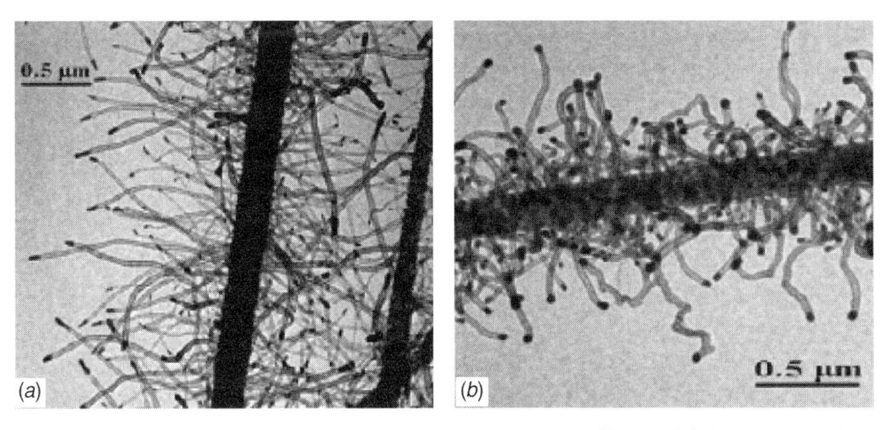

Figure 6.15 TEM images of composite carbon nanofibers: (*a*) long nanotubes formed at 850°C; (*b*) curved nanotubes formed at 700°C. Reprinted with permission from Hou and Reneker (2004). Copyright 2004. John Wiley & Sons.

These geometries are particularly appealing for filtration-type applications, in sensors, and in highly effective composite design. But, at least for the present, the fabrication routes for these more exotic nanofiber-based constructs remain too expensive for high-volume applications. Most have yet to be moved out of the laboratory into pilot-plant scale before their viability in volume applications can be meaningfully discussed.

7

BIOMEDICAL APPLICATIONS OF NANOFIBERS

As is clear from the majority of patents devoted to the topic, biomedical applications remain the most intensely researched application area for electrospun nanofibers. Nanofibers are expected to contribute in diverse emerging medical areas such as organogenesis, genomic medicine, high-throughput screening, rapid bedside clinical tests, and smart wound dressings. Two particularly promising biomedical research areas are focused on nanofiber-based three-dimensional scaffolds for tissue engineering and the design of nanofiber devices for controlled delivery of pharmaceuticals. The success of polymer nanofiber mats in scaffolding applications is primarily because their size range closely matches the structural features present in body tissue environments, especially with regard to their high porosity and readily tailorable surface chemistry. In addition to the biodegradable organic nanofibers commonly used for such applications biocompatible inorganic nanofibers (such as titanate nanofiber mats) have also been studied. Also, most synthetic polymers are bioinert while several classes are biodegradable in the human body. This chapter reviews recent work on the use of nanofibers in these two key biomedical application areas. The reader is also directed to several excellent recent reviews of the literature for more detailed information (Li, W.-J., et al. 2005a; Lim et al. 2004; Pham et al. 2006; Smith and Ma 2004; Xu et al. 2004b).

7.1 DRUG DELIVERY APPLICATIONS

Conventional delivery of a drug in successive doses results in a blood (or other tissue phase) concentration profile of the drug that fluctuates over the duration of therapy. Therefore, over significant durations, the concentrations may exceed the recommended maximum value C_{max}, with the risk of biotoxicity, or fall below the minimum effective concentration, C_{min}, limiting the therapeutic effect. To derive the highest therapeutic value, an optimum concentration C $\{C_{min} < C < C_{max}\}$ should be maintained in the body tissue over the full duration of treatment. With controlled delivery techniques, the bioavailability of the drug is designed to be close to this optimum value throughout the therapy. It also minimizes potential side effects, as the amount of drug that needs to be administered is relatively lower in the controlled-release mode. These advantages often outweigh the drawbacks of higher costs associated with controlled-delivery devices and possible discomfort where implants are involved.

In designing polymer scaffolding in tissue engineering, the ability to incorporate growth factors and other bioactive agents to be released over a period of time into the growing tissue is desirable (Pham et al. 2006). In nanofiber applications such as wound dressings or artificial skin, the controlled release of antibiotic substances locally may help the healing process (Katti et al. 2004). Excluding active systems such as osmotic pumps, polymer-based delivery systems may be diffusional or chemically controlled (via bioerosion of the matrix, or the biodegradation of linkages that bind the drug to the matrix). The reader is referred to several excellent reviews on the use of polymers for controlled delivery of drugs for a comprehensive discussion of the process (Park and Mrsny 2000; Rathbone et al. 2003; Saw et al. 2006).

The simplest configuration of a controlled release device is where a drug is either dissolved in high concentration or suspended as particles in a monolithic polymer such as a cylindrical polymer fiber.[1] The release of the drug from it may occur via:

1. Diffusive transfer through the polymer matrix to the surrounding tissue.
2. Release of the dissolved or suspended drug due to slow biodegradation or erosion of the surface layers of the fiber.
3. Slow release of covalently bonded drug via hydrolytic cleavage of the linkages.
4. Rapid delivery of the drug due to dissolution of the fiber.

[1]This is contrasted with the "reservoir device" where a highly concentrated reservoir of drug is enclosed by a thick polymer membrane. As with monolithic devices the release is via diffusion across the polymer membrane, but the drug release kinetics are different for the two types of devices.

Although, in principle, polymer-based delivery devices can operate via any of these mechanisms, most of the reported data pertain primarily to mechanism (1). The concept of using a melt-spun water-soluble fiber for the purpose is hardly new [U.S. patent 4,855,326 (1988) and WO 04014304 (2004)]. However, the use of a nanoscale fiber is a fairly new development (Brewster et al. 2004; Katti et al. 2004; Verreck et al. 2003a, 2003b; Zeng et al. 2004a, 2005d). Rapidly dissolving dosage forms based on nanofibers carrying pharmaceutical agents as fillers have been discussed in detail in U.S. patent application # 20060083784A1. Incorporating certain types of drugs into the polymer matrix by blending has been reported to result in an increase in the average fiber diameters (Luong-Van et al. 2006), possibly because of increased conductivity of the spinning solution.

Expressions for the kinetics of diffusion of solutes through polymers have been extensively discussed by Crank (1980). These are derived primarily by assuming that Fick's Law applies; the concentration gradient of the drug inside and outside the polymer matrix is the driving force for diffusive transfer:

$$J = -D(\mathrm{d}c/\mathrm{d}x), \tag{7.1}$$

where $J =$ local flux of the drug ($\mathrm{g/cm^2 \cdot s}$), D is the concentration-independent diffusion coefficient[2] of the drug, and ($\mathrm{d}c/\mathrm{d}x$) is the concentration gradient. Strictly, it is the gradient of the chemical potential rather than ($\mathrm{d}c/\mathrm{d}x$) that is the driving force for diffusion. Experimentally, the fraction of drug released by the polymer matrix (M_t/M_∞) at time t is the quantity most conveniently measured. In a typical drug-release study a drug-laden polymer nanofiber mat sample is placed in a well-stirred reservoir of buffer and the release of drug is monitored spectroscopically over a period of time. The anticipated relationships between (M_t/M_∞) and t for different geometries of monolithic polymer release devices have been derived (Crank 1980). For the simplest slab geometry for short durations,

$$(M_t/M_\infty) = 4(D \cdot t/\pi r^2)^{1/2}, \tag{7.2}$$

where r is the thickness of the slab and D is the diffusion coefficient. The rate of release of drug varies with $t^{1/2}$. Low-density, loosely spun nanofiber mats, however, are better approximated by cylindrical or fiber geometry. For the case of a fiber with a circular cross-section and a diameter of r units, the relevant equations are as follows (Comyn 1985).

[2] D is regarded as a constant here. In most real systems the value of D varies with the concentration of the drug in the matrix and will change with the composition of the extractant (water or buffer solution).

During early stages of the release process where $(M_t/M_\infty) \leq 0.4$, the expression is approximated by

$$(M_t/M_\infty) = 4(D \cdot t/\pi r^2)^{1/2} - (D \cdot t/r^2) \quad (7.3)$$

and

$$\mathrm{d}(M_t/M_\infty)/\mathrm{d}t = 2(D/\pi r^2 t)^{1/2} - (D/r^2). \quad (7.4)$$

As $(M_t/M_\infty) > 0.6$, a more complicated expression applies:

$$(M_t/M_\infty) = 1 - 4/(2.405)^2 \exp\{-(2.405)^2 Dt/r^2\} \quad (7.5)$$

and

$$\mathrm{d}(M_t/M_\infty)/\mathrm{d}t = (4D/r^2) \exp\{-(2.405)^2 Dt/r^2\}, \quad (7.6)$$

where M_t is the amount of drug released at time t, and M_∞ is the total amount released at infinite time. The release kinetics are affected by the crystallinity of the polymer; it is predominantly the amorphous fraction of the matrix that carries the dissolved drug.

The simple kinetics predicted by Fick's second law for diffusive release of a drug from a device having a thin-film geometry appears to apply to nanofiber mats as well. Controlled delivery of heparin anticoagulant from poly(ε-caprolactone) (PCL) nanofibers display Fickian diffusion, and the fraction of drug released increased with square root of time as suggested by equation (7.2) (Luong-Van et al. 2006). Comparison of the observed release rates for heparin from nanofiber mats with the data for release from thin films suggested the diffusion coefficients to be about the same for both the thin film and mat systems.

7.1.1 Drug-Loaded Fibers

The kinetics of release of the drug is controlled by the semicrystalline nature of the polymer as well as by the morphology of the polymer/drug composite. Three basic morphological models for drug-loaded polymers (or polymer particles), first proposed by Kissel et al. (1993) apply to drug-loaded nanofibers as well (Verreck et al. 2003a):

1. Drug dissolved in the polymer matrix at the molecular level.
2. Drug distributed in the polymer matrix as crystalline or amorphous particles.
3. Drug enclosed in the polymer matrix yielding a core of the drug encapsulated by a polymer layer (similar to a reservoir device).

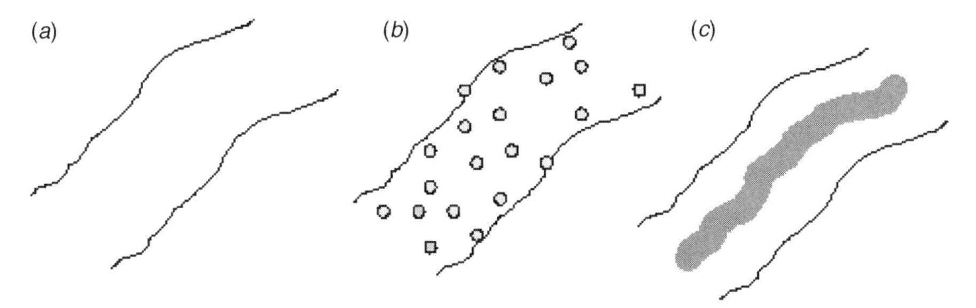

Figure 7.1 Illustration of the three morphological models of drug-loaded polymer nanofibers.

With nanofiber devices the third morphology is obtained by coaxial spinning of core–shell nanofibers rather than from phase separation. The three morphologies are illustrated in Fig. 7.1.

There is little interest in the simpler morphology type 1, as the solubility of the drug in the polymer limits the maximum drug loading possible. As the mass of nanofiber mats in the implant will be small, they can carry only impractically small amounts of most drugs. An example of morphology type 1, however, is given by poly(ethylene oxide) (PEO)/Nabumetone nanofibers (U.S. patent application # 2003/0017208A1). Nanofibers of poly(L-lactide) (PLLA) carrying Rifampin (an antituberculosis drug) also illustrate diffusive release of a drug uniformly dissolved in a polymer. The nanofiber mat in this instance was electrospun from $CHCl_3$: acetone (2 : 1 v/v) at a PLLA concentration of 3.9 wt%. Rifampin was dissolved at a level of 5–100 wt% (based on the polymer) into the spinning solution (Zeng et al. 2003a). Microscopic examination of the nanofibers (average d of ~700 nm) electrospun at the lower drug loadings showed no crystalline inclusions, suggesting the drug was likely completely dissolved in the matrix at that concentration. The diffusive release of Rifampin into a buffer was understandably slow (as PLLA is a glassy polymer) with virtually no release observed for up to 7 h.

The same system at a relatively higher loading of the drug, however, affords an illustration of the morphology type 2. The release of Rifampin in this case was achieved primarily by enzymatic biodegradation of the surface layers. In the presence of proteinase K (a nonspecific serine protease having a very high specific activity) in the buffer, rapid first-order release of Rifampin (~50% released in 7 h) was observed due to slow biodegradation of the polymer matrix (Zeng et al. 2003a). Studies on the delivery of tetracycline hydrochloride from poly(lactic acid) (PLA), poly(ethylene-*co*-vinyl acetate) [P(E-VAc)], or from a 50 : 50 blend of the two polymers (Kenawy et al. 2002, 2003) also illustrate systems of similar morphology. Morphology type 2 covers a majority of the nanofiber-based drug release matrices of interest in the research literature. The high porosity of nanofiber mats (often exceeding ~90%) facilitates rapid removal of any biodegradation products (Jiang et al. 2004b) in bioerodible systems minimizing possible autohydrolysis. Degradation products are known to retard cell proliferation in scaffolding studies (Higgins et al. 2003).

The release kinetics of nanofibers where the drug is uniformly dissolved in the matrix (morphology 1) can be approximated by assuming the mat to be a collection of monodisperse cylinders. With poorly soluble drugs, particularly at higher loadings, however, the nanofibers tend to form beads (Zhang, C. X., et al. 2005a), and the assumption may not always be a reasonable one. Chew et al. (2005) studied the release of protein-stabilized human beta nerve growth factor (β-NGF) from nanofibers of a copolymer of ε-caprolactone and ethyl ethylene phosphate (PCLEEP). A Fickian kinetic expression derived for one-dimensional diffusion from monodisperse perfectly cylindrical matrices under perfect sink conditions (Ritger and Peppas 1987) was used to analyze the data. The following simple model was used to describe the release data.

$$(M_t/M_\infty) = 0.19t^{0.34} \qquad (r^2 = 0.99). \tag{7.7}$$

As a uniform distribution of the drug in the polymer is assumed in Fickian models, good compatibility between the polymer and drug is implied. The value of exponent, however, is lower than predicted ($k \approx 0.45$), possibly due to heterogeneity of fiber diameters in the mat and the semi-crystalline fiber morphology. Hydrophobic drugs such as doxorubicin hydrochloride, for instance, do not disperse well in lipophobic electrospun PLLA fibers (Zeng et al. 2005d).

With drug-laden nanofibers of morphology type 2, the delivery process can be qualitatively described as follows. The drug initially dissolves and saturates the polymer surrounding the particles embedded in the fiber matrix. The dissolved drug invariably reaches the surface layer of the fiber by diffusion and partitions into the aqueous boundary layer at the fiber/buffer interface. Finally, the drug molecules diffuse across the boundary layer into the aqueous medium. Higuchi (1961) and others (Roseman 1972) have derived quantitative expressions that describe the kinetics of the process in terms of a "zone of depletion" of the drug that is obtained at the surface layers as the drug delivery progresses. A detailed treatment of the kinetics of release profiles is beyond the scope of this chapter.

With nanofibers of morphology 2 (i.e., where undissolved drug particulates are suspended in the fiber matrix), the drug release profile typically shows an unexpectedly high initial rate of release, commonly referred to as the "burst phase" (Kenawy et al. 2002; Jiang et al. 2004b; Luu et al. 2003). The period over which this burst release occurs varies with the polymer/drug system, lasting anywhere from several hours to days. Nanofibers saturated with the drug also in some instances do show burst phase behavior (Comyn 1985). But, the rapid dissolution of drug particles embedded, and partly exposed at the surface of the nanofiber likely contributes very significantly to this phenomenon (Kenawy et al. 2002; Zeng et al. 2003a). At lower drug loadings the nanofiber mats generally do not display such a burst phase (Verreck et al. 2003b). X. H. Zong et al. (2002) have speculated that the charged drug species tend to be concentrated near the fiber surface in any event as the jet surface itself is charged. Also, Zeng et al. (2005a), working with poly(vinyl alcohol) (PVA) nanofibers carrying bovine serum albumin, found the burst phase to be eliminated when a thin coating of poly(*p*-xylene) (PPX) was deposited (by chemical vapor deposition) on the surface of the fiber. A drug that is compatible with the polymer or has a high solubility in the polymer is less likely to show burst phase kinetics during early release (Zeng et al. 2005d). However, this initial burst release may not always be undesirable, as in the case of delivery of antibiotics to control infection (Kim, K. S., et al. 2004; Zong, X. H., et al. 2002).

Figure 7.2 (Zeng et al. 2005d) compares the release of doxorubicin (converted to the free base from the HCl form by ammonia) from the biodegradable polymer PLLA in the absence and in the presence of an enzyme that can biodegrade the polymer. The primary release mechanism in this case was bioerosion. In the absence of the enzyme, only a minimal diffusive release

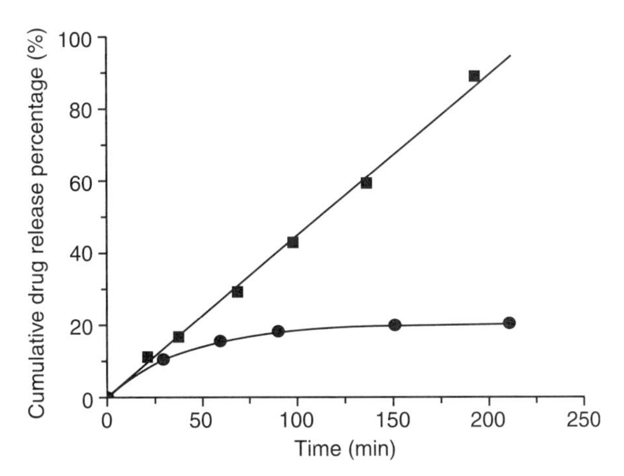

Figure 7.2 Percentage release of doxorubicin base vs time from electrospun nanofibers of poly(L-lactide) (PLLA) in a Tris buffer medium. The concentration of proteinase K was 3.0×10^{-3} mg/ml for the square symbols and zero for the circles. Reproduced with permission from Zeng et al. (2005d). Copyright 2005. Elsevier.

was observed. With proteinase K present, rapid release with the expected zero-order kinetics was obtained (Zeng et al. 2005d). With the more permeable P(LLA-PEG) diblock copolymer nanofibers, however, Xu et al. (2005) found similar release kinetics for doxorubicin even in the absence of a biodegrading enzyme (therefore via diffusion). Release kinetics of the drug are invariably controlled by the morphology of the fiber and particularly by drug–matrix interactions. The latter is illustrated by the release of Ibuprofen from poly(L-glycolide) (PLGA) and PEG-chitosan (graft copolymer) matrices (Jiang et al. 2004b). The release occured rapidly, with most of the drug delivered within a four-day period (incubated in 0.1 M PBS buffer at 37°C). Blending PEG-g-chitosan polymer into the PLGA matrix significantly reduced the release rate, presumably due to non-bonded interactions between the —NH_2 groups in chitosan and the acid moieties in the drug. Covalant bonding of the drug to chitosan via the amine groups further prolonged the release. As seen from Fig. 7.3, however, the general shape of the release curves was the same in each case.

Morphology type 3 has the distinct advantage over the other two in that the drug or biological material (such as protein or DNA), does not come into contact with aggressive spinning solvents, avoiding the possibility of denaturing or other changes that alter their efficacy. Also, long-term contact between the drug and the polymer, which can potentially lead to reaction is avoided. Core–shell nanofibers where the drug is restricted to the core layer have been evaluated for their release profile (Jiang et al. 2005). The kinetic features of drug release for these materials is expected to be qualitatively similar to that for nanofibers where drug particles are distributed in the matrix. With

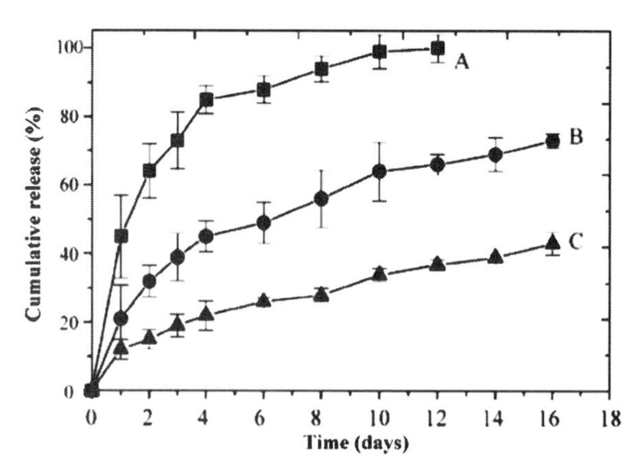

Figure 7.3 The cumulative release of Ibuprofen from three matrices at a loading of 4.45 wt% of drug. (A) PLGA nanofiber mat; (B) PLGA/PEG-*g*-chitosan (70/30) nanofiber mat; (C) same as B except that Ibuprofen is covalently bonded to the polymer ($n = 3$). Reproduced with permission from Jiang et al. (2004b). Copyright (2004). Springer Science and Business Media.

poly(ε-caprolactone) (PCL) shell and a drug core, Huang and Yang (2006) found no burst phase but a smooth delivery profile for Reservatol and Gentamycin sulfate. With bovine serum albumin (BSA) cores with PCL shells, however, a small burst phase was initially observed (Jiang et al. 2005; Liao et al. 2006), but the release profile was still qualitatively similar to that shown in Fig. 7.3. The BSA itself was suggested to act as a porogen, allowing its release from the core via the PCL shell (Zhang et al. 2006c). Table 7.1 lists selected examples of controlled delivery via nanofibers.

7.1.2 Controlled Delivery of Macromolecules

The delivery of peptides and proteins is a particularly demanding application of controlled drug delivery. Electrospun nanofibers designed to deliver proteins, particularly BSA, have been investigated by several researchers (Jiang et al. 2005; Zeng et al. 2005a; Zhang, C. X., et al. 2005b). PVA ($M_w =$ 190,000 g/mol) loaded with FITC-labeled BSA, was electrospun from 5 wt% aqueous solution to obtain 250–300 nm nanofibers (Zeng et al. 2005a) and its release rate into a buffer was studied. The hydrophilic PVA nanofiber matrix showed signs of disintegration in the buffer and the release of protein from the nanofibers showed an initial 2-h burst phase. The release rates were modulated by a hydrophobic surface coating of poly(*p*-xylylene) (PPX) deposited on the surface of the nanofibers. The coating thickness of the PPX was varied in the chemical vapor deposition process using paracyclophane dimer as the precursor. Not only was the burst phase eliminated by this thin film coating, but the kinetics of BSA-FITC release could be controlled by the coating thickness. The effect of changing the PPX shell thickness from 80–100 nm to 250 nm on the release rate is shown in Fig. 7.4.

A similar core–shell construct electrospun (using a coaxial capillary tip as described in Chapter 9) from aqueous poly(ethylene glycol) (PEG) solution shell from poly(ε-caprolactone) (PCL) solutions in $CHCl_3$/DMF (7 : 3 v/v) was used by Jiang et al. (2005) to deliver BSA or lysozyme and also by Zhang et al. (2006b) to deliver BSA-FITC conjugate and luciferase. The thicknesses of the core and shell materials were controlled by adjusting the feed rate of the polymer solutions to the capillary tips. The BSA loading rates used were in the range of 2–5.5%. Microscopy of the nanofibers suggested that initial swelling of the fibers in the PBS buffer compromised the thin shell layer of PCL, giving rise to surface pores. The controlled release of the BSA therefore essentially occurred through these pores and the collapse of empty nanofiber shells was observed microscopically (Jiang et al. 2005). Although some degree of failure of the PCL shell and pore formation was also observed by Zhang et al. (2006c) in BSA/PCL core/shell fibers, their kinetic data supported the release of the agent via a

TABLE 7.1 Nanofiber drug delivery systems from recent literature

Polymer	Solvent System	Drug	Technique	Reference
PEDOT	Electrochemically polymerized on PLGA nanofibers	Dexamethasone	UV (237 nm)	Abidian et al. (2006)
HPMC and PU	C_2H_5OH/CH_2Cl_2	Itraconazole	UV (254 nm)	Brewster et al. (2004)
HPMC and PU		Itraconazole ketanserin	HPLC	Verreck et al. (2003a, 2003b)
PCL	Core/shell ($CHCl_3/C_2H_5OH$)	Resveratrol, gentamycin sulfate	UV (307 nm and 232 nm)	Huang and Yang (2006)
PLGA	THF : DMF (3 : 1)	Cefazolin	—	Katti et al. (2004)
EC		Tetracyclin	UV	Wu et al. (2006)
PLLA PEVA	$CHCl_3$ (14% w/v)	Tetracyclin	UV 360 nm	Kenawy et al. (2002)
PLLA	$CHCl_3$/acet. (2 : 1)	Pacilitaxel doxorubicin HCl	HPLC/UV (227 nm)	Zeng et al. (2005d)
PLLA-PEG	$CHCl_3$ (5–6 wt%)	Doxorubicin HCl	UV (483.5 nm)	Xu et al. (2005)
PVA	Water	Aspirin, bovine serum albumin	UV 295 nm visible 595 nm	Zhang, C. X., et al. (2005b)
PCLEEP	CH_2Cl_2 (2–12%)	β-NGF	ELISA analysis	Chew et al. (2005)
PLGA	DMF (33%)	Cefoxitin sodium	UV 233 nm	Kim, K. S., et al. (2004); Zong, X. H., et al. (2004)
PEO/CH blend	Aqueous	Potassium 5-nitro-8-quinolinolate	Microbiological assay	Spasova et al. (2004)
PCL shell PEO core	$DMF/CHCl_3$ (3 : 1) water	BSA, lysozyme	Protein microassay	Jiang et al. (2005)
PCL	CH_2Cl_2/MeOH (7 : 3)	Heparin (fluorescent labeled)	Fluorescence at 485 nm	Luong-Van et al. (2006)
PVA	Water (10 wt%)	BSA-FITC	Visible 400–500 nm	Zeng et al. (2005a)
PLGA + PEG-g-CH	DMF	Ibuprofen	UV 264 nm	Jiang et al. (2004a)

EC = ethyl cellulose; HPMC = hydroxypropylmethyl cellulose; CH = chitosan; PU = polyurethane.

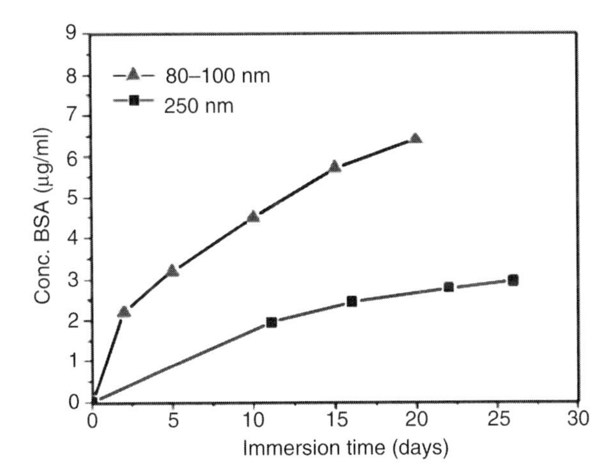

Figure 7.4 Release profile for FITC-BSA into water at 37°C from PPX-coated PVA/BSA nanofibers with different layer thicknesses of PPX. Reprinted with permission from Zeng et al. (2005a). Copyright 2005. American Chemical Society.

diffusion-controlled mechanism. With controlled release of enzymes such as lysozyme or the model compound luciferase (Jiang et al. 2005; Zeng et al. 2005d), the primary protein structure (as ascertained by electrophoresis of the hydrolysate) as well as the bioactivity of the released agent was shown to be intact. Exposure to the high electric fields used or contact with the spinning solvents did not appreciably alter the structure, stability, or bioactivity of these compounds. The core–shell morphology (type 3) might be particularly well suited for protein or DNA delivery as it minimizes burst release and provides closer control of the release rates (Huang et al. 2006; Pham et al. 2006).

Chew et al. (2005) demonstrated the utility of nanofibers electrospun from copolymers of caprolactone and ethyl ethylene phosphate [P(CL-EEP)] in drug delivery. These were electrospun from 2–12% solutions in CH_2Cl_2 and were loaded with low concentrations ($\sim 1 \times 10^{-4}$%) of recombinant human β-NGF stabilized in BSA carrier protein. Nanofibers were spun as aligned mats using a rotating drum collector and their BSA-NGF release characteristics were studied over a three-month period. The released protein was shown to be bioactive (using a neurite outgrowth assay on PC-12 cells).

Controlled delivery of viral or non-viral DNA (lipoplexes or naked DNA) is an important aspect of gene therapy. The potential of nanofiber scaffolds for site-specific delivery of genetic material is beginning to be appreciated. Luu et al. (2003) recently demonstrated for the first time successful incorporation of plasmid DNA into an electrospun nanofiber scaffold. Polymers, PLGA (LA:GA 75:25), and a poly(D,L-lactide)-poly(ethylene glycol) (PLA-PEG-PLA) block copolymer were electrospun from 10–15% solutions in DMF, and the DNA used was pCMVβ plasmid. The pCMVβ plasmid with

7164 base pairs encoding β-galactosidase was extracted from *E. coli.* Up to 80% of the loaded DNA was released from nanofiber mats over a 20-day period, but an initial two-hour burst phase was observed, originating possibly from the DNA localized at the surface layers. The same was incorporated into random PLGA (LA/GA 75/25) copolymer nanofibers spun from DMF solutions (Liang et al. 2005). In this case, however, the plasmids were first encapsulated in a triblock copolymer (polylactide-poly(ethylene glycol)-polylactide), prior to being electrospun as particles in a biodegradable PLGA matrix. Plasmids were again demonstrated to be released into a buffer, apparently structurally intact, and were able to transfect cells *in vitro*. DNA, itself a biopolymer, can also be electrospun into nanofibers. Calf thymus Na-DNA electrospun from aqueous solutions at concentrations from 0.3% to 1.5% yielding nanofibers ($d \sim 50-80$ nm) was reported in 1997 (Fang and Reneker 1997). Salmon testis DNA with 200–900 K base pairs was recently electrospun from water : ethanol (70 : 30) solutions at concentrations of 1.5 wt% (Takahashi et al. 2005). Short nanofibers (~1 μm in length) of diameter *d* of about ~2 nm were obtained from the electrospinning process.

7.2 SCAFFOLDING APPLICATIONS OF NANOFIBERS

An important objective of tissue engineering is to provide an alternative to conventional transplants (ultimately including even entire functional organs) through the development of three-dimensional polymer scaffolds populated by an appropriate mix of cells and tissue. Live tissue comprises of collections of cells arranged in complex geometries within an extracellular matrix (ECM) that is in intimate association with cells. The specific arrangement of cells in the tissue depends on the functional architecture of the organ. The ECM, however, plays a vital role in lending structural integrity to most types of tissue and is generally composed of glycosoaminoglycans and fibrous proteins such as the various types of collagens. In addition to this, ECM also plays a vital role in the transduction of chemical signals that direct tissue development and cellular differentiation.

Polymer scaffolding is essentially a surrogate or a synthetic substitute for the native ECM in the body. As such, nanofiber scaffolding must also provide a three-dimensional environment for cell adhesion and proliferation, guiding growing cells to organize themselves into complex tissue. Cell growth proceeds very differently on two- versus three-dimensional matrices (Cuikerman et al. 2001). For instance, chondrocytes cultured on nanofiber scaffolds retained their phenotype and showed their characteristic shape, while those on polystyrene surfaces were flat and diffused (Li, W.-J., et al. 2003). As it serves as temporary ECM during *in vitro* culture of cells as well as during tissue repair, polymer scaffolding must meet the additional criteria of permeability, high porosity, and high surface area for initial attachment of cells during the seeding stage. A comparison of the adhesion and spreading of human mucosal keratinocytes on nanofibers ($d \approx 80$ nm), microfibers ($d \approx 1100$ nm), and film samples of silk fibroin biopolymer, for instance, illustrates this advantage of the higher surface area (Min et al. 2004a, 2004b). The nanofiber matrices were concluded to be preferred for biomedical scaffolding applications over the other two matrices.

The main characteristics of scaffolding materials or synthetic ECM can be readily anticipated:

1. The material in contact with the host body tissue should not elicit any undesirable immune or tissue responses. The polymers selected and their products of biodegradation should not interfere with the physiology of the body tissue about it.
2. The material used should ideally biodegrade once initial tissue growth has taken hold in the implant (avoiding the need for a second invasive surgery to remove it). Cell growth, proliferation, and differentiation *in vivo* should ideally occur over the timescale of biodegradation of the scaffold.
3. Scaffolding topology should be conducive to attachment or proliferation of cells and its pore-size distribution must match the requirement of the cells being cultured on it. A biomaterial such as trabecular bone, for instance, is 50–90% porous. In vascular grafts the effective pore diameter for cell ingrowth was reported to be in the range of 20–60 μm (von Recum et al. 1996).
4. The mechanical characteristics of the scaffolding material must match those of the tissue with which it will interface. The choice of material therefore varies with the type of tissue in question; thus, scaffolds based on a single material but seeded with appropriate cellular components still cannot always be expected to be generic replacements for all tissue types. In small-diameter vascular grafts, for instance, compliance mismatch between the implanted graft and host artery has been reported to be a major factor in graft failure (Kinley and Marble 1980).
5. In addition to mechanical support of cellular components scaffolding may also deliver growth factors (Casper et al. 2005) or other molecules needed by the host cells (Hirano and Mooney 2004). Some of the polymers used in controlled drug delivery applications will therefore be attractive candidates for scaffolding as well.

A consideration of these criteria and the observation that native body tissue components (fibrin, actin, myosin, elastin, collagen) often tends to have a fiber geometry suggest nanofiber mats to be particularly promising materials for the construction of three-dimensional scaffolding. This suggests that cells need to undergo minimal rearrangement as the natural ECM gradually replaces the biodegrading scaffolding. Recent data on cell growth studies show nanofiber mats to not merely provide a physical support for proliferation but also to promote, via their topological characteristics, *in vivo* like organization and even facilitate morphogenesis of tissue on the scaffold (Ma, Z., et al. 2005b; Pan and Jiang 2006; Schindler et al. 2005; Tuzlakoglu et al. 2005). Furthermore, growth factor induced chondrogenesis of marrow-derived human mesenchymal stem cells (hMSCs) growing on nanofiber scaffolds is enhanced relative to that in cell pellet cultures (Li, W.-J., et al. 2005c). Nanofiber mats have a high surface area to volume ratio to support copious cell growth, and exceptionally high porosity to allow the exchange of gases and nutrients to support the growing tissue. The relatively higher specific surface area of nanofibers in comparison to other scaffolding is expected to enhance cell adhesion as well as adhesion-dependent phenomena such as migration during proliferation.

Unlike in native tissue where cell adhesion is mediated by specific interaction with integrins, the attachment of cells to uncoated polymer nanofibers is often the result of nonspecific interactions. Surface adsorption of serum proteins may, however, facilitate cell adhesion. Common ECM proteins such as vitronectin, fibronectin, and collagen are adsorbed on polymer surfaces (Nikolovski and Mooney 2000; Woo et al. 2003) and the high specific surface area of the nanofiber provides ample capacity for such absorption. In recent studies, fibronectin-grafted nanofibrous scaffolds of P(LLA-CL) copolymers, seeded with porcine esophageal epithelial cells, showed promise as a functional esophagus substitute (Zhu et al. 2007). Adhesive protein-coated nanofibers can be an economical route to developing replacement tissue (for instance, in the treatment of esophageal cancer).

Two common approaches to scaffold design are described in the biomedical engineering literature. A tissue-compatible scaffold of very good mechanical characteristics can be derived from decellularized xenogenic tissue seeded with autologous cells (to minimize tissue reactions to the implant). Alternatively, a sterile synthetic scaffold of a biodegradable polymer, can be seeded with autologous or other cells and cultured *in vitro* to a stage where there is sufficient cell attachment for implantation. In either case the scaffolding is surgically implanted into the defect site to promote repair and regeneration of tissue. Scaffolding can be also designed as foams, by salt leaching of polymer/inorganic salt composites (Lee, S. B., et al. 2005), or by self-assembly (Yuwono and Hartgerink 2007).

Electrospun nanofiber scaffolding can be fabricated out of (1) natural polymers, (2) synthetic polymers, or (3) polymer blends of natural or synthetic polymers. Using natural polymers generally assures a certain degree of biocompatibility and biodegradability of the scaffold. Most biopolymers can be readily electrospun into fibers and are beginning to be used in tissue engineering applications (Saw et al. 2006). Comparative cell proliferation studies carried out on nanofiber mats versus microspheres or even three-dimensional braided materials show nanofibers to perform exceptionally well. Figure 7.5 shows an image of rat hepatocyte cells growing on a galactose-grafted film substrate and on a nanofiber mat. While cell contact with the film surface is minimal, integration with the fiber mesh is extensive. Electrospun polymers reported in scaffolding studies are tabulated in Appendix I. The use of electrospun nanofibers in tissue engineering applications has been recently reviewed (Murugan and Ramakrishna 2006; Pham et al. 2006). Most of the reported studies, however, are limited to descriptions of electrospinning the fiber mats, seeding the construct with specific cell types, and observing cell attachment/proliferation on them over a period of time. Qualitative or semi-quantitative data on interactions of cells with the nanofiber scaffolding, generated using microscopic methods or assays, are also sometimes reported. Usually, microscopic studies of cell growth using environmental scanning electron

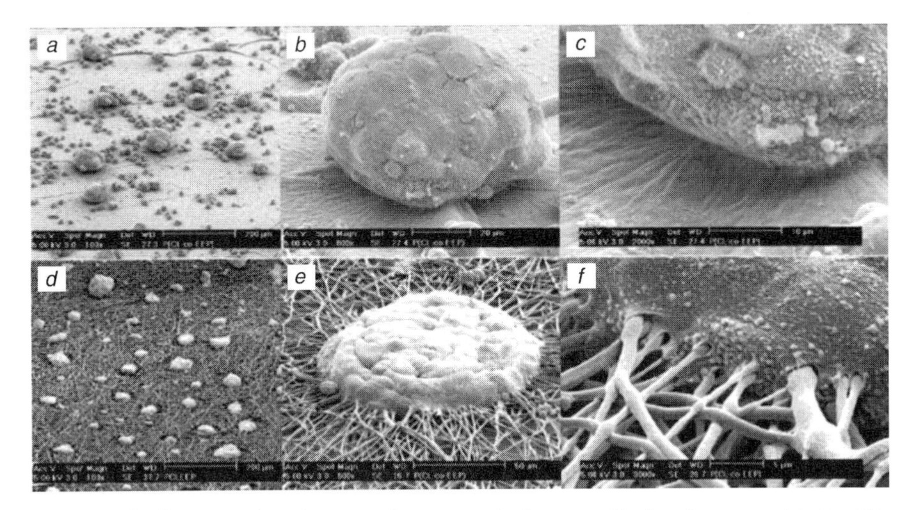

Figure 7.5 Rat hepatocytes growing on galactose-grafted polymer surfaces. The top row of images are for a polymer film surface and the lower row is for a nanofiber mat. Reproduced with permission from Chua et al. (2005). Copyright 2005. Elsevier.

microscopy (ESEM), confocal laser scanning microscopy (CLSM), optical microscopy, and biochemical assays based on DNA and extracellular matrix production (collagen or glycosaminoglycans) are used. Detailed information on the polymer used and characterization of the nanofiber mats, however, is generally lacking or incompletely reported, limiting the value of published research data.

With scaffolding, several key physical parameters need to be quantified:

1. The average diameter of the nanofibers and the distribution of fiber diameters in the scaffolding (Laurencin et al. 1999).
2. The average porosity and the pore-size distribution of the nanofiber mat.
3. An adequate description of cell types used in the experiment.

Porosity is a particularly useful measurement as it determines the kinetics of cell migration to the interior of the scaffold. With scaffolds of high fiber density or where the pores are too small to allow migration of cells, only a superficial layer of seeded cells has been observed to attach onto the scaffold (Boland et al. 2004a; Pham et al. 2006). With hydrophilic fiber mats, however, the possibility of amoeboid movement of cells, pushing against fibers and forcing them apart to create paths for migration, has been suggested (Boland et al. 2004a). Techniques used to emplace the live cells on the scaffolding also appear to play a key role in determining the infiltration of cells into the fiber matrix (Telemeco et al. 2005). A recent development in this regard is simultaneous electrospraying of live cells and electrospinning of the nanofiber (both from positively-charged tips and the fiber/cell construct collected on a rotating mandrel maintained at a negative potential (Stankus et al. 2006). Infiltration and copious integration of the cellular materials with the nanofiber is critical in obtaining a three-dimensional scaffold.

7.2.1 Natural Biopolymers

A variety of biopolymers, such as cellulose (Kim, C. W., et al. 2005), DNA (Fang and Reneker 1997; Luu et al. 2003; Takahashi et al. 2005), gelatin (Huang et al. 2004; Ki et al. 2005; Kim, H.-W., et al. 2005; Li, Z. Y., et al. 2006b; Li, M. Y., et al. 2006; Ma, Z. W., et al. 2005a; Zhang, Y. Z., et al. 2004, 2005b, 2006b), alginate (Wayne et al. 2005), hyaluronic acid (Ji et al. 2006b; Yoo et al. 2005), dextran (Jiang et al. 2004a) and starch (Pavlov et al. 2004a), have been electrospun successfully into nanofibers. However only a few of these have obvious scaffolding applications and even fewer have been studied in any detail. These will be briefly discussed below.

7.2.1.1 Collagen and Elastin Type I and type III collagen[3] are the most abundant structural materials in mammalian ECM (comprising approximately one-third of the protein in human body) and are also among the most frequently used biopolymer scaffolding materials. Type I collagen occurs in native tissue as fibrils of up to several hundreds of nanometers in diameter. In common with other biopolymers, the isolation of collagen from tissue results in structural changes that limit its usefulness. Acid-soluble collagen type I, usually derived from tendons, is a gel of very limited structural integrity at physiological pH (Okano and Matsuda 1998). Blending the collagen with another polymer often improves its integrity and it can then be spun into useful nanofiber materials. In any event, collagen by itself cannot be readily electrospun from aqueous solutions unless blended with another natural (Zhong et al. 2005) or synthetic polymer (Huang et al. 2001a, 2001b). This is also true of other natural polymers, where electrospinning is only possible (or at least the nanofiber quality obtained is greatly improved) by blending; examples include casein electrospun with PEO (Xie and Hsieh 2003), chitosan with silk (Park, W. H., et al. 2004; Spasova et al. 2004), egg shell membrane protein with PEO (Yi et al. 2004), and chitin with PGA (Park et al. 2006) are examples.

Type I collagen/PEO solutions at concentrations of 1–2 wt% dissolved in 34 mM NaCl have been electrospun successfully by Huang et al. (2001b). Bead-free uniform fibers, average diameter ($d = 50–150$ nm), were obtained with blends where the collagen/PEO ratio was 1 : 2. Collagen can also be electrospun from non-aqueous solutions as well — solutions of collagen in 1,1,1,3,3,3-hexafluoro-2-propanol (HFP) (boiling point 61°C), for instance, can be readily electrospun into fine nanofibers (Venugopal et al. 2005a; Rho et al. 2006). However, even at a concentration of 0.083 g/mL, type I collagen is not completely soluble in HFP but yields a cloudy solution (which, however, could still be electrospun to yield nanofibers of about $d \sim 100$ nm) (Matthews et al. 2002, 2003). Both of the sources (calfskin vs human placenta) as well as the isotypes (type I vs type III) of collagen employed in the study had a direct impact on the structural features of the resulting nanofiber. Thus, collagen type I isolated from skin and placenta yielded different fiber morphologies (Matthews et al. 2002). Human smooth muscle cells and endothelial cells were successfully cultured on these collagen nanofiber scaffolds. Electron microscopy, immunohistochemical studies, laser scanning confocal microscopy, and cell proliferation assays all suggested that the electrospun nanofibrous scaffold promoted good cell attachment and proliferation (Matthews et al. 2002; Xu et al. 2005).

[3]Of the 28 types of collagen, type I collagen is the most abundant form in the human body and is found primarily in tendons and in scar tissue. The type II variety is present in articular cartilage. Type III collagen is found in granulation tissue.

The ability to electrospin tubular scaffolding of various diameters, particularly those smaller than about 5 mm in diameter, is important because of their potential use as arterial substitutes in vascular grafts (He, W., et al. 2005a, 2005b; Miller et al. 2004; Stitzel et al. 2006). The challenge involved is twofold; not only should the diameter and mechanical integrity be adequate, but tissue compatibility also needs to be ensured. Mechanical properties of electrospun nanofibers that match those of vascular tissue have already been achieved (Stitzel et al. 2006). Boland et al. (2004a) successfully fabricated tubular scaffolds by electrospinning type I and type III collagen from 3–10 w/v% solutions of the protein in HFP. Electron microscopy of the nanofibers revealed a 65-nm banding pattern that is also apparent in native collagen fiber (Matthews et al. 2002). In native collagen the bands are due to the regular repeating sequences in its primary structure. Relatively less work is reported on electrospinning type II collagen; Matthews et al. (2003) reported electrospinning of this type isolated from chicken cartilage, as well as collagen types I and III, all from HFP solutions. The morphology of the electrospun collagen nanofibers varied with the isotype of collagen. A detailed description of electrospinning of collagen and elastin is given in a recent publication (Li, M. Y. et al. 2005).

Both collagen and elastin make up the protein content in native blood vessels. It is the presence of elastin in the vessel wall that imparts flexibility and compliance to the tissue (Faury 2001). Native elastin, however, is crosslinked and therefore cannot be electrospun (see Fig. 7.6). Attempts at electrospinning blends of collagen and soluble elastin fractions are, however, reported in the literature (Buttafoco et al. 2005, 2006). An 81 kDa recombinant protein having the repeat structure for the key peptide sequence in elastin[4] (Val-Pro-Gly-Val-Gly)$_4$(Val-Pro-Gly-Lys-Gly) (Huang et al. 2000) as well as an acrylate-modified recombinant protein (Val-Pro-Gly-Val-Gly)$_4$ (Val-Pro-Gly-Lys-Gly)$_{39}$ (Nagapudi et al. 2002) have been successfully electrospun. The first of these was electrospun from aqueous solutions at concentrations of 10–15 wt% to obtain nanofibers 400–450 nm in average diameter (Huang et al. 2000). The electrospinning of both collagen and elastin is usually facilitated by adding NaCl to the solution (42.5 mM), presumably because of increased conductivity. Buttafoco et al. (2006), however, suggest the salt also to induce hydrophobic interactions in or between the protein molecules to facilitate fiber formation. Spinning collagen/elastin/PEO blends yielded nanofiber mats with values of d ranging from 220 nm to 3000 nm (Huang et al. 2000).

[4]The amino acid sequence (Val-Pro-Gly-Val-Gly) occurs frequently in the primary structure and is responsible for the β-turns in the secondary structure of elastin.

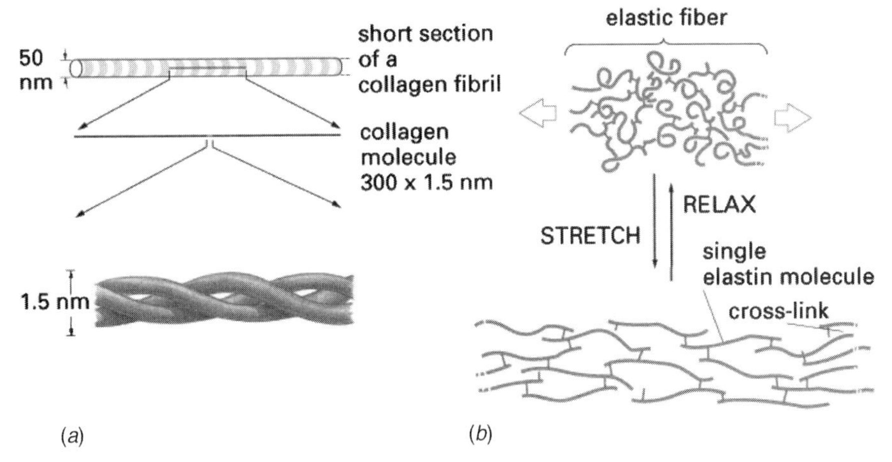

Figure 7.6 (*a*) Collagen is a triple helix of three extended protein chains wrapped around one another. Crosslinked collagen in the extracellular matrix forms collagen fibrils of low extensibility and very high tensile strength. (*b*) Elastin polypeptide chains are crosslinked together to form elastic fibers. Each elastin molecule reversibly uncoils into an extended conformation when the fiber is stretched. Reprinted with permission from Alberts et al. (2003). Copyright 2003. Garland Science.

As with collagen, the soluble fraction of elastin isolated from bovine *ligamentum nuchae* could also be electrospun in blends with PEO (Buttafoco et al. 2006). Unlike collagen, however, even at a blend ratio of protein to PEO ($M_w = 8 \times 10^6$ g/mol) as high as 5 : 1, elastin blends still yielded continuous nanofibers. Collagen/elastin (1 : 1) nanofiber mats were prepared by electrospinning blends of the two biopolymers and were stabilized by crosslinking with *N*-(3-dimethylaminopropyl)-*N*-ethylcarbodiimide hydrochloride (EDC) and *N*-hydroxysuccinimide (NHS). This treatment afforded materials with a high thermal stability ($T_d = 79°$C) and insolubility without affecting the original morphology (Buttafoco et al. 2006). The water-soluble PEO fraction can be extracted from the blend mats to yield a collagen/elastin nanofiber mat that is closer in composition and in physical integrity to vascular tissue. A particularly impressive finding in this study was the intimate mixing of the two proteins achieved during electrospinning as evidenced by their very similar morphology under electron microscopy.

As scaffolding performance invariably depends on the surface chemistry of fibers, synthetic polymer nanofibers coated with a layer of collagen can also be used in place of bulk collagen nanofibers (Dunn et al. 1997; He, W., et al. 2006; Venugopal et al. 2005a) as scaffolding. Human coronary artery smooth muscles cells (hSMC) cultured on collagen-coated nanofibers proliferated well and were more firmly attached to biopolymer scaffolds relative to the control nanofibers of poly(ε-caprolactone) (PCL). Evidence of cell migration towards the interior of the nanofiber mat and organization into tissue was also observed. Human coronary endothelial cells growing on aligned nanofibers of P(LLA-CL) coated with a layer of collagen were found to grow along the direction of nanofiber alignment and showed an elongated morphology (He, W., et al. 2006) that resembled endothelial cells *in vivo* under blood flow. He and colleagues fabricated a core–shell nanofiber with a core of PCL (385 ± 82 nm) and shell of collagen (64 ± 26 nm) using coaxial spinning (He, W., et al. 2005a, 2005b). Human dermal fibroblast proliferation on the core–shell nanofibers was found to be 32% higher (over a six-day period of exposure) relative to that obtained with PCL nanofibers. Even with the partial coating of collagen achieved by soaking PCL mats in collagen solutions overnight (He, W., et al. 2006), the fibers performed better relative to PCL controls. Surface modification by plasma treatment has also been shown to improve coating effectiveness by improving the wetting of hydrophobic polymer surfaces by the collagen solution (He, W., et al. 2005a, 2005b, 2006).

7.2.1.2 Fibrinogen Fibrinogen is a glycoprotein of molecular weight 340 kDa (with ~3000 amino acid residues), synthesized in the liver and generally present in plasma at a concentration of about 200–400 mg/dL. The molecule consists of six chains linked by 29 disulfide bonds and is often denoted by $(A\alpha\beta\gamma)_2$ as the protein is a dimer. Being the principal protein involved in clotting of blood, there is considerable interest in electrospinning fibrinogen. It not only yields the fibrin material that constitutes the clot but also acts as a cofactor in platelet aggregation at the site. Conversion of the protein to fibrin during clot formation is facilitated by thrombin (and Ca^{2+}), which cleaves the fibrinopeptides A and B from α and β chains, exposing the N-terminal polymerizable sites in the lysine residue, that react to form the clot. These sites usually lie buried within the tertiary structure of the native protein and are generally inaccessible for reaction. Clearly, the process of electrospinning does not perturb the tertiary structure of fibrinogen to an extent to expose the lysine residues, as the clotting characteristics of fibrinogen do not appear to be altered by the process. Studies on fibrinogen nanofibers are encouraged by the potential for developing hemostatic wound dressings based on such fibers.

Nanofibers of human and bovine fibrinogen with an average diameter of ~80 nm were electrospun from HFP and minimal essential medium or Earle's salt solution mixed in the ratio of (9 : 1), at protein concentrations of 0.083, 0.125, and 0.167 g/mL. Fine fibers with $d = 80 \pm 20$ nm, 310 ± 70 nm, and 700 ± 110 nm were electrospun at the different concentrations (Wnek et al. 2003). Fibrin fibers that occur in clots typically show diameters in this same size range ($d \sim 80$–90 nm).

Another study investigated electrospinning of bovine fibrinogen using the same solvent system at concentrations of the protein ranging from 80 mg/mL to 140 mg/mL (McManus et al. 2006). Characterizing the mats included an evaluation of the average fiber diameter *d* as well as mat porosity, which in turn controls the permeability of the mat. Both the fiber diameter (McManus et al. 2006; Wnek et al. 2003) as well as mat porosity (McManus et al. 2006) were found to be a linear functions of concentration. Typical values for average porosity were 54% and 59% for fiber mats spun from HFP at concentrations of 110 mg/mL and 130 mg/mL, respectively; the corresponding surface area to volume ratio for the samples were 16,800 and 14,200 cm^2/cm^3. Typically, higher levels of porosity can be obtained in nanofiber mats by shortening the duration of collection. Samples electrospun from solutions of concentration >100 mg/mL had sufficient structural integrity to be mechanically tested, and the moduli of PBS buffer-hydrated electrospun fiber scaffolds were in the range of 0.30–0.58 MPa.

7.2.1.3 Silk Silk fibroin is a protein produced by many insect and arachnoid species (e.g., *Nephila clavipes*) (Ayutsede et al. 2005, 2006; Jin et al. 2002, 2004; Khil et al. 2003; Kim, S. H., et al. 2003; Kim, K. H., et al. 2005; Lee, K. H., et al. 2005; Li, C. M., et al. 2005, 2006; Min et al. 2004a; Ohgo et al. 2003; Park et al. 2004; Seidel et al. 1998; Stephens et al. 2005; Sukigara et al. 2003, 2004; Wang, M., et al. 2006; Zarkoob et al. 2004). Commercially, silk is, however, derived from cocoon silk (from the silkworm *Bombyx mori*) as natural fibers 10–20 μm in diameter. Biomedical interest in silk is mainly due to its exceptionally good mechanical properties (Ohgo et al. 2003). Spiders extrude an aqueous liquid-crystalline fibroin protein through their spinnerets and the liquid hardens on exposure to air forming a highly crystalline oriented mesofiber. The primary structure of the two proteins that constitute silk has been elucidated (Hinman and Lewis 1992); the drag-line spider silk is made of fibroin ($M_w \approx$ 20,000–300,000 g/mol), a protein that is 42% glycine and 25% alanine. The structurally simple amino acids without any side chains allow close chain packing, resulting in a high degree of crystallinity that is primarily responsible for the high strength and tensile modulus of the material. These proteins for the most part exist as antiparallel β-sheet crystallites in the native silk. Recombinant dragline spider silk analogs that mimic silk from *Nephila clavipes* have been successfully expressed in *E. coli* host and electrospun into nanofibers from HFP (Stephens et al. 2005).

Native silk is readily isolated from the crude silk on cocoons. Typically, crude degummed silk is dissolved in 50% aqueous $CaCl_2$ followed by dialysis against distilled water. The regenerated silk sponge from this treatment is dissolved in formic acid (98–100%) at a concentration of 9–15 wt% for electrospinning (Ayutsede et al. 2005). Others have used the ternary solution system $CaCl_2$/ethanol/water (1 : 2 : 8 molar) at 70°C for 6 h followed by dialysis. Electrospinning was carried out in formic acid solution in this case as well. Wang and colleagues reported a somewhat complicated but environmentally benign aqueous electrospinning process that included post-treatment of the bicomponent nanofibers produced (Wang, M., et al. 2006). Sukigara et al. (2003, 2004) studied the different variables involved in the electrospinning of silk from 10–15% solutions in formic acid solution to conclude that concentration was the most important parameter in producing continuous ($d \sim 100$ nm) nanofibers. Figure 7.7 shows an SEM image of a silk fibroin nanofiber mat showing uniform bead-free fibers spun from formic acid. Nanofiber diameter distribution obtained by SEM image analysis for a similar mat but spun from a 19.5 wt% solution of silk in formic acid is shown in Fig. 7.8.

Isolated silk fibroin has been readily electrospun into nanofibers by several researchers (Jin et al. 2002, 2004; Khil et al. 2003; Kim, S. H., et al. 2003; Kim, K. H., et al. 2005; Lee, K. H., et al. 2005; Li, C. M., et al. 2005, 2006; Min et al. 2004a; Park, W. H., et al. 2004; Seidel et al. 1998; Wang, M., et al. 2006). Degummed crude dragline silk can be directly dissolved in HFP at ambient temperature (a very slow process taking weeks in the case of silk from *Bombyx mori*) and electrospun at a concentration of 0.23–1.2 wt%. Fine nanofibers (most frequent diameter $d \approx 100$ nm and a range of diameters from 8 nm to 200 nm) were obtained (Zarkoob et al. 2004). These diameters are an order of magnitude smaller than those for natural silk fibers and approach the dimensions of small subfibers in natural silk.

Figure 7.7 SEM image of silk fibroin nanofibers spun from 12% formic acid illustrating the small fiber diameters obtained. Reproduced with permission from Min et al. (2004b). Copyright 2004. Elsevier.

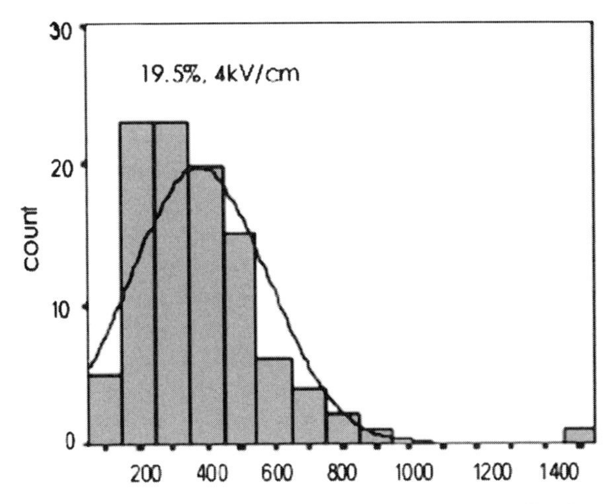

Figure 7.8 Fiber diameter distribution of electrospun silk fibroin nanofiber from SEM image analysis. A 19.5% solution of silk in formic acid was used in electrospinning. Fiber analysis was based on SEM images of the mat. Reproduced with permission from Sukigara (2003). Copyright 2003. Elsevier.

In a later study (Sukigara et al. 2004), the average fiber diameter was correlated to the magnitude of the electric field and concentration using a second-order polynomial equation. At low silk concentrations the changes in fiber diameter were found to be more responsive to changes in the electric field. As nanofibers are formed rapidly during electrospinning, their degree of crystallinity is generally quite low (Deitzel et al. 2001a). The as-electrospun silk nanofibers therefore have a predominantly random coil structure as opposed to the β-pleated crystalline structure typical of the native protein (Stephens et al. 2005). On annealing the fibers, crystallinity slowly develops, as seen from X-ray diffraction measurements (WAXD). The rate of random coil to β-sheet transition in nanofibers spun from HFP was studied by time-resolved IR spectroscopy by L. Jeong et al. (2006). The FTIR absorption band at 1624 cm^{-1} and that at 1663 cm^{-1} are indicative of the β-sheet and random coil conformations of the protein respectively and allow the crystalline transition to be followed spectroscopically. The transition can be accelerated by exposing the material mats to solvent vapor. Methanol vapor and aqueous methanol treatment of the mats facilitate the change in morphology within a 10-min period of exposure, as confirmed by ^{13}C NMR studies (Min et al. 2004a). Raman spectroscopy has also been used similarly to study conformational features (Ayutsede et al. 2005).

In recent research, recombinant hybrid silk fibers that incorporate both the highly crystalline domains of native silk fibroin from silkworms and the amorphous domains from a different variety of silk (from a wild species of silkworm *Samia cynthia ricini*) were electrospun from hexafluoroacetone (HFA) solution into nanofibers ($d \sim 100$ nm) (Ohgo et al. 2003). The latter species produces a silk of different primary structure and mechanical properties. The structure of the hybrid silk was [Gly-Gly-Ala-Gly-Ser-Gly-Asp-Gly-Gly-Gly-Tyr-Gly-His-Gly-Tyr-Gly-Ser-Asp-Gly-Gly-(Gly-Ala-Gly-Ala-Gly-Ser)$_3$]$_6$. Changes in crystallinity on post-treatment of silk fibers in methanol was monitored by ^{13}C CP/MAS NMR spectroscopy.

7.2.1.4 Chitin/Chitosan Chitin is the second most abundant biopolymer in nature (after cellulose) and is a water-insoluble biodegradable polysaccharide with excellent mechanical properties (Bhattarai et al. 2005; Geng et al. 2005; Li and Hsieh 2006; Min, B.-M. et al. 2004c; Noh et al. 2006; Ohkawa et al. 2004a; Park, K. E., et al. 2006; Park, W. H., et al. 2004; Subramanian et al. 2005). It occurs in insects, fungi, and is the principal component of exoskeleton of shellfish, making it an inexpensive product in fairly good supply. Chitin is an unbranched polymer of *N*-acetyl-D-glucosamine. Structurally, it is a cellulose analog where the hydroxyl group at the C-2 carbon has been replaced by an acetamido (—NH(C=O)CH_3) group. Chitin is readily deacetylated into chitosan, also a biodegradable, easily processable biopolymer. Chitosan, in particular, has been successfully used as a scaffolding material for cartilage, nerve, and liver cells. It is therefore a natural candidate polymer for scaffolding applications.

Noh et al. (2006) electrospun chitin from HFP (3–6% w/w) solutions to obtain continuous nanofibers ($d \approx 50$–460 nm). High-molecular-weight commercial chitin of degree of deacetylation of about 8% was degraded by gamma irradiation into a sample of $M_w \approx 91{,}000$ (g/mol) to be electrospun. Cell growth studies on the scaffolds seeded with human keratinocytes and fibroblasts showed the nanofibers to promote enhanced cell attachment and spreading compared to commercial microfibers of the material (Noh et al. 2006). Partially depolymerized chitin electrospun from HFP into nanofiber

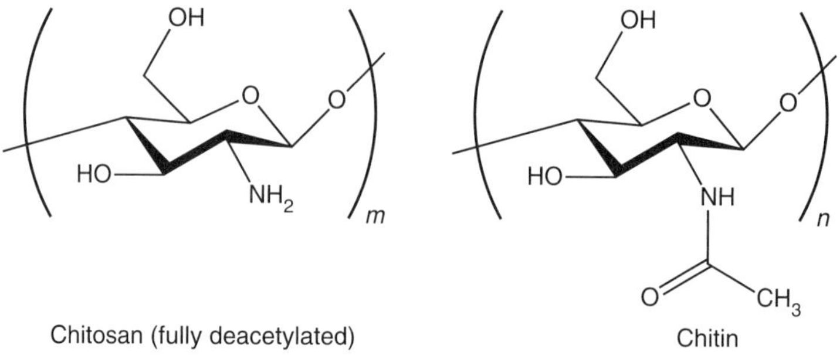

Chitosan (fully deacetylated)

Chitin

Scheme 7.1

mats where most fibers have $d < 100$ nm could be easily deacetylated with 40% NaOH to yield chitosan nanofiber mats (Min, B.-M., et al. 2004c).

Chitosan is obtained by deacetylation of chitin and has an average molecular weight between 50,000 and 1,000,000 (g/mol) depending on its source; its degree of crystallinity (50–90%) depends upon the degree of deacetylation. Chitosan has been wet-spun into microfibers using conventional methods (Hirano and Mooney 2004), but is difficult to electrospin by itself as it forms a viscous gel at very low concentrations (1–3 wt% in dilute acid, for instance) due to its polyelectrolyte nature (Bhattarai et al. 2005). However, the low molecular weight polymer was electrospun from a 7 wt% solution of chitosan (molecular weight 106,000 g/mol) in 90% (v/v) aqueous acetic acid (Geng et al. 2005) and from trifluoroacetic acid (TFA) (Ohkawa et al. 2004a) to obtain bead-free nanofibers. A higher molecular weight chitosan (200,000 g/mol) was electrospun from TFA or a TFA : CH_2Cl_2 (80 : 20) mixture by Ohkawa et al. (2004a) or from HFP (Min, B.-M., et al. 2004c). Chitosans ranging in M_v from 21×10^4 to 18×10^5 g/mol have also been electrospun from TFA recently (Ohkawa et al. 2006). It is likely that the TFA protonated the amine functionalities and disrupted the intermolecular forces responsible for the very high viscosity of chitosan in dilute solutions. An electro-assisted wet-spinning process where chitosan is spun into a coagulating solution of alkaline ethanol has recently been reported (Lee, C. K., et al. 2006).

Blends of chitosan with synthetic polymers have been electrospun by several research groups (Bhattarai et al. 2005; Duan et al. 2004; Park, K. E., et al. 2006; Park, W. H., et al. 2004; Subramanian et al. 2005). Chitosan/PEO blends in ratios of 2 : 1 and 5 : 1 were electrospun from 6 wt% solutions in dilute acetic acid into nanofibers with $d = 80–180$ nm (average 124 ± 19 nm) (Duan et al. 2004). Electrospinning a 1 : 1 mix of the polymers yielded a mat where the fiber diameters were bimodally distributed; nanofibers were interspersed with microfibers of relatively larger diameter. Using spectroscopic and XPS techniques, the larger microfibers were for the most part found to be composed of PEO only. Others (Bhattarai et al. 2005), working with the same blend, obtained similar results using acetic acid/DMSO mixtures containing 0.3% Triton X-100 as the solvent. Using a spinning collector disk it was possible even to obtain aligned nanofibers of the chitosan/PEO blends (Subramanian et al. 2005). Water-soluble polymers such as PVA (Mincheva et al. 2005; Ohkawa et al. 2004a), poly(vinylpyrollidone) (PVP) (Ignatova et al. 2007) or polyacrylamide (Mincheva et al. 2005) can be substituted for PEO in the blends to obtain chitosan or chitosan derivative nanofibers. The bicomponent fibers electrospun from chitosan/PVA blends (3 wt% solutions in 2% (v/v) acetic acid) have values of d in the range of 20–100 nm. Removing the PVA content from the nanofibers by post-extraction with alkaline solution yielded a porous chitosan nanofiber (Li and Hsieh 2006). Blends of chitosan with other biopolymers have also been electrospun successfully. For instance, Park and colleagues electrospun silk fibroin/chitosan blends with up to a maximum of 30% chitosan from formic acid solutions to obtain nanofibers ($d = 130-430$ nm) (Park, W. H., et al. 2004). Collagen/chitosan blends were successfully electrospun from HFP/TFA (90/10 v/v) mixed solvent at a concentration of 6–10 wt% by Z. Chen et al. (2007).

Chondrocytes (Bhattarai et al. 2005; Subramanian et al. 2005) and osteoblasts (Bhattarai et al. 2005) were shown to grow well on chitosan blend nanofibers. In a comparative study, cell viability on polymer film, nanofiber mat and PS surfaces has been reported. Chondrocyte cell proliferation (10 days of culture) on the chitosan/PEO scaffolds was 81% of that obtained for tissue-culture-grade polystyrene surfaces while that for film geometry was about 56% (Subramanian et al. 2005).

7.2.1.5 Poly(3-hydroxybutyrate-co-3-hydroxyvalerate) (PHBV)

PHBV is a bacterial polyester copolymer with molecular weights ranging between 400,000–750,000 g/mol and is now commercially produced with hydroxyvalerate contents of 7 and 22 percent (molar). It is hydrolytically degradable and is also biodegradable (Picha and Schiemenza 2006). The polyester can be electrospun from several solvents (Choi, J. S., et al. 2004; Fang et al. 2004; Han et al. 2004; Ito et al. 2005; Lyoo et al. 2005). Electrospun nanofibers of the polyester was used as a scaffold to successfully culture chondrocytes derived from rabbit ears (Lee, I. S., et al. 2004). Poly(3-hydroxybutyrate) (PHB) $M_w \sim 300{,}000$ g/mol, PHBV (HV ~ 5%) $M_w \sim 680{,}000$ g/mol, and their 50/50 w/w blend were recently electrospun from 14 wt% $CHCl_3$ solution (Sombatmankhong et al. 2007). Human osteoblasts and mouse fibroblasts seeded on the mats proliferated well, and again afforded better support for cell growth compared to film substrate.

7.2.2 Synthetic Polymers

Using synthetic polymers for tissue engineering applications is advantageous because of the uniform chemical composition and consistency in the quality materials obtained from commercial sources. Also, a polymer with mechanical properties that best matches a particular scaffolding application can be readily selected or even designed. However, as most synthetic polymers are virtually nonbiodegradable, the available choices of synthetic polymers for scaffolding are somewhat limited. Electrospun biodegradable scaffolds reported in the literature are for the most part based on ε-poly(caprolactone) (PCL), poly(lactide) (PLA), poly(glycolide) (PGA),

TABLE 7.2 Properties of common biodegradable polymers used in biomedical applications

Polymer	T_m (°C)	T_g (°C)	Modulus (MPa)	Degradation[a]
Poly(glycolic acid) (PGA)	225–230	35–40	7	6–12 months
Poly(L-lactide) (PLLA)	173–178	60–65	2.7	>24
Poly(DL-lactide) (PDLLA)	Amorphous	55–60	1.9	12–16
Polycaprolactone (PCL)	58–63	−(65–60)	0.4	>24
85/15 poly(lactide-*co*-glycolide)[b]	Amorphous	50–55	2.4	5–6
75/25 poly(lactide-*co*-glycolide)[b]	Amorphous	50–55	2.0	4–5
65/35 poly(lactide-*co*-glycolide)[b]	Amorphous	45–50	2.0	3–4
50/50 poly(lactide-*co*-glycolide)[b]	Amorphous	45–50	2.0	1–2

[a]*In vivo* degradation rate also depends on the molecular weight of the polymer, fiber diameter and mat thickness.
[b]Properties as well as the degradation rates vary with copolymer composition.
Source: Lu and Chen 2004; Liang et al. 2007.

or copolymers composed of the repeat units of these. A summary of their characteristics is given in Table 7.2.

7.2.2.1 Polyglycolides (PGA) Ring-opening polymerization of the glycolide monomer yields high-molecular-weight PGA that generally has a degree of crystallinity of about 30–55%, a glass-transition temperature just above room temperature and a relatively high melting point (Boland et al. 2001; Gao et al. 1998; Park, K. E., et al. 2006; You et al. 2005a, 2005b, 2006b; Zong, X. H., 2002, 2003a). Because of its semicrystallinity, only highly fluorinated organic solvents such as HFP dissolve the polymer. Electrospun nanofiber mats of PGA have high strength and modulus and are readily biodegradable in the body (Zong, X. H., et al. 2003a). Nanofibers, in fact, tend to biodegrade even faster than thin films of the same material (Zong, X. H., et al. 2002). Sutures made from the material, for instance, are fully embrittled following weeks of implantation and are completely absorbed in as short a time as a few months (depending on the M_w of the polymer). Glycolide has been copolymerized with other monomers to reduce their crystallinity and to increase the flexibility or extensibility of the resulting nanofibers. Copolymers with the lactide have been widely studied (Berkland et al. 2004; Bini et al. 2004; Jiang et al. 2004b; Kim, H. S., et al. 2005; Zong, X. H., et al. 2002, 2003a, 2003b).

Poly(L-glycolide) (PLGA) nanofiber mats of random copolymers of lactide (L) and glycolide (G) have mechanical properties that are in the same range as those of the tissue they are intended to replace (Li et al. 2002; Luu et al. 2003; Shin, H. J., et al. 2006). The tensile modulus (MPa) of these mats are, for instance, two to three times higher than that of skin or cartilage, and the porosity of scaffolding is particularly high. Table 7.2 shows the properties of different copolymers of PLGA. Boland et al. (2001) electrospun PLGA nanofibers ($d = 0.15 - 1.5\ \mu m$) from hexafluoroisopropanol (HFIPA) (1/7 to 1/20 wt%) using a rotating drum collector to obtain aligned fibrous scaffolds that consequently showed anisotropy in their mechanical properties. Tubular constructs of PGA copolymers electrospun onto a rotating mandrel have been used for *in vivo* regeneration of neural tissue (Bini et al. 2004). Nanofibers of PGA were shown to support growth and proliferation of human bone marrow mesenchymal stem cells (hMSC) (Li, W.-J., et al. 2002). Over the first week of growth, the seeded MSC population increased fivefold on the nanofiber scaffold. Stem cells seeded on the fiber mat maintained their shape and their growth was guided by the fiber orientation.

7.2.2.2 Polylactide (PLA) PLA is a relatively hydrophobic polymer because of the methyl group in its structure and therefore inherently slower biodegrading compared to PGA. The homopolymer (PLLA) of the naturally occurring isomer of lactide monomer L-lactide is semicrystalline, has a high modulus, high strength, and is about 37% crystalline with a glass-transition temperature of 60–65°C. It biodegrades slowly (taking months or even years to break down). Copolymers of the racemic mixture of D and L forms (PDLLA) on the other hand are noncrystallizable and therefore tend to biodegrade much faster compared to PLLA. Random copolymers of L-lactide and DL-lactide with reduced, controlled levels of crystallinity and higher rates of biodegradation can also be synthesized. PDLA is obtained by polymerizing a blend of the D and L forms of the monomer.

Nanofibers of crystalline polymers, however, generally do not display a high degree of crystallinity in the as-spun condition. Rapid drying of the solution does not allow crystallite formation during electrospinning, and yields a lower percentage crystallinity compared to that obtained in slower crystallization of the same polymer from melt or solution (Bognitzki et al. 2001a). Zong and colleagues (Zong, X. H., et al. 2002), using WAXD techniques, determined that the as-electrospun poly(glycolide-*co*-lactide) nanofibers were essentially amorphous but with a high degree of chain orientation. On annealing at 55°C for 24 h, two crystalline peaks at 2θ values of 16.4° and 18.7° (associated with a pseudo-orthorhombic unit cell of the α-form of PLLA crystals) developed. (This general tendency of crystalline polymers to yield nanofibers of low crystallinity was discussed in Chapter 5.)

Relatively few studies have investigated PLA nanofibers as a scaffolding material; most have been on its copolymers, particularly with glycolide (Bini et al. 2006; Bognitzki et al. 2001a; Boland et al. 2006; Gu et al.

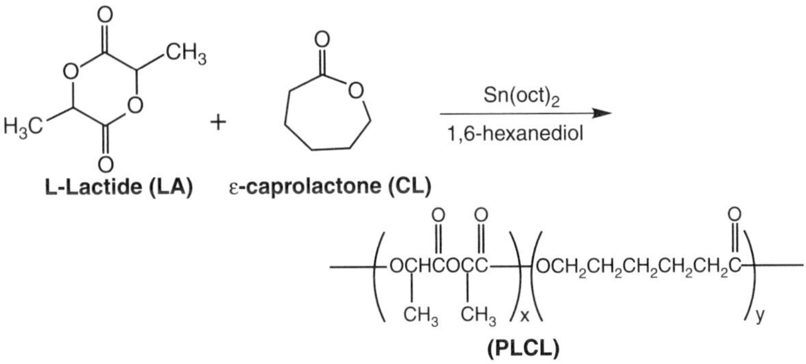

L-Lactide (LA) + ε-caprolactone (CL) $\xrightarrow[\text{1,6-hexanediol}]{Sn(oct)_2}$ (PLCL)

Scheme 7.2

2005; He, W., et al. 2006; Jeun et al. 2007; Kim, K. S., et al. 2003; Spasova et al. 2007; Yang, F., 2004, 2005; Zong, X. H., et al. 2002). F. Yang et al. (2004, 2005) found neural stem cells to successfully grow and even show a degree of orientation on nanofiber mats of PLLA electrospun from CH_2Cl_2/DMF (70 : 30) solvent. When compared to two-dimensional scaffolding of comparable porosity, a twofold improvement in osteoblast attachment to the scaffolds was obtained (Woo et al. 2003). This was partly a result of the higher surface capacity of nanofibers to adsorb serum proteins. It has been suggested that chondrocytes are more biocompatible with nonwoven PLA compared to nonwoven PGA (Sittinger et al. 1996). Copolymers of lactide with caprolactone (P[LA-CL]) was electrospun into nanofibers by Kwon et al. (2005) who reported human umbilical vein endothelial cells (HUVECs) to adhere and proliferate on mats with the average fiber diameters ranging from 300 nm to 1.2 μm. Mo et al. (2004) came to a similar conclusion but using muscle cells and endothelial cells.

7.2.2.3 Poly(ε-caprolactone) (PCL) PCL is an aliphatic polyester with a semi-crystalline morphology synthesized by the ring-opening polymerization of ε-caprolactone. The polymer biodegrades slowly over a period of about two years in the body, yielding ε-hydroxycaproic acid as the primary metabolite. The homopolymer can be easily electrospun from CH_2Cl_2 or THF : DMF (1 : 1 v/v) mixture (Li, W.-J., et al. 2005b, 2005c). A limitation of the homopolymer is its low melting point of only 63°C. To enhance the rate of biodegradation and to modify the mechanical properties, copolymers of ε-caprolactone (CL) are therefore often used in scaffolding. For example, copolymers of ε-caprolactone with DL-lactide tend to biodegrade at comparatively faster rates. Copolymers can be electrospun from a range of common solvents such as $MeOH : CHCl_3$ (3 : 1 v/v) mixtures

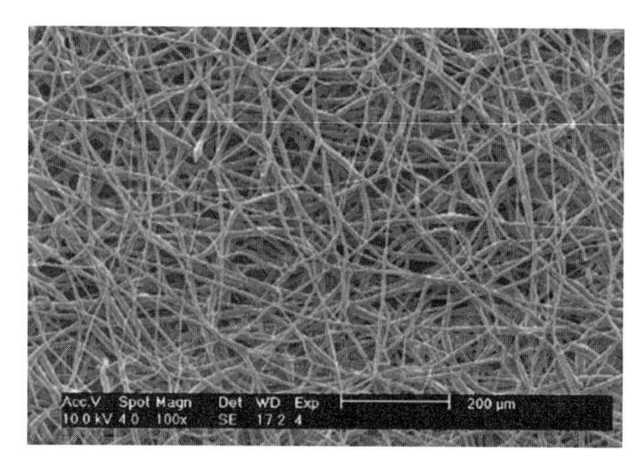

Figure 7.9 Nanofiber mat of poly(ε-caprolactone) (PCL) prior to seeding with cells, demonstrating the high degree of porosity. Average fiber diameter $d = 400 \pm 200$ nm. Reproduced with permission from Yoshimoto et al. (2003). Copyright 2003. Elsevier.

(Venugopal et al. 2005a), DMF : $CHCl_3$ mixtures (Bölgen et al. 2005), $CHCl_3$ (Yoshimoto et al. 2003), or acetone (Reneker et al. 2002), to obtain continuous defect-free nanofiber mats.

PCL and its copolymers can be readily electrospun (Bölgen et al. 2005; Hsu and Shivkumar 2004a; Jeun et al. 2005; Li, W.-J., et al. 2003; Ma, Z. W., et al. 2005a; Zhu et al. 2007). Figure 7.9 shows a nanofiber mat of PCL electrospun from a 10 wt% solution in $CHCl_3$. Mats were seeded with rat MSCs harvested from femoral bone marrow (Yoshimoto et al. 2003). The cell suspension ($\sim 4 \times 10^6$ cells per 50 mL) was centrifuged, the supernatant discarded, and the cell pellet resuspended in 200 μL of osteogenic

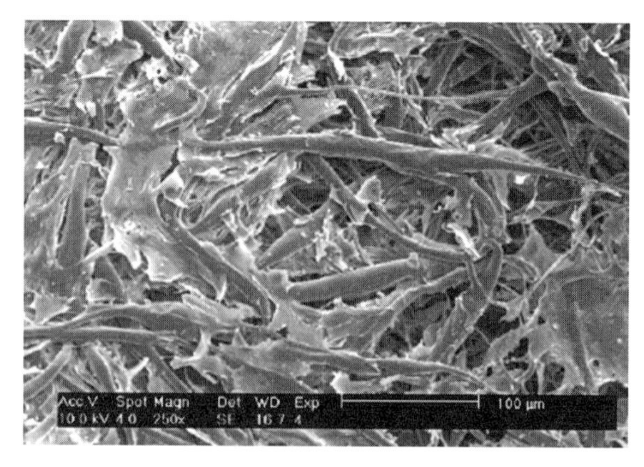

Figure 7.10 The same fiber mat as in Fig. 7.9, seeded with MSCs from the bone marrow of rats, after one week of growth. Multiple layers of osteoblast-like cells adhering to the surface of the fibers are seen. Reproduced with permission from Yoshimoto et al. (2003). Copyright 2003. Elsevier.

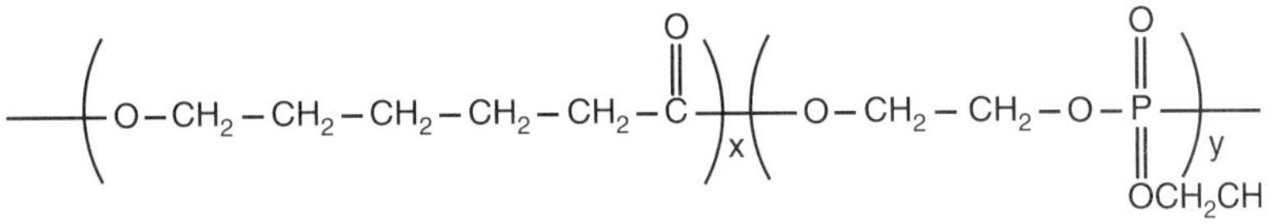

PCLEEP, x+y=100

Scheme 7.3

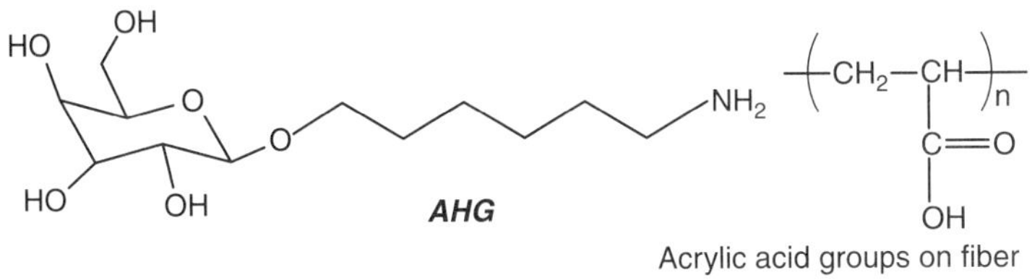

AHG

Acrylic acid groups on fiber

Scheme 7.4

differentiation medium for seeding experiments. Figure 7.10 illustrates the adhesion and proliferation of layers of MSCs. Biocompatibility of the coplymer [P(LL-CL)] with LL : CL ratio of 3 : 1 electrospun from a 5% acetone solution was investigated by Mo and colleagues (Mo and Weber 2004; Mo et al. 2004). Smooth muscle cells and endothelial cells grew and proliferated well on the scaffold.

Unlike with cast films, nanofibers of lactide copolymers do not show the well-defined peak at $2\theta \approx 16.7^\circ$ generally associated with crystalline domains of the L-lactide sequences. This lack of crystallinity in the nanofibers was demonstrated in electrospun copolymers of lactide and glycolide such as P[LA-GA] (10 : 90) (Zong, X. H., et al. 2003a, 2003b) as well as for the homopolymers PLLA and PDLA (Zong, X. H., et al. 2002). In the latter study, shrinkage and loss of porosity of copolymer scaffolds on incubation in buffer was also studied; these properties are particularly important in tissue engineering, as changes in dimensional stability and porosity of implanted scaffolds may seriously impair their performance. Copolymer composition controls the percent crystallinity as well as the crystalline morphology developed in annealed nanofibers and therefore their biodegradation rate as well. This was illustrated by the dependence of hydrophilicity and biodegradation rates on the structure in triblock copolymers (PLA-*b*-PEG-*b*-PLA) of different copolymer compositions (Kim, K. S., et al. 2003).

Copolymer with an LLA : CL ratio of 50 : 50 is elastomeric. The elastomeric copolymer was recently electrospun from HFP into a tubular scaffold (Inoguchi et al. 2006) that showed wall-thickness-dependent compliance when tested under pulsatile pressure. Several [P(LLA-CL)] copolymers (100/0, 74/26, 50/50, 31/69, 0/100) were also electrospun from 4–11 wt% solutions in CH_2Cl_2 to obtain nano- and microfabric scaffolds (Kwon et al. 2005). The fabrics were characterized with respect to mat porosity and density in addition to the average fiber diameter d. A much higher degree of cell proliferation for human umbilical vein endothelial cells (HUVEC) was obtained with the smaller ($d \approx 300$ nm or 1200 nm) fibers as opposed to mats of the larger $d \approx 7$ μm fibers. The same research group also reported (Kwon and Matsuda 2005) electrospinning of [P(LLA-CL)] copolymer blends with type I collagen from HFP solutions to obtain nanofibers with $d \approx 120$–520 nm. The mean fiber diameter decreased as the ratio of collagen to [P(LLA-CL)] ratio was increased. However, TEM images show collagen to be phase-separated into spherical domains within the nanofiber matrix to form a dispersed phase within the fibers. Cell infiltration into synthetic biodegradable nanofiber mats [PGA, PLA, and copolymer of (GA/LA)] of comparable porosity and average pore dimensions, implanted in rats, was poorer compared to collagen type I nanofiber mats electrospun from HFP solution (Telemeco et al. 2005). Infiltration in this case was measured over a 7-day period in terms of the accumulation of endothelial or interstitial cells within the bulk of the cylindrical scaffolds implanted.

Cell interaction with nanofibers depend also on the chemical nature and hydrophilicity of their surface (Kim, K. S., et al. 2003). Acid treatment of the PGA nanofibers ($d \approx 220$ nm) with concentrated HCl significantly increased the proliferation of rat cardiac fibroblasts relative to that on the untreated nanofiber controls (Boland et al. 2004b). The improved biocompatibility was attributed to the increased surface hydrophilicity due to partial acid hydrolysis of fiber surface.[5] Modification of the nanofiber surface to encourage interaction with cells is an important technique in scaffold design. The biodegradable copolymer poly(ε-caprolactone-*co*-ethyl ethylene phosphate) [P(CL-EEP)] electrospun from a 21.5% acetone solution yielded nanofibers with $d \approx 350$–1500 nm. The surface of the fiber was galactosylated by treatment with acrylic acid/UV followed by reaction with 1-*O*-(6-aminohexyl)-D-galactopyranoside (AHG) to obtain a galactose ligand surface density of 66 nmol/cm^2. Galactose ligands mediate hepatocyte adhesion to the surface via glycoprotein receptor interaction. Hepatocytes cultured on functionalized scaffolds were similar in physiological functions to those cultured on a two-dimensional substrate, but had a distinctive integrated fiber-spheroid morphology (Chua et al. 2005).

In order to improve scaffolding performance of nanofibers via surface modification (and to improve their mechanical integrity), fillers are sometimes used in nanofibers intended for scaffolding. For instance, composite nanofibers of

[5]Improving biocompatibility by partial hydrolysis of polyesters in scaffolding is well known (Gao et al. 1998). Hydrolytic pretreatment of the amorphous domains may also reduce the lowering of local pH in the scaffolds during early biodegradation, encouraging cell proliferation (Boland et al. 2004b).

PCL with 25% or 75% (w/w) of calcium carbonate were seeded with osteoblasts (Fujihara et al. 2005). Although both samples showed good cell attachment in the first five days after seeding, thereafter the scaffold with higher level of $CaCO_3$ showed reduced attachment. Hydroxyapatite (HAP), the major inorganic component of bone and teeth, is also a particularly well-studied filler.[6] Ito et al. reported the incorporation of HAP on the surface layers of electrospun microbial polyester poly(3-hydroxy butyrate valerate-*co*-3-hydroxyvalerate) (PHBV). Improved hydrophilicity of the scaffold was shown to enhance porosity and biodegradability of the matrix but no significant improvement in cell adhesion was observed (Ito et al. 2005). Biomimetic bone matrices based on electrospun nanofibers of gelatin/HAP (Kim, H.-W., et al. 2005) and PCL/HAP (Thomas et al. 2006) have been explored. Silk fibroin/PEO nanofiber scaffolds with HAP particles were investigated for their ability to sustain and proliferate human-bone-marrow-derived MSCs (Li, C. M., et al. 2005, 2006). The highest calcium deposition and upregulation of the transcripts levels observed in the study were only achieved when both HAP and BMP-2 (bone morphogenetic protein) were present in the nanofiber scaffold.

In implanted scaffolding, too high a rate of degradation may potentially impair cell proliferation, and too slow a rate may slow down the integration of the implant into the surrounding tissue. Selecting a polymer material with a compatible rate of degradation is therefore important. Tuan's group (Li, W.-J., et al. 2005a, 2006) studied hydrolytic degradation of various poly(α-hydroxyester) nanofiber mats in PBS over 42-day period of exposure. Based on previously published data based on shorter duration studies taken together (Li, W.-J., et al. 2006; You et al. 2005b; Zong, X. H., et al. 2003b), common biodegradable synthetic polymers can be arranged as follows (in ascending order of their relative biodegradability):

$$\text{PCL} < \text{PLA} < \text{PGA} < \text{PDLLA} < \text{PLGA}(85:15) < \text{PLGA}(75:25) < \text{PLGA}(50:50).$$

However, these rates of breakdown are invariably dependent on the percent crystallinity, crystalline morphology, the average molecular weight of the polymer as well as on fiber diameter and mat thickness (You et al. 2005b).

Polymers that are not biodegradable in physiological environments have also been electrospun and evaluated as scaffolds for tissue engineering applications. With only limited interest in these polymers, relatively few examples are available in the literature. Some recent examples include

[6]Hydroxyapatite [$Ca_5(PO_4)_3(OH)$] can be made into an inorganic nanofiber mat by electrospinning an inorganic precursor mixture of a calcium salt, triethoxy phosphine, and a polymer, followed by calcinations (Wu et al. 2004).

endothelial cells grown on poly(ethylene terephthalate) (PET) nanofibers grafted with polyacrylate functionalities (Ma, P. X., et al. 2004; Ma, Z. W., et al. 2005d), smooth muscle cells grown on polystyrene nanofibers (Baker et al. 2006), human vascular endothelial cell growth on polyurethane nanofiber surfaces electrospun on wet-spun PCL vascular scaffolds (Williamson et al. 2006), and fibroblasts or kidney cells grown on polyamide nanofiber scaffolds (Schindler et al. 2005).

7.2.3 Scaffolding with Stem Cells

A particularly exciting recent development is the successful culture of stem cells on biodegradable nanofiber scaffolding. The utility of embryonic and adult stem cells[7] and their potential applications in regenerative medicine are well known (Lin 1997; Tuan et al. 2003). These are able to differentiate in response to chemical cues yielding different types of tissue such as adipose, cartilage, bone or muscle. Ideally, these are pluripotent, quiescent, undifferentiated cells that can be triggered by chemical and physical cues to undergo lineage-specific differentiation. In tissue such as bone, consisting of several different types of cells patterned in specific geometries, the use of stem cells for tissue engineering can have a decided advantage. Using autologous tissue on a biodegradable scaffold can avoid immunorejection of scaffolds. Although a range of engineered tissue phenotypes derived from stem cells, particularly MSCs, is available, a corresponding range of scaffolds capable of accommodating and optimally nurturing them are yet to be developed (Xin et al. 2007).

Ideal scaffolding not only supports cell growth but also controls their differentiation into a multiphase tissue supporting different cell types self-organized appropriately. Adult trabecular bone, bone marrow, muscle, and fat tissue as well as chord blood (or amniotic fluid) contain MSCs or "stem-like" multipotent progenitor cells that can potentially differentiate into a host of different tissue types such as cartilage, bone, muscle, tendon, and ligament under appropriate physiological conditions. Inclusion of autologous stem cells in biodegradable nanofiber scaffolding to facilitate their development and differentiation into multiple cell types in a lineage-specific manner is particularly attractive.[8] This is an emerging research area with only a very limited number of published reports.

[7]The term "stem cell" is used generically and somewhat loosely in this discussion to include true pluripotent stem cells, progenitors of limited potency (monopotent and bipotent) perhaps determined by the tissue they reside in (Tuan et al. 2003).

[8]Multiphase tissue can be bioengineered by combining tissue types grown separately *in vitro* from stem cells, and subsequently combined into a single implantable construct.

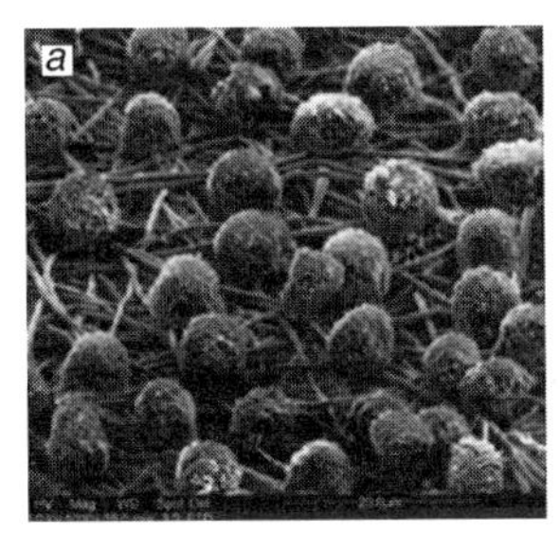

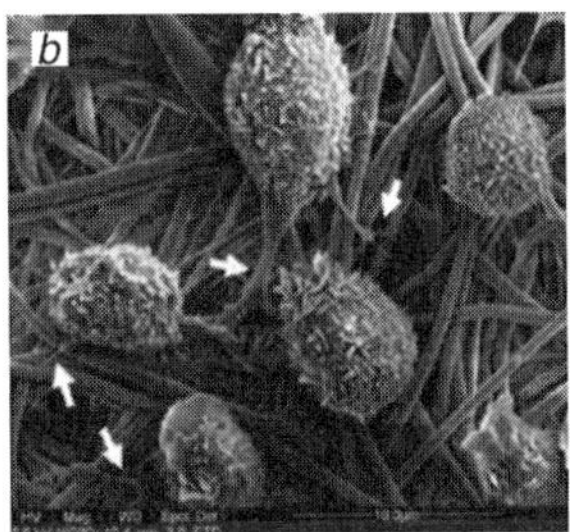

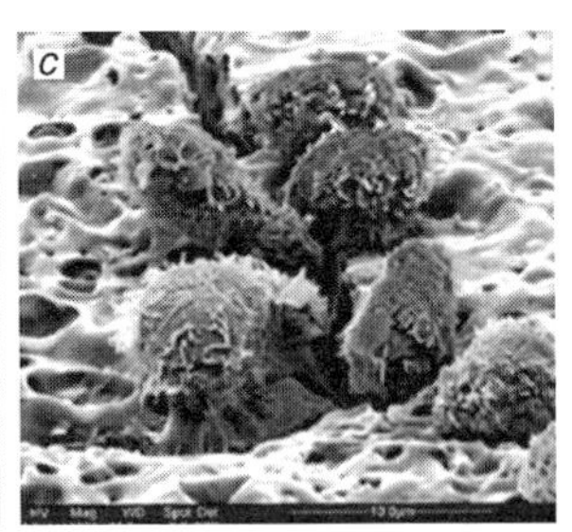

Figure 7.11 (*a*) Human umbilical chord-blood HSPCs growing on electrospun PES nanofibers (surface modified by amination) after 10 days of growth. (*b*) An enlarged view of the cells showing filopodia from the cells interacting with fibers (see white arrows). (*c*) The same cells growing on surface-aminated PES thin film. Reproduced with permission from Chua et al. (2006). Copyright 2006. Elsevier.

A comparison of the effectiveness of nanofiber mats and thin films in supporting stem cell growth was recently carried out using surface-functionalized poly(ethersulfone) (PES). Acrylic acid was first photografted onto the surface of nanofibers (average $d \approx 529$ nm) and the —COOH functionalized mats were then conjugated with ethylene diamine (EtDA) using a two-step carbodiimide crosslinking reaction. In studies using hematopoietic stem/progenitor cells (HSPCs), aminated nanofibers (56.2 ± 12.6 mmol/cm^2 coverage) were shown to support a relatively higher degree of cell attachment and expansion of cells *ex vivo* (Chua et al. 2006). Figure 7.11 shows images of HSPCs growing on film and a nanofiber surface; in the latter case filopodia emanating from the cells are seen to interact with the fiber.

Studies on the growth of adipose-derived stem cells on nanofibers of PLLA copolymer for neural tissue engineering (Yang, F., et al. 2004), human mesenchymal stem cells (hMSCs) on PCL for cartilage repair (Li, W.-J., et al. 2005c), MSCs derived from neonatal rats on PCL for bone tissue formation (Shin et al. 2004b; Yoshimoto et al. 2003), and neural stem cells on PLLA (Yang, F., et al. 2004), all on nanofiber scaffolding, have been reported in the literature. The Table 7.3 lists polymer nanofibers successfully demonstrated as scaffolding to support the growth of stem cells.

Successful differentiation of cell types can be established by assays that show upregulated gene expression and in some instances by histochemical studies of the seeded cells supporting the development of appropriate cell types. Li and colleagues (Li, W.-J., et al. 2005b) demonstrated gene expression consistent with differentiation of hMSCs from a single source into adipogenic, chondrogenic, and osteogenic lineages when cultured *in vitro* on PCL nanofiber scaffolding. TGF-β1 directed chondrogenesis of the progenitor cells showed upregulated expression of aggrecan and Col II transcripts and downregulated expression of ColX. Chondrogenesis yielded

TABLE 7.3 Some biodegradable polymer nanofiber scaffolds supporting stem cell growth

Polymer	Solvent	Fiber d (nm)	Stem Cell[a]	Reference
PLLA	CH_2Cl_2	400–4000	hMSC	Boudriot et al. (2005)
PLLA	CH_2Cl_2/DMF (70/30)	150–350	NSC	Yang, F., et al. (2004)
Silk fibroin	LiBr solution	700 ± 50	hBMSC	Jin et al. (2004)
PCL	$CHCl_3$ (10%)	400 ± 200	rMSC	Yoshimoto et al. (2003)
PCL	THF : DMF (1 : 1)	500–900	hBMSC	Li, W.-J., et al. (2005c)
PCL	THF : DMF (1 : 1)	700	hMSC	Li, W.-J., et al. (2005b)

[a]hMSC, human mesenchymal stem cells; hBMSC, human bone marrow stromal cells; rMSC, mesenchymal stem cells from bone marrow of rats; NSC, neural stem cell line.

cartilage-like morphologies, containing chondrocytes surrounded by abundent cartilagenous ECM (Li, W.-J., et al. 2003, 2005b). With adipogenic cells, histochemical examination showed oil-like globules (in addition to the evidence from gene expression) suggestive of morphogenesis (Kang et al. 2007). Similar data are available for differentiation of hMSCs from bone marrow cultured *in vitro* on PCL nanofiber scaffolding (Li, W.-J., et al. 2005c). Particularly interesting is the potential biodegradable nanofiber scaffolding seeded with neural stem cells to repair damage to the nervous system. A multipotent neural stem cell line (C-17) not only supported the differentiation of the stem cells into neurites, but also promoted cell adhesion (Yang, F., et al. 2004). Although general observations such as the relationship between fiber characteristics and cell proliferation patterns have been reported (Badami et al. 2006), research has not progressed to a point to allow a full assessment of the clinical potential of this approach.

7.3 OTHER APPLICATIONS

7.3.1 Wound Care Applications

Skin, the largest organ in the body, provides the first line of defense against infection. Damage to skin due to trauma (particularly burns), although capable of self-repair, often requires immediate medical intervention depending on severity. Wound dressings serve to protect the wound bed from contamination or infection and remove from the bed any exudates generated during the healing process. Natural wound healing initiates with platelet adhesion, vascular constriction (to limit blood loss) and leucocyte mobilization at the site of wound, resulting in inflammation and clot formation. Debridement by lytic enzymes removes any dead tissue, allowing the repair phase to set in with scab formation, fibrillar development, and epithelialization.

This is followed by a maturation phase where laying down and crosslinking of collagen ECM fibers slowly occurs. A desirable wound dressing will therefore have high mechanical integrity, good gas exchange capabilities, will be non-occluding, and be sufficiently sorbant for exudate control. Although the presence of some amount of exudate promotes rapid wound healing and maintains a moist wound surface, lowering the rate of infection, excessive exudate needs to be removed from the bed (Matsuda et al. 1993; Simpson et al. 2006). Patents describing the direct electrospinning or nanofibers onto the surface of the skin and wounds have been issued.[9]

Nanofiber mats of selected polymers possess most of these characteristics as discussed above under scaffolding and controlled release applications (Ji et al. 2005a). These same biodegradable polymers used in scaffolding applications therefore also find use as wound dressings (Błasińska et al. 2004a; Jin et al. 2002; Katti et al. 2004; Khil et al. 2003). Some of these have been shown to also support the proliferation of keratinocytes (Min et al. 2004a, 2004b, 2004d; Rho et al. 2006). The desirable characteristics of biodegradable mats (seeded appropriately with dermal fibroblasts) in wound care applications have resulted in these being suggested as dermal substitute (or artificial skin) (Venugopal et al. 2006). However, electrospun synthetic polymers including nonbiodegradable polymers (Kenawy et al. 2002, 2003; Khil et al. 2003) have also been suggested for the same application.

Collagen nanofiber mats are reported to have particularly good wound healing properties (Huang, L., et al. 2001a, 2001b). In wound dressing applications, high mechanical integrity of the nanofiber mats is important. Collagen can be electrospun into porous fiber mats that can be crosslinked to increase its dimensional stability for use as a dressing. Vapor or liquid phase reaction of the as-spun mat with glutaraldehyde followed by deactivation of any residual aldehyde groups by immersion in 0.1 M glycine yielded a mat of tensile strength ($\sim$10 MPa) comparable to commercial tissue regeneration membranes, but with a slight reduction in porosity (Rho et al. 2006). The collagen scaffold coated with either collagen type 1 (uncrosslinked) or laminin generally show enhanced cell attachment and spreading compared to uncoated mats (Rho et al. 2006). The early stages of wound healing were accelerated when these were used as a wound dressing in rat models.

Some of the synthetic polymers used in scaffolding application are also likely to be good candidates for wound dressings. Combining the controlled delivery capabilities of such scaffolds with the gauze-like qualities of conventional dressings has been exploited (Huang and Yang 2006; Huang and

[9]D. J. Smith, C. Mello, D. H. Reneker, A. T. McManus, H. L. Schreuder-Gibson, and M. S. Sennett (2004). "Electrospun fibers and an apparatus therefor." US patent # 6753454, issued on June 22, 2004.

Zhang 2005; Ignatova et al. 2007). A simple example is a nanofiber mat where the PVA nanofibers carry Ag nanoparticles that impart an antibacterial characteristic to the fiber mat (Hong, K. H., et al. 2006; Hong, K. H. 2007). Katti et al. (2004) suggested the use of a PLAGA (LA and GA 50 : 50) nanofiber mat with up to 30% (w/w) of the drug cefazolin (antibiotic) as a wound dressing. With porous nanofiber mats of PLGA or dextran/PLGA scaffolds, the dermal fibroblasts were reported to assemble into dense multilayer structures mimicking a dermal microstructure (Pan and Jiang 2006). Medical-grade polyurethane ($M_w = 110{,}000$ g/mol) electrospun from 25% w/v DMF/THF solvent mixtures has also been evaluated as a wound dressing material in guinea pig models. These dressings were found to promote faster healing compared to a commercial permeable wound dressing membrane also made of polyurethane (Khil et al. 2003). Sun et al. (2005) found that polystyrene nanofiber mats seeded with keratinocytes, endothelial cells, and skin fibroblasts and cultured at an air–liquid interface self-organized, mimicking the native epidermal–dermal arrangement.

7.3.2 Immobilized Bioactive Moieties on Nanofibers

The high specific surface area of nanofibers allows surface loading of the fiber with bioactive compounds to yield effective biocatalytic materials (Da Silva et al. 2004; Wu, X. H., et al. 2006). Bioactive molecules may be attached physically or chemically to the fiber surface; lipase can be adsorbed on to poly(acrylic acid)-grafted cellulose nanofibers (Chen and Hsieh 2004) or covalently linked to cellulose nanofibers (Wang and Hsieh 2004), PAN nanofibers (Lu et al. 2007) or to a P(AN-MA) nanofiber surface (Ye et al. 2006). Using nanofiber mats surface-functionalized for biological activity in this manner is beneficial over using native bioactive agent in solution or using them as nanoparticles of equivalent surface area per unit mass. Unlike with homogeneous liquid-phase reactions, the use of nanofibrous catalysts allows facile separation of the catalyst from the reaction mixture. In some instances, where the native bioagent is not well solubilized in the reaction medium, the molecule tethered to a nanofiber surface can yield surprisingly high rates of reaction. Comparison of the catalytic activity of film-immobilized and nanofiber-immobilized lipases, for instance, shows the latter to have a relatively higher stability as well as higher catalytic activity (Nakane et al. 2005).

The approach is illustrated by α-chymotrypsin functionalized polystyrene nanofibers. The catalytic effectiveness of nanofibers with 27.4% of surface coverage by a monolayer of enzyme was reported to be 65% of that of the native enzyme when aqueous *n*-succinyl-ala-ala-pro-phe *p*-nitroanilide

(SAAPPN) was used as the substrate. In organic solvents, however, the nanofiber-enzyme displayed activities that were three orders of magnitude higher compared to its native counterpart (Jin et al. 2002). Others have reported 5.6% loading of the same enzyme on silk fibroin nanofibers to retain 90% of the original enzyme activity (Lee, K. H., et al. 2005). Kim, B. C., et al. (2005) used a novel two-step approach to achieve loadings that are even higher than that corresponding to the maximum monolayer coverage of the nanofibers. As demonstrated with chymotrypsin, seed enzyme was first attached to nanofibers of PS and poly(styrene-co-maleic anhydride) [P(S-MA)], followed by a glutaraldehyde (GA) crosslinking step to bind additional enzyme molecules and aggregates from the solution onto the covalently attached seed enzyme molecules on the nanofiber. Compared to a nanofiber simply coated with a layer of the enzyme, these nanofibers with a higher loading of chymotrypsin aggregates showed a ninefold increase in enzyme activity.

Wang and colleagues studied the functionalization of composite nanofibers of poly(acrylonitrile-*co*-acrylic acid) [P(AN-AA)] and multiwalled carbon nanotubes (MWCNTs) by attaching catalase (hydrogen peroxide oxido-reductase from bovine liver) onto the carbon nanotube surface. The activity of the catalase was reported to increase by about 42% with increasing MWCNT content in the composite nanofiber and was attributed to promotion of electron transfer via charge-transfer complexes formed by carbon nanotubes (Wang, Z.-G., et al. 2006).

Table 7.4 shows selected recent studies on electrospun nanofibers that have attached surface biofunctionalities.

TABLE 7.4 Some examples of surface functionalization of polymer nanofibers with enzymes

Polymer	Functionality	Loading	Substrate[d]	Reference
PS	α-Chymotrypsin	1.4% w/w	SAAPPN[a]	Jia et al. (2002)
PS-*co*-maleic acid	α-Chymotrypsin	Aggregate	SAAPPN[a]	Kim, B. C., et al. (2005)
Silk fibroin	α-Chymotrypsin	5.6% w/w	BTPNA[b]	Lee, K. H., et al. (2005)
Cellulose acetate	Triacylglycerol lipase	~1.8%	Olive oil	Wang and Hsieh (2004)
PVA	Lipase		Citranellol[c]	Nakane et al. (2005)

[a] *n*-succinyl-ala-ala-pro-phe *p*-nitroanilide.
[b] *N*-benzoyl-D,L-tyrosine-*p*-nitroanilide hydrochloride.
[c] Esterification of citranellol by acetic acid.
[d] Substrate used to test efficacy of enzyme.

7.4 FUTURE DIRECTIONS

Advances in biomedical applications of nanomaterials require a multidisciplinary research orientation with a focus on both the engineering aspects of nanofiber mats as well as their biological interactions at the implant site. W.-J. Li et al. (2005a) suggest two promising directions for future research: bioactivation and incorporation of controlled release functionality into polymers.

The possibility of stem cells on nanofiber scaffolds undergoing multilineage differentiation is beginning to be experimentally demonstrated (Li, W.-J., et al. 2005b). Successful cell proliferation and growth is mediated by specific binding sequences such as the RGD containing the motif (Arg-Gly-Asp) for integrin binding. Bioactivation of nanofiber surfaces by incorporating such motifs is a promising strategy towards high-efficiency bioactive materials. Controlled delivery of factors such as TGF-β1, insulin-like growth factor (IGF), platelet-derived growth factor (PDGF), fibroblast growth factors (FGFs), bone morphogenic protein (BMP-2), vascular endothelial growth factors (VEGFs) and other angiogenic stimulators via the scaffold itself to the growing tissue can regulate cell proliferation and differentiation in a controlled manner. Recent patents also suggest the use of nanofiber mats to deliver nitric oxide (NO) in wound care and other medical applications [WO 04094050 (2004) and WO 06096572 (2006)]. Inclusion of such biological cues in addition to the physical cues (in terms of porosity and fiber morphology) in the scaffolds is already being explored. This has been demonstrated with TGF-β1 release from PCL scaffolds facilitating chondrogenesis in hMSC on the scaffold (Li, W.-J., et al. 2005c). However, ideally, multiple growth factors need to be released at different rates at different durations and the design of such devices will be challenging. It is perhaps a challenge that can be met using multicomponent fiber mats. Particularly intriguing is the report by Lee and colleagues (Lee, S.-W. and Belcher 2004) that M-13 virus could be electrospun by itself and as composite nanofibers of PVP/M-13, could still infect bacterial hosts after re-suspension in a buffer. Anti-streptavidin M-13 bacteriophage (virus), possessing an engineered peptide sequence, was used with or without conjugation with R-phycoerythrin as the basic building block to fabricate micro- and nanoscale fibers. It is conceivable that scaffolding that releases DNA in a controlled manner can concurrently transfect the cells proliferating on the scaffold, affording an additional design dimension in tissue engineering.

A review of the literature suggests materials selection criteria for scaffold design needs to be guided by both the chemical nature of the polymer and the physical characteristics of the nanofiber mat (Dong, W., et al. 2006). The various polymers electrospun into nanofibrous and evaluated as tissue engineering materials are compiled in Appendix I. However, in cell–nanofiber interactions as well as in controlled release modalities, nanofiber mat characteristics play a crucial role in determining functionality. Parameters such as the fiber diameter distribution, fiber surface morphology and the porosity (fraction of free space in the construct) as well as pore size distribution are particularly important in determining their performance.

8

APPLICATIONS OF NANOFIBER MATS

Filtration is the leading nonbiomedical application of electrospun nanofibers, with products containing layers of nanofibers already in the marketplace. It is also one of the areas where nanofibers are likely to make a significant and lasting impact. This chapter will discuss the application of nanofiber mats as air filters, emphasizing their advantages over conventional fiber mat microfibers. Another emerging application area for nanofibers is in sensors for sense chemical and biological agents including the growing number of toxic industrial chemicals (TICs). Their high specific surface area is a particular advantage when using nanofibers in this application area. In nearly all application areas the fragility of organic nanofibers and their temperature sensitivity limits their use. In catalysis applications, for instance, the substrate gas streams that need to be processed or the liquid-phase reactions that need to be catalyzed are at relatively high temperatures. The use of inorganic nanofibers, including metal and metal oxide nanofibers, can therefore be of particular value in such applications. This chapter reviews these application areas as well with illustrations from recent electrospinning literature.

8.1 INTRODUCTION TO AIR FILTRATION

The hazard to human health from exposure to air-borne dust is primarily due to the smaller particles with aerodynamic diameters of less than a few

Science and Technology of Polymer Nanofibers. By Anthony L. Andrady

micrometers. Inhaled small particles including nanoparticles are well known to lodge deeper in the lungs causing asthma-like symptoms and other complications. The simplest means of removing these aerosols from air is by filtration. Filters are used in diverse applications including personal masks for inhalation protection, air cleaning of industrial effluents, in electronic equipment and in maintaining clean room manufacturing environments. Despite the popularity of the approach and relatively simple engineering construction of filters, their aerosol collection efficiencies are not easily predictable from fundamental considerations because of the inability to precisely measure the detailed properties of the filters. However, the collection mechanisms for conventional fiber-based filters are well established and assuming the individual mechanisms to be additive, semi-empirical theoretical equations can be developed using statistical fits to experimental data. A well-designed filter needs to be of desired efficiency in removing particles from the air stream and have other characteristics such as the dust loading capacity and durability suitable for particular applications. The required efficiency can simply be achieved by making the filter thicker. However, the filter might then be of little use because invariably the pressure drop obtained increases with thickness (unless the solidity is reduced to compensate for this, but then only at the expense of increased particle penetration). Therefore, a balance needs to be maintained between the efficiency in particle removal and the pressure drop associated with the filter.

Nanofiber mats, because of the small fiber diameter, are very well suited for air filtration providing that sufficiently low solidity and mechanical strength can be obtained. The range of particle sizes that need to be separated from an air stream can be quite broad, ranging from environmentally relevant large particles such as PM-10 (particulate matter with aerodynamic diameter smaller than 10 μm), all the way down to engineered nanoparticles only several nanometers in diameter. The dynamics of nanoparticles suspended in a gas is very different from that of microparticles; the behavior of nanoparticles tends to be dominated by Brownian motion, and they behave more like molecular gases diffusing through the air. Filtration models, however, address only the simpler case of spherical particles. Engineered nanoparticles such as nanotubes as well as biological aerosols such as bacteria can have an aspect ratio greater than unity and therefore their filtration will be incompletely described by such models. In filtering bioaerosols in military and homeland security applications the filtration efficiencies based on conventional models must be regarded as being approximate estimates.[1] Although very

[1]The penetration of spherical polystyrene latex particles was higher than that for *M. chelone*, a rod-shaped bacterium, in tests carried out on N95 respirators, although the two particles are of comparable aerodynamic size (Qian et al. 1998).

useful for understanding filter performance and in designing of filtration approaches, filter models are no substitutes for experimental data in specific applications. Testing the filters with aerosols representative of the actual application is critical before filters are placed in service. Furthermore, in practice, nonparticulate constituents in the stream such as volatile organic compounds (VOCs) or moisture may affect the electrostatic charge, or increase the fiber diameters by swelling the polymer fibers, further affecting their predicted performance. In the case of personal protection filters the "fit factor" or the degree of face-fitting attained by a respirator very significantly affects the protection afforded (Han 2002; Janssen et al. 2002).

The removal of particles by a fibrous filter relies on five main mechanisms, of which four are discussed here[2]:

1. *Interception.* A particle approaching the fibers' surface to a distance equal to or less than its radius (r_{particle}), or within the contact range of the fiber, without crossing a flow streamline tends to adhere to and deposit on the surface. It is an important mechanism for particles larger than about 100 nm, and its effectiveness depends on the ratio of particle diameter to the fiber diameter. If the interfiber distance or the porosity of the mat is smaller than the particle diameter d_p, then a special case of interception or particle sieving occurs, but this is usually not an important mechanism in air or gas filtration.
2. *Impaction.* The path of airflow results in a curvature of streamlines in the vicinity of a fiber. Because of their inertia (at high enough particle velocities), particles cross the streamlines, impact on the nanofibers, and are deposited on them. This mechanism too is normally important for particles larger than a few hundred nanometers.
3. *Diffusion.* Smaller particles exhibit Brownian motion and collide into fibrous media by random movement and are collected on it. As these particles diffuse randomly they traverse distances far exceeding their diameters. Therefore attachment and collection can occur whether airflow streamlines bring a particle within a single diameter of a fiber or not. Lower air velocities increase the removal of small particles by diffusion because they spend more time in the vicinity of a fiber. This mechanism usually dominates filter performance for particle diameters smaller than 100 nm.
4. *Electrostatic attraction.* If either the particles or the fiber (or both) possesses a surface charge, electrostatic interaction can occur. Charged particles are attracted to and are retained on the surface of a

[2]Gravitational deposition is also sometimes included in this list. The trajectories of the particles deviate from the streamline course due to gravity and are intercepted by nanofibers. With nanoparticles, the gravitational contribution is very small.

fiber that carries the opposite surface charge; those with similar charges are repelled. Particles in air streams may acquire charges from a number of mechanisms, including triboelectric effects (friction), which allow them to be electrostatically collected on fiber surfaces. The deposition mechanism in this case is similar to that in electrostatic precipitators used in air cleaning, with the exception that a corona discharge is used to charge the particles. Electrostatic augmentation by using an external electric field on the filter mat can enhance such attraction by induction. Electrostatically charged polymer fiber filters that retain their charge over the long term are commercially available.

The importance of these different mechanisms (Fig. 8.1) to overall filtration is a function primarily of the average particle size (Ensor et al. 2003). Moderate- to high-efficiency performance can be qualitatively understood considering only the combined efficiency of interception, E_I and of diffusion, E_D (assuming impaction to contribute negligibly at practical air velocities through the filter). With larger particles, E_I is large, but falls off as the size decreases, but at very small particle sizes, E_D is large and falls off as particle size increases. This results in the typical efficiency curve for a high-efficiency filter where a minimum in efficiency (or a most-penetrating particle size, MPPS) occurs at particle dimensions where E_I and E_D cross over (Fig. 8.2).

The efficiency of a filter depends on fiber diameter d_f (or the dimensionless projected area S) and the single fiber efficiency E_S (Davies 1973):

$$\text{Efficiency } E = 1 - \exp[-(E_S S)] \tag{8.1}$$

where

$$E_S = E_I + E_D,$$

$$S = 4La/[\pi(1 - a)d_f],$$

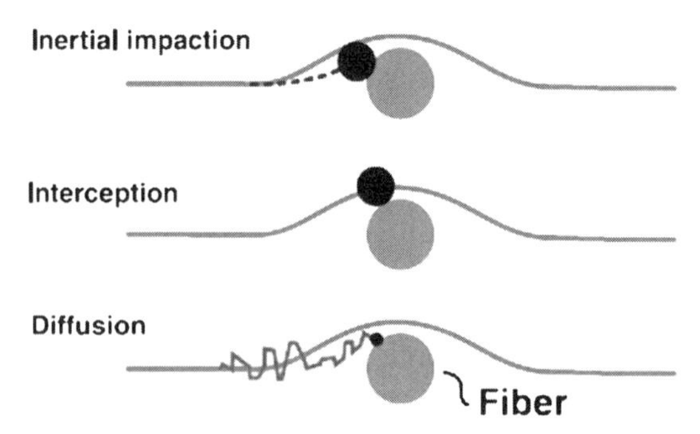

Figure 8.1 Illustration of three mechanisms involved in air filtration. (Courtesy of DHHS (NIOSH), Publication No. 2003-136.)

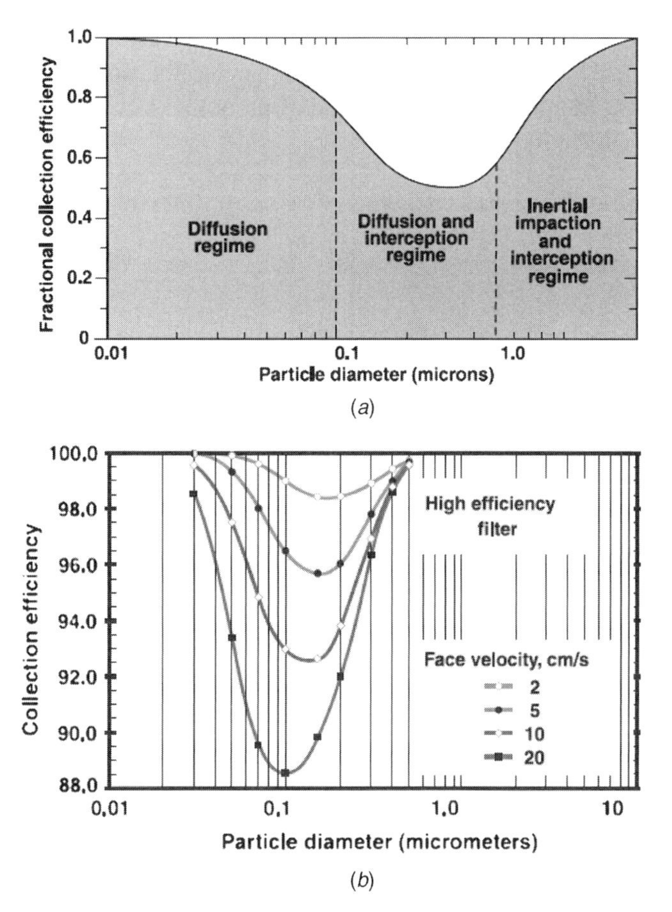

Figure 8.2 (*a*) Theoretical expectation of efficiency of an air filter vs particle size. (*b*) The change in particle collection efficiency on fiber diameter at different face velocities. (Courtesy of DHHS (NIOSH), Publication No. 2003-136.)

and L (mm) is the length or depth of the filter media in the direction of air flow, a is the volume packing density or solidity of the filter media, and d_f (mm) is the fiber diameter. The single fiber efficiency E of collecting monodisperse particles under low-velocity flow is the sum of E_I and E_D. The efficiency due to the diffusional mechanism is given by (Matteson and Orr 1987)

$$E_D = 1.6125[(1 - a)/F_K]^{1/3} P_e^{-2/3}, \quad (8.2)$$

where F_K is a hydrodynamic factor that depends on the volume packing density and P_e is the Peclet number, which is a measure of diffusional deposition and is given by

$$P_e = (1 \times 10^6) U(d_f/D). \quad (8.3)$$

Here, U is the face velocity in m/s and D is the diffusion coefficient of the particle (m^2/s). The Peclet number decreases with fiber diameter, illustrating the advantage of nanofibers in this mode of particle deposition. The diffusion coefficient is given by

$$D = \frac{kTC_c}{3\pi\eta d_p}, \tag{8.4}$$

where k is the Boltzmann constant, T is absolute temperature, d_p is particle diameter (μm), η is viscosity, and C_c is the Cunningham slip correction factor, which depends on the ratio of the mean free path of air molecules to the radius of the particle.

One of the assumptions of continuum fluid dynamics is that the velocity of the fluid at the surface of the particle surface is zero. However, as the size of the particles approaches molecular dimensions, this assumption becomes increasingly unreasonable. The first slip correction factors were developed by Cunningham to take this into account. The correction factor C_c is given by

$$C_c = 1 + \frac{\mathrm{Kn}}{2}\left[2.514 + 0.800\exp\left(-1.1\frac{1}{\mathrm{Kn}}\right)\right], \tag{8.5}$$

where Kn is the Knudsen number ($2\lambda/d_p$) and λ is the mean free path of the gas molecules.

The interception efficiency E_I was shown to be given as follows (Liu and Rubow 1986)

$$E_I = [(1 - a)/\varepsilon F_K]\,[N^2/(1 + N)], \tag{8.6}$$

where $1/\varepsilon$ is an empirical factor to correct for filter inhomogeneity and $N = d_p/d_f$. In this equation, it is assumed that both N and a are small. The efficiency of a filter with multiple fiber diameters is then simply the sum of the single fiber efficiencies:

$$ES = \Sigma E_{Si} S_i. \tag{8.7}$$

The impact of nanofibers on filtration is based on how these basic mechanisms are modified when the fiber diameters are reduced. As the fiber diameter decreases the filtration efficiency increases because diffusion and interception mechanisms are enhanced (equations (8.2) and (8.6)). Not surprisingly, nanofiber mats easily outperform conventional high-efficiency media in filtration of submicrometer (and larger) particles from air or gas streams (Park and Park 2005).

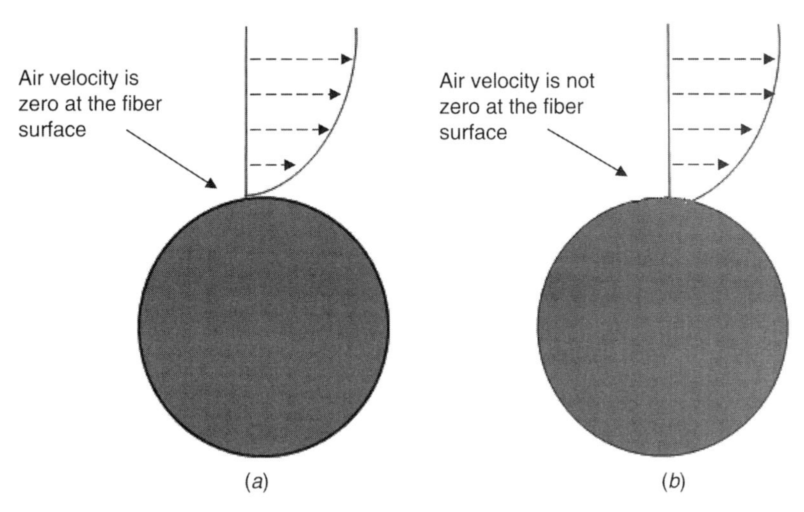

Figure 8.3 Schematic illustrating (*a*) nonslip flow and (*b*) slip flow of air at the fiber surface. Redrawn from Graham et al. (2002).

An additional consideration is that for fibers of nanoscale dimensions the gas stream itself cannot be treated as a continuum and its molecular nature needs to be taken into account in the same way as with the nanoscale particles described above. As the value of Kn becomes significant, the continuum assumption of gas at the surface of the fiber no longer applies (Devienne 1958). The velocity of gas at the fiber surface is no longer assumed to be zero, resulting in "slip flow." This results in a reduced drag force at the surface relative to the slip-free flow (Graham et al. 2002), and in air filtration. This translates into a lower pressure drop (Fig. 8.3). Also, the associated increased contact of air (and therefore the particles) with the fiber surface results in higher efficiencies of impaction, interception, and diffusion. The increased filtration efficiency results from relatively more air and therefore more particles flowing closer to the fiber surface. Given that $\lambda_f \approx 66$ nm, a fiber of $d = 60$ nm (assuming that a fiber of this diameter could be successfully made into a filter mat), leads to values of $\mathrm{Kn} \approx 2$; nanofibers in air filters are therefore likely to operate in the so-called transition range between the free molecular flow regime (Devienne 1958) of $\mathrm{Kn} > 10$ and continuum flow where $\mathrm{Kn} < 1$.

Although several mathematical treatments model the gas flow in filters under free molecular flow conditions, their detailed discussion is beyond the scope of this chapter. In order for a filter to perform in the free molecular regime the ambient pressure would need to be much less than atmospheric or the fibers would need to be on the order of 10 nm. However, these models generally suggest that (1) the pressure drop Δp in the regime varies with gas viscosity, velocity, and filter mat thickness as it does in the continuum region; (2) $\Delta p \sim a$ (where a is the packing density of the filter[3]), meaning that as the porosity of the nanofiber mat increases the pressure drop Δp will decrease; (3) $\Delta p \sim 1/d_f$ in this region (where d_f is the nanofiber diameter); and (4) at constant temperature, Δp decreases linearly with decreasing pressure p (Pich 1987).

8.1.1 Nanofiber Filter Performance

The filtration efficiency of different fibrous filters needs to be compared at the same basis weight[4] of the filter medium. Although only limited data have been published, these all agree that nanofiber mats afford superior filter performance compared to conventional filters (Ensor et al. 2006; Graham et al. 2002; Tsai et al. 2002) and that filter efficiency improves as fiber diameter decreases (Park and Park 2005). Electrospun nanofibers of poly(ethylene oxide) (PEO) (spun from 10% in isopropanol/water solution) having average fiber diameters in the range of 100–500 nm, for instance, showed filtration efficiencies comparable to high efficiency particulate air (HEPA) filters at a basis weight of only 16 g/m^2 (Tsai et al. 2002). As the fiber diameter is decreased, the effectiveness of removal of submicron particles by interception also increases significantly. The resulting shift in the most penetrating particle size (MPPS) has been modeled (Podgórski et al. 2006).

A practical measure of filter efficiency must take into account both the penetration of particles across the filter, P_t (i.e., the ratio of particles upstream of the filter to that exiting the opposite surface of the filter downstream) and the pressure drop ΔP across the filter at a given face velocity. Note that $P_t = 1 - E$. A figure of merit (FoM) for the filter can then be defined as

$$\text{FoM} = -\log(P_t)/\Delta P. \tag{8.8}$$

Numerically, the value of FoM is the gradient of the log-linear plot of penetration vs the pressure drop at constant face velocity. Various phenomena, such as the magnitude of the slip correction factor discussed in the previous sections, tend to increase the FoM as the average fiber diameter is decreased. Ensor et al. (2006) reported the FoM for nanofiber mats to be

[3]In Chapter 5, where porosity of nanofiber mats is discussed, the quantity a is the same as $\Pi - 1$, where Π is the total porosity of the mat.

[4]Basis weight is the weight per unit area of the material. It is related to the porosity of the material Π (see Chapter 5). It can be shown that $\Pi = [1000 \cdot \rho_{pol} \cdot$ thickness of mat (mm)]; basis weight (g/sq $\cdot$ m).

TABLE 8.1 Filtration efficiencies for electrospun nanofiber mats of polysulfone compared to commercial HEPA and ULPA filter paper

Filter Type	Designation	ΔP (Pa)	Penetration	FoM (kPa^{-1})[a]
HEPA	Lydall 213	349	3.0×10^{-5}	13
HEPA	HV 5433	311	1.7×10^{-4}	12
ULPA	Lydall 252	510	8.4×10^{-7}	12
Nanofiber	PSU 55	100	7.5×10^{-6}	52
	PSU 60	147	$\sim 5 \times 10^{-7}$	48
	PSU 65	42.3	5.7×10^{-3}	53
	PSU 66	18.7	4.3×10^{-3}	127
	PSU 74	60.5	1.3×10^{-4}	64
	PSU 77	48.3	4.2×10^{-6}	113
	PSU 89	31.9	2.9×10^{-4}	111

[a]Tested with 300 nm KCl particles at a face velocity of 5.33 cm/s.
Source: Ensor et al. 2006.

much higher than that for HEPA and ULPA (ultra-low penetration air) filters, when tested under the same conditions (Table 8.1).

The improvement in performance reported for these filters even in these preliminary experiments is very significant. Despite the unavoidable scatter in data due to variation in packing density and filter thickness, the performance of nanofiber filters appears to well surpass that of the commercial fiberglass utility filters. Optimistically, the best data in the table suggest as much as an order of magnitude improvement in FoM for nanofiber mats over HEPA filters. The relative contributions of slip and electrostatic effects to the reported filter performance have not been totally assessed and may depend on the conditions used to electrospin each filter mat. It has been estimated that the contribution of electrostatics to the FoM is about 20%.

8.1.2 Filters with Nanofibers

The effectiveness of incorporating a nanofiber layer on conventional filter media has been recognized for some time in the air filter industry. Typically, these applications have been for dust removal and the media has demonstrated robustness for a wide range of applications. Industry leaders such as Donaldson Company (Minneapolis, MN 55431) routinely manufacture air filters with a nanofiber component in them (Graham et al. 2002). Nanofibers are included in commercial filters to serve two purposes:

- When applied as a thin layer over conventional media nanofibers enhance their filtration efficiency.

- Layers of nanofiber on the surface improve surface dust loading characteristics of the media (Kalayci et al. 2006).

In the latter function, nanofibers prevent larger dust particles from clogging the conventional filter media during surface loading and enable facile cleaning of filters by means of backward pulsing or vibration. Essentially, the larger particles are retained by the nanofiber mat and do not reach the media. This prescreening by nanofibers lets a fraction of very small particles through to the conventional microfibers where the efficient Brownian diffusion mechanism removes them. Experimental data on meltblown nanofiber filters tested as a layer over conventional microfiber filter media, found improvements in the FoM of 2.7 times (for $d = 1.18$ μm) (Podgórski et al. 2006).

The solidity (rather than its overall thickness) is a key parameter defining efficiency of a filter. Increasing the solidity of the filter results in an increase in the pressure drop as the drag increases, and also an increase in

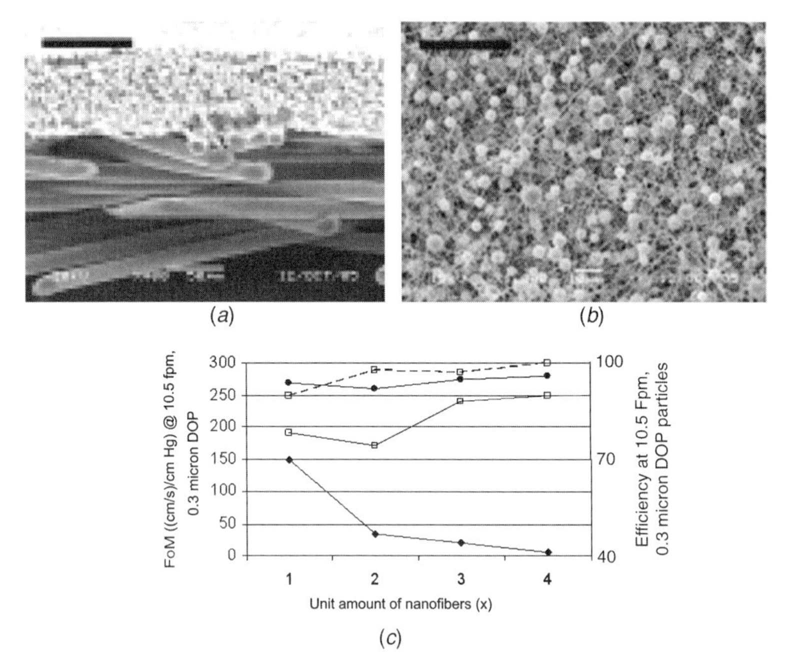

Figure 8.4 (*a*) SEM image of a layer of nanofibers with spacer particles on a layer of conventional media. (*b*) Topographic view of spacer particles in the nanofiber layer. Reproduced with permission from Kalayci et al. (2006). Copyright 2006. The Filtration Society. (*c*) The change in FoM and efficiency as a function of nanofiber content. Redrawn from Kalayci et al. (2006). Filled symbols denote FoM and open symbols the efficiency. In each set, the mat with glass spheres show the higher values.

the single-fiber efficiencies E_I and E_D. Kalayci and colleagues concluded that at least in the vicinity of the most penetrating particle size, the former increases at a relatively faster rate and the FoM of the filter decreases with increasing solidity (Kalayci et al. 2006). However, this may be reversed at larger particle sizes. A practical way to decrease solidity and increase the permeability (and capture efficiency as well) is to use a spacer material to ensure that nanofiber layers do not mat together into a membrane. The spacer, usually beads, keeps the individual nanofibers separate preventing their fusion into a membrane. The effect of the spacer particles in improving filter performance is illustrated in Fig. 8.4 (Kalayci et al. 2006).

8.2 NANOFIBER SENSORS

With increasing emphasis on homeland security and accidental release of toxic industrial chemicals (TICs), there is a continuing demand for better, and more sensitive chemical and biological sensors. Available sensors for the most part do not function particularly well as advanced-warning devices — they provide warning at levels of agent that necessitate immediate mitigation and first responder activity. Their sensitivity to low-vapor-pressure chemicals (such as explosives residues) needs considerable improvement both in limits of detection and in response times (Wang, X. Y., et al. 2002a).

TABLE 8.2 Examples of nanofiber-based chemical sensor materials recently reported in the literature

Class	Parameter Measured	Nanofiber	Analyte	Reference
Gravimetric	Mass of analyte	PAA/PVA	NH_3	Ding et al. (2004b, 2005c)
Amperometric	Electric current	PEO/$LiClO_4$	Humidity	Aussawasathien et al. (2006)
Conductometric	Electrical resistance	PANi/PEO/CSA	NH_3	Liu et al. (2004); Virji et al. (2004)
		PANi/PS/GOX	Glucose	Aussawasathien et al. (2006)
		PANi/PVP/urease	NO_2	Bishop and Gouma (2005)
		Titania	NO_2	Kim, I.-D., et al. (2006)
Optical	Fluorescence	PAA/PM	M^{2+}, DNT	Wang, X. Y., et al. (2002b)
		MATTP/copolymer	M^{2+}	Wan, L.-S., et al. (2006)

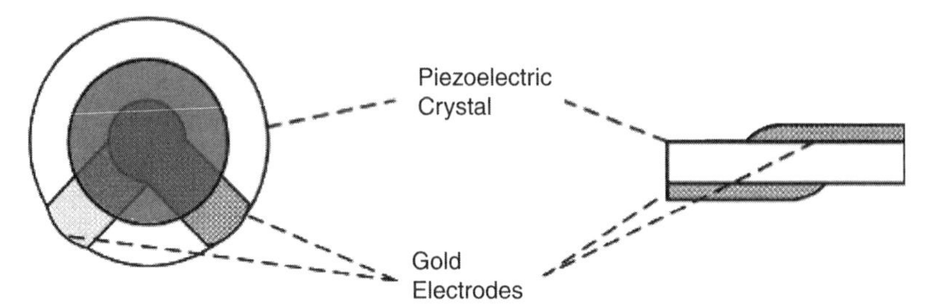

Figure 8.5 Quartz crystal microbalance design.

It is reasonable to expect the sensitivity of a sensor that involves surface interaction with analyte molecules to increase with increasing surface area per unit mass of the sensing material. The high specific surface area of nanofibers therefore suggests the possibility of more efficient and rapid sensor performance, particularly when sensing mechanism is via a surface reaction. Even with bulk reactions the distances over which gaseous agents need to diffuse into the fiber prior to reaction is of the order of tens or hundreds of nanometers. Therefore, nanofiber-based chemo- and biosensors tend to be more effective compared to both microfiber sensors and thin-film-based sensors (Reneker and Chun 1996). Research literature provides support for this expectation as discussed in the illustrative examples given below (Table 8.2). However, the technology of nanofiber-based sensors has not developed to the point where practical sensing devices are available off the shelf.

8.2.1 Gravimetric Sensors

Sorption and surface adsorption of chemical agents by a mat of polymer nanofibers can be detected gravimetrically using quartz crystal microbalance (QCM) detectors (Ding et al. 2004b, 2005c; Kwoun et al. 2001). The mass sensor is essentially a circular quartz crystal with thin metal electrodes deposited on either side of it. The resonant frequency of the piezoelectric crystal in an AC field depends on the crystal characteristics as well as the mass of material deposited on its surface. The Sauerbrey equation[5] predicts a linear relationship

[5]The equation relates the change in oscillation frequency of the crystal (Δf) to the change in mass of the crystal (Δm) in air as follows: $\Delta f = -K\,\Delta m$. The constant K depends on the area of crystal, its resonant frequency, density and transverse wave velocity of quartz. The deposited mass is essentially treated as an extension of the thickness of the crystal. Therefore, the measurement works best with rigid, evenly distributed mass that results in a value of $(\Delta f/f) < 0.05$.

between the change in frequency ΔF (Hz) of oscillation of the crystal and the change in mass Δm (ng) due to material deposited as an even layer on it.

Blends of poly(acrylic acid) (PAA) ($M_w = 90{,}000$ g/mol) with poly(vinyl alcohol) (PVA) ($M_n = 66{,}000$ g/mol) can be directly electrospun from aqueous solution on to a grounded QCM device from aqueous solution (Fig. 8.5). The coated QCM detector might then be used as a reactive gas detector. The acidic PAA in the blend readily reacts with NH_3 and therefore act as a chemical sensor for the gas (Ding et al. 2004b). The PAA nanofiber (400–600 nm)-coated QCM devices were shown to be four times more sensitive to ammonia relative to comparable thin-film-coated devices (Ding et al. 2005b), allowing parts per billion (ppb) level measurements to be carried out at appropriate humidities. Modified piezoelectric crystal sensors might also be used for the purpose. Kwoun and colleagues used a thickness shear mode (TSM) resonator micro viscoelastic sensor with electrospun poly(L-lactic acid-*co*-glycolic acid) (PLGA) copolymer nanofibers ($d \sim 500$ nm) as the sensing material (Kwoun et al. 2001) in the detection of benzene. The same sensor showed different responses on exposure to water and propanol.

8.2.2 Conductivity Sensors

Electrical resistance of a nanofiber mat varies on interaction with chemical and/or biological agents. A volatile solvent sorbed by the nanofibers, for instance, increases the fiber volume due to swelling and will therefore alter its electrical resistance. This phenomenon has been already exploited in electronic-nose technology employing arrays of polymer films. However, unlike with a thin film, very large surface area in a relatively small amount of nanofiber can lead to a higher sensitivity of detection. Organic polymers generally have to be rendered conducting by the addition of a suitable additive in order to be used as conductive sensors. For example, humidity sensors based on a PEO nanofiber carrying 1 wt% of $LiClO_4$ showed a sensitivity six times that of comparable thin-film material (Aussawasathien et al. 2006).

Nanofibers of conducting polymers such as those spun from polyaniline (PANi) are particularly well suited for use in conductivity sensors. Polymers commonly explored in sensor research include PANi, polythiophene (PT), poly(3,4-ethylenedioxythiophene) (PEDOT), and polypyrrol (PPy), sometimes with other constituents, to enhance their sensing capability. Some of these such as PANi (Norris et al. 2000; Zhu et al. 2006a) and PPy (Kang et al. 2005; Nair et al. 2005) (as well as their blends), can be readily electrospun into nanofibers. Polyaniline is somewhat unique in that its doped state can be controlled by the pH of the medium, allowing it to exist either as the emaraldine base or as the salt (Scheme 8.1). Simple acids and bases (ammonia) interconvert the two forms of the polymer, which display a large difference in their electrical conductivities.

Therefore, both basic and acidic analytes can be successfully detected by PANi films and nanofibers. This was illustrated by the PANi–PEO blend nanofiber (or nanowire) device described by Liu et al. (2004). Nanofibers made by scanned-tip electrospinning of a droplet of the blend solution were deposited across a pair of gold electrodes. The spinning solution of the polymer blend was doped with 10-camphosulfonic acid (CSA). The resulting nanowire detector of protonated PANi nanofibers reacts with NH_3, changing its electrical resistance (Fig. 8.6) and displayed a threshold detection of 0.5 ppm of NH_3. In the figure, normalized resistance $[(R/R_o) - 1]$ is plotted as a function of the analyte concentration.

Virji and colleagues reported the use of PANi nanofibers for detection of acids as well as NH_3 (Virji et al. 2004). In both instances the nanofiber

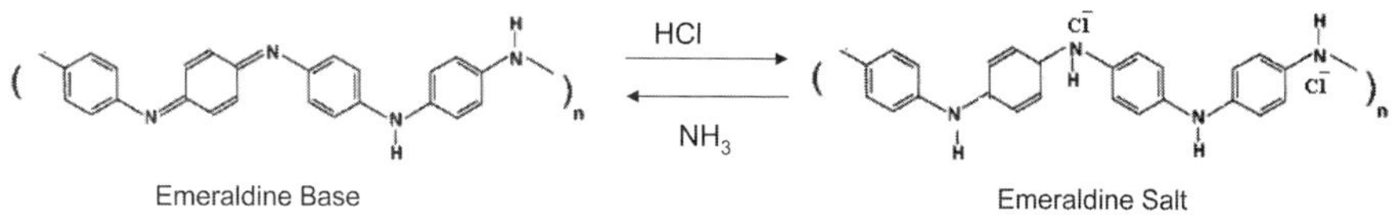

HCl

NH_3

Emeraldine Base

Emeraldine Salt

Scheme 8.1

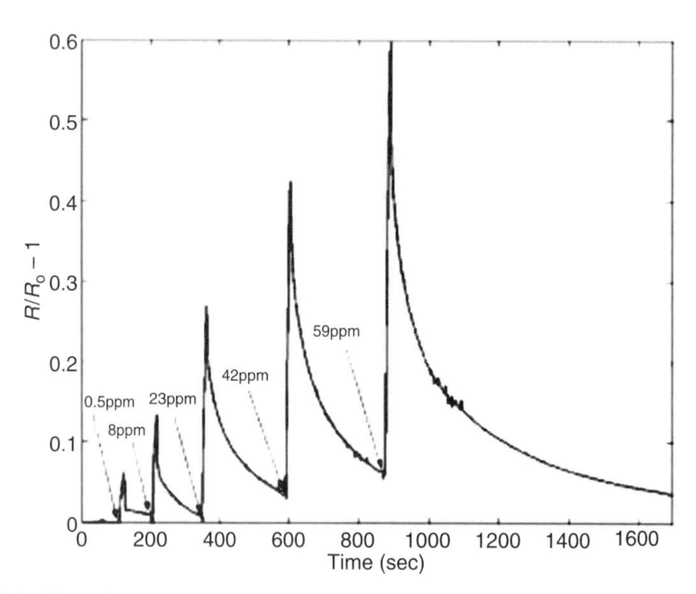

Figure 8.6 The change in electrical resistance of PANi nanowires (doped with CSA) on exposure to NH_3 in air at different concentrations. A 2–4 min recovery time was allowed after each 10 s exposure to the ammonia mixture. Reprinted with permission from Liu et al. (2004). Copyright 2004. The American Chemical Society.

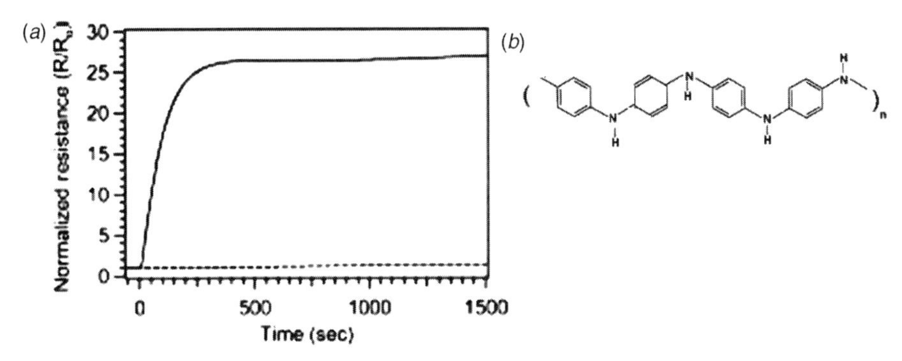

Figure 8.7 (*a*) A comparison of the response of nanofibers (~300 nm) (solid line) and thin film (broken line) to 3 ppm of hydrazine. (*b*) The leucoemaraldine form that is responsible for the change in electrical resistance. Reproduced with permission from Virji et al. (2004). Copyright 2004. The American Chemical Society.

sensors were shown to be more sensitive than the conventional thin-film sensors. Also, only the nanofiber sensors ($d = 300$ nm) detected hydrazine via the rapid reaction with the emeraldine base (converting it into the leuco-form polymorph, as illustrated in Fig. 8.7). Both humidity (Bishop and Gouma 2005; Liu et al. 2004) and temperature (Pinto et al. 2005) affect the electrical properties of PANi, requiring either maintaining these variables constant (in laboratory experiments) or compensating for these in the analysis of data (in operating devices).

Glucose oxidase immobilized on the surface of CSA-doped PANi/polystyrene (PS) nanofibers can function as a glucose sensor (threshold 1–2 mM glucose) (Aussawasathien et al. 2006). In this system the H_2O_2 generated by the reaction of glucose with the glucose oxidase enzyme is detected by cyclic voltametry. Again, the nanofibers were found to be markedly more sensitive to the analyte relative to thin-film sensors with the same chemistry. PANi/poly(vinyl pyrollidone) blends containing urease enzyme have been demonstrated successfully as NO_2 sensors (Bishop and Gouma 2005; Bishop-Haynes and Gouma 2007). Inorganic nanofibers of TiO_2 (anatase) on Pt electrodes (formed by electrospinning a composite nanofiber onto the electrode followed by calcinations) have been shown to have exceptional sensitivity, with a possible detection limit down to 1 ppb of NO_2 (Kim, I.-D., et al. 2006).

Increasing the available surface area of nanofiber mats used in sensor applications is generally desirable. Chapter 9 summarizes the various approaches to electrospin porous nanofibers. Patel and colleagues introduced porosity into nanofibers using a leaching method by adding glucose (up to 40 wt%) into the tetramethyl orthosilicate/PVA solution (Patel et al. 2006). On leaching out the glucose 7–35 Å pits were obtained on the silica nanofibers as surface features; the horseradish peroxidase (HRP) immobilized on the porous fibers were demonstrated to have high activity (for conversion of H_2O_2 to water). The activity was assessed to be four-fold of that from conventional non-templated silica samples.

8.2.3 Optical Sensors

Detecting very low levels of analytes optically usually requires resorting to an efficient fluorescence measurement. For instance, a high-quantum-efficiency pyrene methanol (PM) fluorophore might be used for the purpose; electron-deficient analytes such as metal ions quench the PM fluorescence. Nanofibers of PM-functionalized poly(acrylic acid) (PAA) (see Scheme 8.2) were electrospun and the nanofiber mats were used to detect Fe^{3+} and 2,4-dinitrotoluene (DNT) in solution (Wang, X. Y., et al. 2002b). Sensor performance for such systems is quantified using the Stern–Volmer equation,

$$I_o/I = 1 + K_{SV}[Q], \tag{8.9}$$

where I_o and I are the fluorescence intensities obtained with the system in the absence and in the presence of the quencher (analyte) respectively, and $[Q]$ is its concentration. The value of K_{SV} is a measure of the efficiency of quenching (it is the product of the luminescence decay time of the fluorophore in the absence of the quencher and the bimolecular quenching rate constant). The values of K_{SV} obtained with Fe^{3+}, Hg^{2+}, and DNT were 2 to 3 orders of magnitude higher than that obtained previously for similar thin-film sensors.

Another polymer-attached fluorophore that binds metal ions is the porphyrin moiety (5-(*p*-methacrylamidophenyl)-10,15,20-triphenylporphyrin (MATPP) (Scheme 8.3). It can be polymerized into a copolymer that is then electrospun into nanofibers ($d \sim 300$ nm) (Wan, L.-S., et al. 2006). Metal conjugation shifts the fluorescence emission wavelength of these

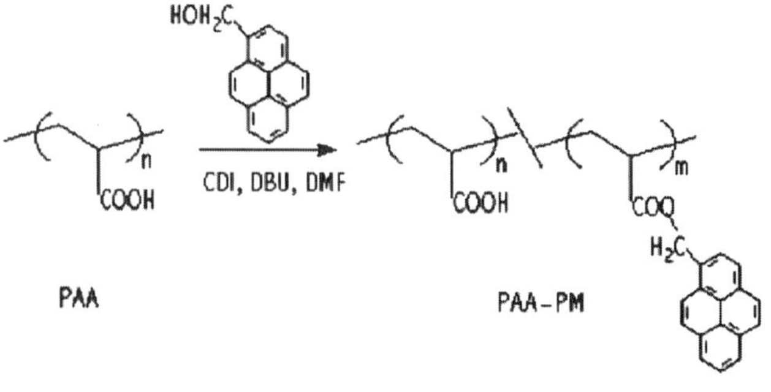

HOH₂C
COOH
n
CDI, DBU, DMF
COOH
n
COO
m
H₂C
PAA
PAA–PM

Scheme 8.2 Chemical structure for PAA–PM polymer.

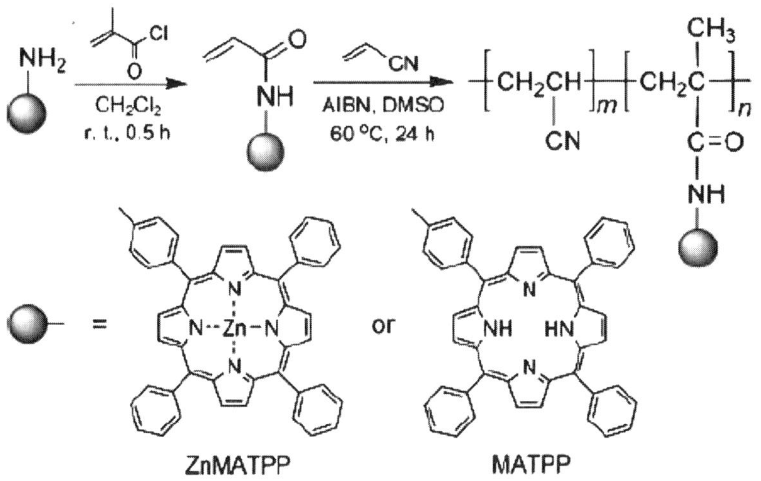

Scheme 8.3 Molecular structure of the copolymer containing porphyrin and the MATPP moiety with and without Zn^{2+} conjugation (Wan, L.-S., et al. 2006).

luminescent nanofibers; adding Zn^{2+} to the porphyrin polymer results in a shift of the emission peak fluorescence λ_{max} from 607 nm to 651 nm (excitation with $\lambda = 420$ nm) (Wan, L.-S., et al. 2006).

8.3 INORGANIC NANOFIBERS

Nanofibers that are made of inorganic materials such as inorganic polymers, metals, or metal oxides are of interest in semiconductor or electronic materials, photonic materials, aerospace applications, catalysis, and high-temperature sensor applications. Unlike organic polymer nanofibers, they are temperature resistant and in general have higher mechanical integrity, but have the same high specific surface area typical of nanofibers. This allows inorganic nanofibers to be used over a wide range of operating temperatures. These are usually generated in a two-step reaction involving the electrospinning of a precursor nanofiber (which is usually semiorganic) followed by a post-treatment, usually calcination to generate the inorganic material.

8.3.1 Sol–Gel Chemistry

Inorganic fibers are electrospun from sol-gel systems generally prepared by the hydrolysis of a metal alkoxide $[MOR]_n$ in acidic alcohol solvents. Although base catalyzed hydrolysis is also possible an acid medium is more likely to yield an electrospinnable solution. A "sol" in this context refers to a dispersion of colloidal particles in a liquid and "gel" to an interconnected loose polymeric network formed by self-assembly of the sol. Typically, alkoxides of silicon, titanium, and tin can be used in sol–gel reactions and some of these can be electrospun, usually in the presence of a template organic polymer. The chemistry involved might be illustrated using reactions of tetraethyl orthosilicate (TEOS). TEOS and alcohol/water in the presence of an acid catalyst (0.01 M HCl) can undergo three possible reactions — hydrolysis of the alkoxide into a silanol, water condensation, and alcohol condensation:

$$(EtO)_3\text{-Si-OEt} + \text{H-O-H} \rightarrow (EtO)_3\text{-Si-OH} + EtOH$$

$$(EtO)_3\text{-Si-OH} + (EtO)_3\text{-Si-OH} \rightarrow (EtO)_3\text{-Si-O-Si-}(OEt)_3 + H_2O$$

$$(EtO)_3\text{-Si-OH} + EtO\text{-Si-}(OEt)_3 \rightarrow (EtO)_3\text{-Si-O-Si-}(OEt)_3 + EtOH \qquad (8.10)$$

The reaction will continue, with progressively longer sequences of the reaction product being formed, with the reactant mixture eventually reaching the sol colloid state. The kinetics and extent of reaction depends on the molar fraction of reactants and the reaction conditions (particularly the temperature and pH of the medium). Therefore, sol-gel systems of different degrees of complexity are obtained by altering the reaction conditions. The presence of an organic acid, usually acetic acid or a mineral acid, helps the hydrolysis proceed rapidly and yields a viscous electrospinnable reaction product. The condensation reactions shown above can also lead to some cyclic byproducts that may subsequently undergo ring-opening reactions to yield a gel. The gel formed is generally blended with an organic polymer prior to electrospinning. For instance, calcium phosphate sol precursor and PVA solutions electrospun into nanofibers and calcined at 600°C yielded hydroxyapatite, a biocompatible scaffolding material (Dai and Shivkumar 2007).

The electrospun nanofiber mats are calcined at high temperature to remove the organic components and, in some cases, to convert the inorganic content into the required oxide species. Metal oxide nanofibers can also be prepared by electrospinning blends of a polymer with an organic metal salt dissolved in the spinning solution, followed by calcination of the nanofiber. Table 8.3 summarizes the different chemistries used to synthesize inorganic nanofibers from recent literature; the list is not intended to be comprehensive but illustrates the general approach used in synthesis of inorganic nanofibers.

8.3.2 Oxide Nanofibers

A wide range of oxide nanofibers have been reported in the literature prepared using either a procedure based on sol–gel chemistry or via calcination of a polymer nanofiber carrying a salt of the metal of interest, such as the acetates

TABLE 8.3 A selection of metal oxide nanofibers reported in the recent electrospinning literature

Oxide	Precursor	Alcohol	Acid/Catalyst	Polymer	Reference
$PbTiO_3$	Lead acetate/$Ti(OCH_4)_4$	Ethanol	Acetic acid	PVP	Lu et al. (2006)
V_2O_5	Vanadium oxide isopropoxide	Ethanol	—	PVAc	Viswanathamurthi et al. (2003a)
	Vanadium oxide isopropoxide	Propanol	CF_3COOH	B-76[b]	Macías et al. (2005)
	Vanadium chloride	Ethanol	HCl	P-123[a]	Macías et al. (2005)
Nb_2O_5	Niobium ethoxide	Ethanol	Acetic acid	PVAc	Viswanathamurthi et al. (2003b)
	Niobium chloride	Ethanol	HCl	P-123	Macías et al. (2005)
	Niobium butoxide	Butanol	HCl	P-123	Macías et al. (2005)
TiO_2	Tetrabutoxy titanate	Ethanol	HCl	P-123	Madhugiri et al. (2004)
	Titanium isopropoxide	Ethanol	Acetic acid	PVP	Watthanaarun et al. (2005)
	Titanium isopropoxide	—	Acetic acid	PVAc	Ding et al. (2004a)
	Titanium isopropoxide	Ethanol	HCL	P-123 B-76	Macías et al. (2005)
PdO	Palladium acetate	—	—	PC	Viswanathamurthi et al. (2004a)
SnO_2	Dimethydineodecanoate tin	—	—	PEO	Wang, Y., et al. (2004)
	Tin(IV) chloride	Isopropanol	—	PEO	Wang, Y., et al. (2005, 2007)
		Ethanol	—	PVAc	Dharmaraj et al. (2006)
Ta_2O_5	Tantalum chloride	Ethanol	HCl	P-123 B-76	Macías et al. (2005)
	Tantalum pentoxide	Ethanol	Acetic acid	PVAc	Dharmaraj et al. (2006b)
	Tantalum butoxide	Propanol	—	F-127[c]	Macías et al. (2005)
$TaNbO_7$	$TaCl_5$, $NbCl_5$, and $Nb(OC_4H_9)_4$	Ethanol	HCl	B-76	Macías et al. (2005)
$VTiO_2$	VCl_3 and $Ti(OC_3H_7)_4$	Ethanol	Acetic acid	P-123	Macías et al. (2005)
	$VO(OC_3H_7)_3$, $(OC_4H_9)_4$	Ethanol	Acetic acid	B-76	Macías et al. (2005)
$MgTiO_3$	MgOEt, titanium isopropoxide	2-methoxy ethanol	DMF	PVAc	Dharmaraj et al. (2004a)

(Continued)

TABLE 8.3 *Continued*

Oxide	Precursor	Alcohol	Acid/Catalyst	Polymer	Reference
ZnO	Zinc acetate	—	—	PVA	Viswanathamurthi et al. (2004b); Wu and Pan (2006); Yang, X. H., et al. (2004)
NiO/ZnO	Nickel oxide/zinc oxide	Ethanol	Acetic acid	PVA	Shao et al. (2004g)
NiO	Nickel acetate	—	—	PVA	Guan et al. (2003c); Shao et al. (2004d)
NiO/ZnO	Nickel acetate + zinc acetate	—	—	PVA	Shao et al. (2004g)
$NiTiO_3$	Nickel acetate + titanium isopropoxide	Ethanol	HNO_3	PVAc	Dharmaraj et al. (2004b)
$NiFe_2O_4$	Nickel ethylhexano-isopropoxide	Isopropanol	Acetic acid	PVP	Li et al. (2003a)
SiO_2	Tetraethyl orthosilicate	Ethanol	HCl	PEO	Kataphinan et al. (2003)
		Ethanol	H_3PO_4	PVA	Shao et al. (2003)
		Ethanol	Acetic acid	PVP	Zhang, G., et al. (2005)
ZrO_2	Zirconium oxychloride	—	—	PVA	Shao et al. (2004a)
Mn_2O_3, Mn_3O_4	Manganese acetate	—	—	PVA	Shao et al. (2004b, 2004e)
CeO_2	Cerium nitrate	—	—	PVA	Yang, X. H., et al. (2005)
Co_3O_4	Cobalt acetate	—	—	PVA	Shao et al. (2004f)
Nb_2O_5	Niobium ethoxide	Ethanol	Acetic acid	PVAc	Viswanathamurthi et al. (2003b)
$BaTiO_3$	Barium acetate + titanium isopropoxide	Ethanol	Acetic acid	PVP	Yuh et al. (2005)
Fe_2O_3	Iron(III) chloride	—	—	PVA	Shao et al. (2004h)
Fe_3O_4	Iron(III) chloride + iron(II) chloride	—	—	Copolymer	Wang, M., et al. (2004b)
Al_2O_3/ZnO	Aluminum nitrate + zinc acetate	—	—	PVA	Lin et al. (2007)
Al_2O_3	Aluminum dibutoxide ethyl acetoacetate	Ethanol	HCl	Photopolymer	Larsen et al. (2003)

[a] Pluronic P-123 is a copolymer of ethylene oxide and propylene oxide $(EO)_{20}$-$(PO)_{70}$-$(EO)_{20}$ manufactured by BASF.
[b] B-76 is the abbreviation for Brij 76, a copolymer $C_{18}H_{37}$—$(OCH_2CH_2)_{10}$—OH (ICI Americas, Inc.).
[c] Pluronic F-127 (a non-ionic difunctional block copolymer surfactant terminating in primary hydroxyl groups.)

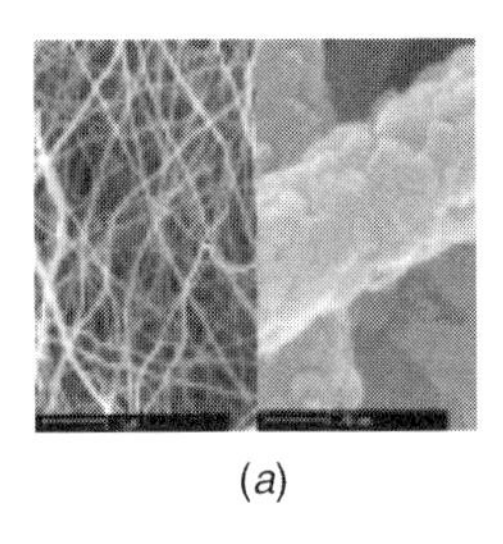

(*a*)

(*b*)

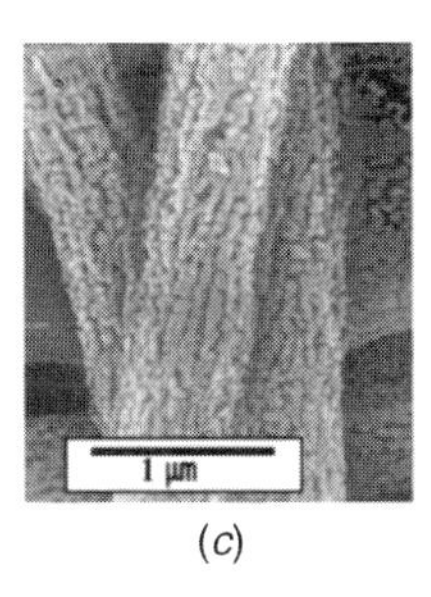

(*c*)

Figure 8.8 (*a*) SEM images of titanium dioxide nanofibers showing surface irregularities after calcination. Reproduced with permission from Ding et al. (2004a). Copyright 2004. The Korean Fiber Society. (*b*) SEM image of palladium oxide nanofibers showing surface irregularities after calcination. Reproduced with permission from Viswanathamurthi et al. (2004a). Copyright 2004. Elsevier. (*c*) SEM image of niobium oxide nanofibers showing surface irregularities after calcination. Reproduced with permission of Viswanathamurthi et al. (2003b). Copyright 2003. Elsevier.

of Cu, Co, Zr, Mn, Al, Zn, mixed into aqueous solutions of PVA or PVP (Table 8.3). The solutions can be electrospun to yield well-formed defect-free nanofibers comparable in morphology to those spun from polymer solution without any additive. Calcination of these yields inorganic nanofibers that are generally smaller in diameter relative to that of the original as-spun fibers and generally have a rougher surface (Fig. 8.8). Calcination has to be carried out at a high enough temperature to completely degrade the organic moiety; with the zinc acetate/PVA system a temperature of $\sim$480°C was needed (Wu and Pan 2006). The progress of calcination can be monitored by thermal analysis of the nanofibers.

Inorganic nanofibers often have a rough surface appearance. In $PbTiO_3$ mixed-oxide nanofibers, the uneven surface morphology is exaggerated into a "necklace-like" morphology (Lu, X. F., et al. 2006), with the nanofibers having "beads on a string" appearance. Atomic force microscopy (AFM) has been effectively used to study the gross features of inorganic nanofibers (Dharmaraj et al. 2006b; Viswanathamurthi et al. 2003b; Wang et al. 2004a) and to image the surface irregularities. As discussed in Chapter 6, however, the effectiveness of AFM in representing surface features depends upon instrument variables as well, particularly the material, geometry, and dimensions of the probe tip.

Inorganic nanofibers are usually characterized by wide angle x-ray diffraction (WAXD) to establish the crystallinity of the fibers, transmission electron micrograph (TEM) images to demonstrate the crystallite morphology, and Fourier transform infra-red (FTIR) spectroscopy to confirm the surface chemistry. A few the studies have also employed Raman spectroscopy (Madhugiri et al. 2004) and X-ray photoelectron spectroscopy (Wang, Y., et al. 2005) for the purpose.

With alumina-borate oxide nanofibers, both x-ray diffraction (XRD) data and FTIR data have been used to establish changes in morphology at different calcination temperatures (Dai et al. 2002). Nanofibers were electrospun from 10 wt% PVA solutions containing dissolved aluminum borate. At the lower calcination temperatures of 1000°C and 1200°C, $Al_4B_2O_9$ and $Al_{18}B_4O_{33}$ nanofibers were formed; these are unstable but can be decomposed into the stable α-Al_2O_3 form when calcined at 1400°C. XRD data also show the as spun vanadium isopropoxide/PVAc nanofibers to be essentially amorphous, with the crystalline V_2O_5 phase emerging only after calcination at 500°C (Viswanathamurthi et al. 2003a). AFM images show the as-spun smooth fibers to develop surface irregularity on calcination. In organic titanate sol–gel incorporated into P-123 (a water-soluble polyether polymer), XRD data on the nanofibers calcined at lower temperatures (400 or 600°C) showed predominantly anatase structure. The data showed *d*-spacings of 3.54, 2.38, and 1.89, corresponding to (101), (004), and (200) reflections of anatase (Madhugiri et al. 2004). Also, Raman spectra of commercial anatase matched well with those of the inorganic nanofibers. With samples calcined at 900°C, however, the XRD data were consistent with the evolution of the rutile phase of titania. The peaks corresponding to reflections (101) and

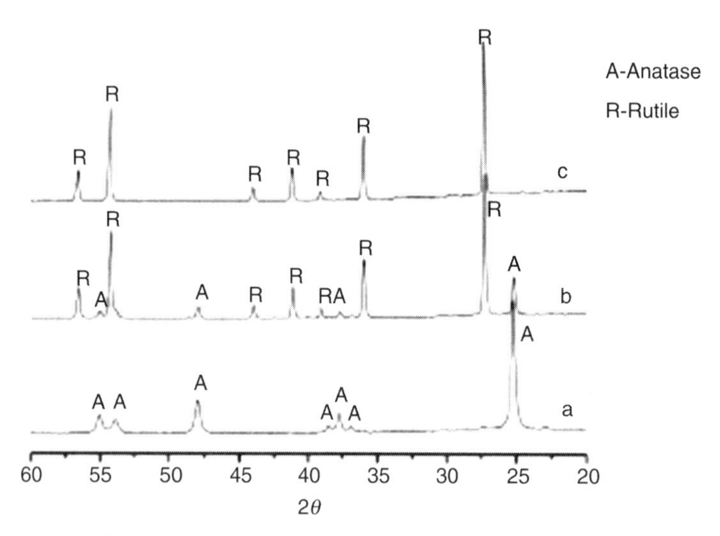

Figure 8.9 WAXD patterns for nanofibers of PVAc with 63.5 wt% of titanium isopropoxide calcined at three different temperatures: (*a*) 600°C; (*b*) 800°C; and (*c*) 1000°C. Reproduced with permission from Ding et al. (2004a). Copyright 2004. The Korean Fiber Society.

(200) decreased to be replaced by new reflections corresponding to (110), (101), (200), (111), and (210) associated with the rutile form of titanium dioxide. Nanofiber morphology, however, remained intact, despite the change in crystalline morphology of the fiber. Ding and colleagues also reported on the anatase–rutile transformation in nanofibers spun from PVAc solutions, as shown in the XRD data shown in Fig. 8.9 for nanofibers calcined at different temperatures (Ding et al. 2004a). A gradual transition of the crystalline morphology was evident in the data.

Fourier transform infra-red spectroscopy has also been used in sol–gel chemistry to confirm that the gel-forming reactions have progressed as expected. As reported, information on the evolution of phases obtained from WAXD correlates well with that from FTIR data (Dharmaraj 2004a). Work by X. H. Wu et al. (2006) on ZnO nanofibers and Viswanathamurthi et al. (2003b) on Nb_2O_5 nanofibers illustrate the use of IR spectra to monitor the reaction. In these studies either PVA or PVAc polymer solutions were used and the as-spun nanofibers containing the precursors showed their characteristic IR absorption bands in the spectrum of the mats. On calcinations these bands disappeared indicating an inorganic matrix; also, no peaks characteristic of water were detected, indicating a completely dry product.

9

RECENT DEVELOPMENTS IN ELECTROSPINNING

Innovative modifications of the basic spinning apparatus and methodology used in electrospinning allows a wide range of fiber and mat morphologies to be produced. Many interesting variations of the basic electrospinning process have been described in the literature over recent years. These include novel electrode arrangements, the use of AC voltage to drive the process, reactive electrospinning, unusual collector geometries, unique tip designs, vibrating tip designs and the use of different spinning environments. Most of these are intriguing scientific phenomena that help better understand the complexities of the process but invariably remain laboratory curiosities. A few however, show promise in terms of extending the range of applications of nanofibers. These include process and material changes that result in unusual surface morphologies in nanofibers and complex mat structure. Although the full range of applications that best exploit these new developments are yet to be developed, the emerging innovative applications of nanofibers in biomedical, sensor, electronic, and other areas will likely be enabled or enhanced by these recent advances in several key techniques.

9.1 NANOFIBERS WITH SURFACE POROSITY

The unique advantage of nanofiber constructs is that they combine high specific surface area with a permeable and easily handled mat structure.

The surface area per unit mass in fiber mats increases as the average fiber diameter *d* (nm) is reduced. However, there are practical processing limits to electrospinning solutions of very low concentrations to reduce *d* (nm) [as well as to reducing *d* (nm) in post-processing treatments (e.g., by drawing of the preformed fiber mats, Zong et al. 2005)]. Changing the surface morphology of individual fibers either by creating pits, pores, or bumps on their surface or by altering their circular cross-section by regular or irregular patterning of the surface offers an alternative or complementary method of increasing the specific surface area of nanofibers. Electrospinning is one of the few techniques[1] available to create nanoscale topologies in a controllable manner on soft materials. Nanofibers of high specific surface area are especially attractive in applications such as in catalysis, in the controlled delivery of bioactive agents via nanofiber matrices, and in sensor technology where rapid reaction kinetics are desirable. Changes in surface morphology can also have an impact on the optical, electrical, and permeability properties of nanofiber mats. Several approaches to modifying the surface morphology of nanofibers discussed in the electrospinning literature are reviewed here.

9.1.1 Extraction of a Component from Bicomponent Nanofibers

When two or more thermodynamically immiscible polymers (e.g., polystyrene (PS) and poly(bromo styrene)) are dissolved in a common solvent, concentration-dependent phase separation occurs in solution. Thermal analysis of such polymer blend samples shows first- and second-order transitions at temperatures characteristic of their component polymers, indicating immiscibility. [In the case of fully miscible polymers, differential scanning calorimetry (DSC) tracings, for instance, yield a single glass transition signal that occurs at a temperature lying between those for the pure components (Guo 2003)]. Cast polymer films and nanofibers electrospun from solution also show the same phase-separated morphology. As the homogeneous ternary solution evaporates during casting or spinning, the polymer concentration rapidly increases and the solution undergoes glass transition and/or crystallization, leading to solidification. In electrospinning, phase separation occurs during rapid solvent evaporation and the phase morphologies obtained in the spinning jet are therefore generally finer compared to those in cast films, where drying occurs at a much slower rate.

[1]Lithographic methods (near-UV photolithography or electron beam lithography) can also yield topologies at this size scale. The latter technique does not need a mask and can produce array features of 30–40 nm in dimension (Norman and Desai 2006).

The millisecond timescale in which phase separation sets in during drying of the fiber is far too short for "coarsening" of morphology, and the finer phase morphology leads to a narrow distribution of the dimensions of separated phases in the fiber. Phase geometry can be controlled by changing the ratio of the polymers and the concentration of the spinning solution. Post-treatment of phase-separated nanofibers by extraction with a solvent that selectively removes one of the polymers leads to well formed surface pores or other complex geometries on the fiber.

Changing the composition of the polymer blend generally does not alter the average pore[2] dimensions obtained in this technique significantly, because phase separation is kinetically controlled. However, changes in blend composition alter the density of pores obtained on the fiber surface (You et al. 2006b). Spinodal phase separation, which sets in as the solvent evaporates, may lead to a co-continuous phase morphology (as opposed to a matrix-dispersed structure), as observed in the case of nanofibers poly(D,L-lactide) (PDLLA)/poly(vinyl pyrrolidone) (PVP) blend and PLLA/PVP blend nanofibers electrospun from CH_2Cl_2 (Bognitzki et al. 2001b). With these blends phase domains of dimension 300 nm to 1 μm have been reported. Another example is the electrospinning of a 1 : 1 blend of gelatin (GT) and poly(ε-caprolactone) (PCL) from 2,2,2-trifluoroethanol (TFE) solution to yield a bicomponent nanofiber (Zhang et al. 2006). These nanofibers showed phase separation in AFM, and high-resolution field emission scanning electron microscope (FESEM) images. The nanofibers extracted by PBS buffer solution at 37°C over a two-week period to dissolve away the GT component yielded a highly porous construct resulting in a ~240% increase in the accessible specific surface area in a fiber of average $d \approx 800$ nm (Fig. 9.1). This corresponds to as much specific surface area as in a smooth fiber in the diameter range of 200–300 nm. There is considerable room for even further increase in area by both reducing fiber diameters and by using higher volume fractions of the extractible polymer in the blend. A similar approach was used with blend nanofibers of polyacrylonitrile (PAN) and PVP, with the PVP component was leached out from the nanofibers by aqueous extraction (Li and Nie 2004) to yield nanopores as fine as 30 nm. Other systems such as a blend of nanofibers of poly(glycolide) (PGA) with PLLA spun from 1,1,1,3,3,3-hexafluoro-2-propanol (HFP), with the latter component removed using chloroform (You et al. 2006b), and a poly(vinyl alcohol) (PVA)/chitosan blend of nanofibers,

[2]The terms "porosity" and "pore" are used in this chapter to mean the features on the nanofiber itself as opposed to the interstitial porosity of mats dealt with in Chapter 5. Note that porous nanofibers (with surface pores on the fiber) are electrospun into porous mats with "interstitial pores" as well.

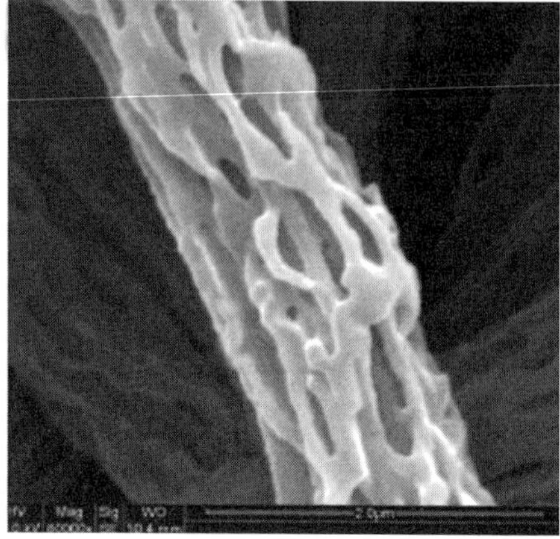

Figure 9.1 PCL/GT nanofibers as spun $d \sim 900$ nm (magnification $\times$40,000) and the high surface area nanofiber after extraction with buffer to remove gelatin $d \sim 800$ nm (magnification $\times$80,000). Reprinted with permission from Zhang et al. (2006a). Copyright 2006. Institute of Physics.

with the PVA phase removed by aqueous alkali (Li and Hsieh 2006), have also been reported.

A variation of the approach uses crosslinking to insolubilize one component in the blend of nanofibers prior to extraction of the second polymer. Blends of poly(vinyl cinnamate) (PVCi) and the biodegradable polymer poly(3-hydroxy-butyrate-*co*-3-hydroxyvalerate) (PHBV) electrospun from $CHCl_3$ solution are examples of this approach. The bicomponent nanofiber mats were irradiated under a high-pressure Hg vapor lamp to photocrosslink only the PVCi phases (the process being initiated by the UV-B radiation). Chloroform (also used to electrospin the fiber) was then used to extract the PHBV component, yielding a nanofiber of high porosity (Lyoo et al. 2005). With a blend of nano-fibers of polyetherimide (PEI)/PHBV, the PHBV component was removed by thermal degradation (instead of by dissolution), leaving the thermally resistant PEI as a porous nanofiber (Han et al. 2004).

The detailed topology of the porous fiber, consisting of interconnected volumes and channels produced by leaching or annealing away of one component, depends on the phase morphology of that component.

9.1.2 Phase Separation During Electrospinning

Under appropriate electrospinning conditions nanofibers of a single polymer, with their surface patterned into closely spaced regular or irregular pores, are obtained. Coupled with the already small average fiber diameters, these surface features can potentially yield remarkably high specific areas, with values in the range of 100–1000 m^2/g at times surpassing even that of conventional porous materials such as silica gel (400 m^2/g) (Megelski et al. 2002). These nanoporous morphologies are often obtained in electrospinning a single polymer dissolved in a particularly volatile solvent, but only under specific process conditions. Most literature references to porous nanofibers describe fibers of micrometer-scale dimensions, yielding pores in the range of several hundred nanometers. The average pore size obtained depends on the process conditions and polymer/solvent system used in the electrospinning.

Polymers such as polycarbonate (PC), poly(vinyl carbazole) (PVCz), and PLLA electrospun from particularly volatile solvents such as CH_2Cl_2 illustrate this phenomenon, with surface pores ranging in size from 100–250 nm (Bognitzki et al. 2001a). Polystyrene (PS) electrospun from highly volatile tetrahydrofuran (THF) also resulted in porosity amounting to as much as much as 20–40% in surface area depending on fiber diameter (Megelski et al. 2002). Phase separation is believed to play a key role in the formation of pores. Thermodynamic instability in the system caused by evaporative cooling and an increase in polymer concentration during drying of the jet are believed to be the driving forces behind the phase separation that leads to a porous morphology. With rapid extension of the jet and concurrent solvent evaporation, thermally-induced phase separation (TIPS) occurs primarily at the surface of the jet and results in a solvent-rich and a solvent-lean phase being created. The solvent-lean phase ultimately solidifies into the nanofiber matrix, while solvent-rich domains invariably dry up to form the surface pores. Localization of the pores on the surface of the fiber is expected because evaporative cooling primarily occurs at the surface; this is in fact observed experimentally (Casper et al. 2004; Megelski et al. 2002). Thermal analysis often reveals a signature of this rapid structure formation during electrospinning (Bognitzki et al. 2001a).

Pore formation can therefore often be suppressed by reducing the volatility of solvent to retard evaporative cooling that initiates TIPS (Han et al. 2005). For instance, PLLA when electrospun from less-volatile $CHCl_3$ instead of from CH_2Cl_2, did not result in pore formation (Bognitzki et al. 2001a). Similarly, the porosity on a micro-textured surface of PS nanofibers was reduced and finally eliminated when the less volatile dimethyl formamide (DMF) content in the electrospinning solvent mixture DMF/THF was gradually increased from 10/90 to 100/0 (Megelski et al. 2002). Electrospinning cellulose triacetate solutions in CH_2Cl_2 yielded porous nanofibers with pore sizes in the range of 50–100 nm. Spinning from a solvent mixture of CH_2Cl_2/ethanol (90/10) also yielded porous nanofibers, but with larger pores in the size range of 200–500 nm. With solvent mixtures of higher ethanol content (CH_2Cl_2/ethanol: 85/15 or 80/20) no such porosity was obtained (Han et al. 2005).

Phase separation in this case likely proceeds via spinodal decomposition (with the cooling solution passing through the binodal curve of the phase diagram to enter the metastable region) as suggested by the interconnectivity observed between the surface pores (Megelski et al. 2002). Figure 9.2 shows an environmental scanning electron microscope (ESEM) image illustrating the porosity in nanofibers of PC spun from $CHCl_3$ (~4 wt%) solutions (Kim, G. M., et al. 2005b). Note that the pores are elongated in the axial direction due to rapid extension of the jet during the nucleation/growth stage of their formation. It is not easy to study the incidence of bulk porosity (as opposed to these surface markings) in nanofibers microscopically, except from fracture images. However, in the PMMA/closite clay nanofibers electrospun from $CHCl_3$, Kim and colleagues reported observing pores in the bulk of the fiber as well as on the surface (Kim, G. M., et al. 2005a).

TIPS can be encouraged by forced cooling of the spinning jet and is useful when the available solvents for the polymer of interest are not volatile enough to result in substantial evaporative cooling. Lowering the temperature of the collector to control fiber morphology has been reported (Kim, C. H., et al. 2006). Freezing the spinning jet (polymer and solvent) by electrospinning into a liquid nitrogen collector followed by drying under vacuum yielded nanofibers with high levels of bulk porosity (McCann et al. 2006). Several polymers, including PS, PCL, poly(vinylidene chloride) (PVDC), and poly(acrylonitrile) (PAN) electrospun from less volatile DMF or

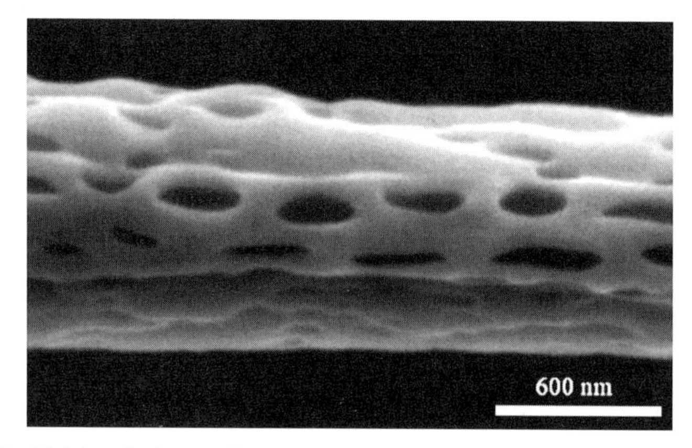

Figure 9.2 Field emission environmental SEM image of PC nanofibers electrospun from chloroform, showing porosity due to phase separation. Reprinted with permission from G. M. Kim et al. (2005b). Copyright 2005. Elsevier.

dimethylacetamide (DMAC) solvents still yielded highly porous fibers using this technique. With PAN nanofibers the porosity increased the specific surface area of the $d \sim 1\ \mu m$ nanofiber nearly threefold and the porosity was distributed throughout the bulk of the polymer. In contrast, electrospinning from highly volatile solvents generally results in porosity localized at the nanofiber surface, although both techniques rely on the TIPS phenomenon.

An alternative explanation of pore formation during electrospinning is based on the observation that condensation of moisture on the surface of drying polymer films exposed to volatile vapors leads to the formation of so-called "breath figures" (Srinivasarao et al. 2001). Electrospinning polymer solutions in an environment rich in water vapor also obtains the same effect. In this case, minute droplets of moisture from the spinning environment may nucleate and condense on the surface of an evaporatively cooling jet. The immiscible water droplets amount to hard spheres embedded in the wet surface and leave imprints on the electrospinning jet. Nucleation and growth of droplets is a relatively slower process compared to the rate of phase separation and the pores are therefore expected to be localized at the surface of the fiber. As the mechanism still depends on cooling of the jet surface, nanofiber porosity is expected to be suppressed when a less volatile spinning solvent is used or when the fraction of less volatile component of a solvent mix is increased lowering the rate of evaporation. However, the irregularly shaped heterogeneous surface pores obtained in electrospun nanofibers appear to be quite different in their morphology from the ordered arrays of hexagonal pores typically observed in breath figures on polymer films (Megelski et al. 2002). Some of these differences, however, might be due to the differences in geometry between the fiber and film, and the fact that fiber extension occurs concurrently with pore formation during electrospinning (compared to the slow drying of a film). Figure 9.3 shows an image of a typical porous fiber obtained (Casper et al. 2004).

Electrospinning PS from THF (35 wt% solution), Casper and colleagues obtained flat ribbon-like fibers with a cross-section of about 15 μm (Casper et al. 2004). On varying the relative humidity in the spinning environment, different average pore sizes were obtained on the nanofiber surface. The pore distribution was only slightly affected by percentage humidity, but the surface density of pores increased at the higher humidity levels. Their data are summarized in Table 9.1.

Increasing the average molecular weight of polymer (with the weight percent of polymer in solution held at 35 wt%) resulted in relatively larger, less uniformly shaped surface pores. Pores on nanofibers spun from higher molecular weight samples of PS were also found to be relatively larger and deeper compared to those spun from lower molecular weight polymer

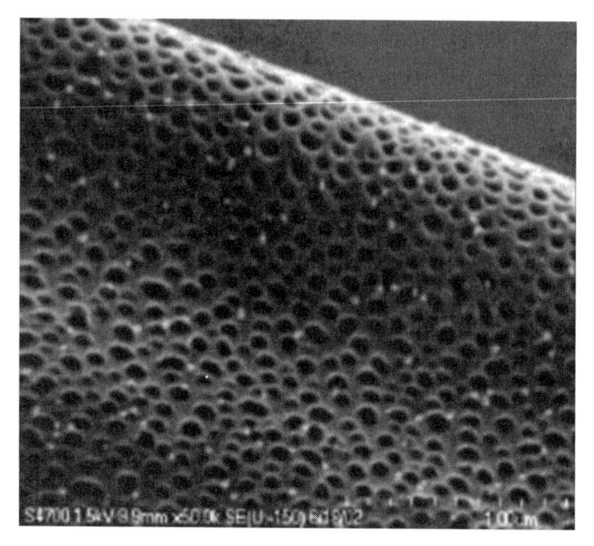

Figure 9.3 FESEM image of the porosity on a polystyrene nanofiber (M_w = 171,000 g/mol) electrospun in an environment with 50% humidity. Reprinted with permission from Casper et al. (2004). Copyright 2004. American Chemical Society.

(Casper et al. 2004); atomic force microscopy (AFM) was used to estimate the depth of the pores. This mechanism, however, should be insensitive to the molecular weight of the polymer and is inconsistent with the observed dependence of average pore size on molecular weight. But, increasing the average molecular weight of the polymer will result in a more viscous jet surface, impairing the diffusion of water molecules into the jet. This might be expected to leave a shallower imprint on the more viscous higher molecular weight polymer nanofibers. The data in fact show the opposite trend, suggesting the process leading to porosity to be far more complex and possibly involving some degree of TIPS in addition to the breath figure effect.

TABLE 9.1 Effect of humidity in the spinning environment on the porosity of electrospun nanofibers of PS

	Range of Pore Sizes (nm)	Most Frequent Pore Size (nm)	Range of Pore Sizes (nm)	Most Frequent Pore Size (nm)
Humidity (%)	Average M_w = 190,000 g/mol		Average M_w = 560,900 g/mol	
31–38	60–190	85	150–650	250
40–45	90–230	115	150–600	350
50–59	50–270	115	100–850	300
60–72	50–280	135	200–1800	350

Source: Casper et al. 2004.

TABLE 9.2 Selected examples of core–shell nanofibers

Core Material	Shell Material	Reference
PMMA (6–10% in DMF[a])	Polyaniline (PANi)[b]	Dong et al. (2004)
Drug[c] (H_2O)	PCL ($CHCl_3$: EtOH) (3 : 1)	Huang, Z. M., et al. (2006)
Mineral oil	PVP + sol–gel precursor	McCann et al. (2005)
Gelatin	PCL (TFE)	Huang et al. (2005)
PEO	Polysulfone	Sun et al. (2003)
PEO	Poly(dodecylthiophene)	
BSA/PEG (1 : 10) 44% in H_2O	PCL (30% w/v in $CHCl_3$: DMF(3 : 7))	Jiang et al. (2005)
PANi (10–16% $CHCl_3$)	PC or PS (16% THF)	Wei et al. (2005)
PVP (6% in EtOH : H_2O (8.5 : 1.5)	MEH–PPV (2.5 mg/mL) MEH–PPV + PHT blend	Li et al. (2004a)
Oil or glycerin	Sol–gel precursor	Loscertales et al. (2004)

[a]5–20 wt% (based on PMMA) of tetrabutylammonium chloride was added to solution.
[b]Not fabricated by coaxial spinning. [c]Core was not polymeric, but either resveratrol (in EtOH) or gentamycin sulfate (aqueous). PMMA, poly(methyl methacrylate); PCL, poly(ε-caprolactone); PVP, poly(vinyl pyrrolidone); MEH–PPV, poly(2-methoxy-5-(2-ethylhexyloxy)-1,4-phenelyenevinylene); PHT, poly(3-hexylthiophene); TFE, 2,2,2-trifluoroethanol.

At the lowest average molecular weights of PS investigated, only beads could be electrospun, but interestingly even these showed surface porosity, with both large and small pores (nanopores) on their surfaces. However, the smallest of the pores (26–71 nm) observed on beads were not observed on nanofibers (Casper et al. 2004); this suggests possible coalescence of these nanopores on the fiber surface during rapid fiber extension. Findings from this study are also consistent with the explanation based on TIPS, while a contribution from a moisture-induced breath figure mechanism cannot be entirely ruled out. Table 9.2 summarizes recent examples of porous and other high-surface-area nanofibers from the literature. Porous beads are not unique to electrospinning in humid environments, but have also been observed in electrospinning from a volatile solvent when phase separation and bead formation occur simultaneously. This is illustrated in electrospinning of poly(butylenes succinate) (PBS) (He, J.-H., et al. 2007a), where beads with a complex pore morphology were obtained (Fig. 9.4).

9.2 CORE–SHELL NANOFIBERS

Core–shell bicomponent nanofibers, where a core nanofiber of one polymer is sheathed by the shell of a different polymer, can have interesting applications in several areas. For instance these might be used in the controlled delivery of pharmaceuticals or biological macromolecules (as already

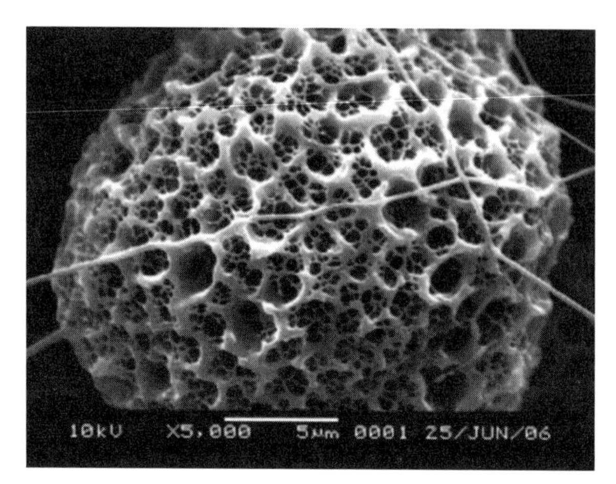

Figure 9.4 Porous beads of poly(butylene succinate) (PBS) ($M_w \approx$ 200,000–300,000 g/mol) from electrospinning of polymer from chloroform (8, 10, 12% wt). Reprinted with permission from J.-H. He et al. (2007a). Copyright 2007. Elsevier.

pointed out in Chapter 7), and may perhaps be used to protect (in the core) labile biomaterials such as enzymes. Where the core material is a conducting polymer such as polyaniline (PANi) and the shell is an insulating polymer, the nanofiber may function as an insulated nanowire for nanoelectronics and sensor applications. The approach is also useful in electrospinning polymers that are generally difficult to electrospin; using a readily electrospinnable polymer for the shell component allows these to be spun into the core. Polymer–polymer core–shell fibers of mesoscale and nanoscale dimensions have been fabricated by techniques other than electrospinning. For instance, electrospun nanofibers can be coated with a second polymer by dipping in a solvent (He, W., et al. 2005a) or by vapor deposition (Caruso et al. 2001).

9.2.1 Coaxial Electrospinning

Coaxial nanofibers with the core polymer sheathed by a layer of the same or a different polymer fabricated using an electrospinning technique was first reported[3] by Sun et al. (2003). Figure 9.5 shows the specially engineered tip of concentric capillaries used for the purpose. Homopolymer core–shell nanofibers (e.g., PEO–PEO fibers with the shell containing a dye to contrast it with the core in optical imaging) as well as two-polymer fibers such

[3] A stable core–sheath complex liquid jet formed using a coaxial capillary tip, however, has been reported in earlier literature (Loscertales et al. 2002).

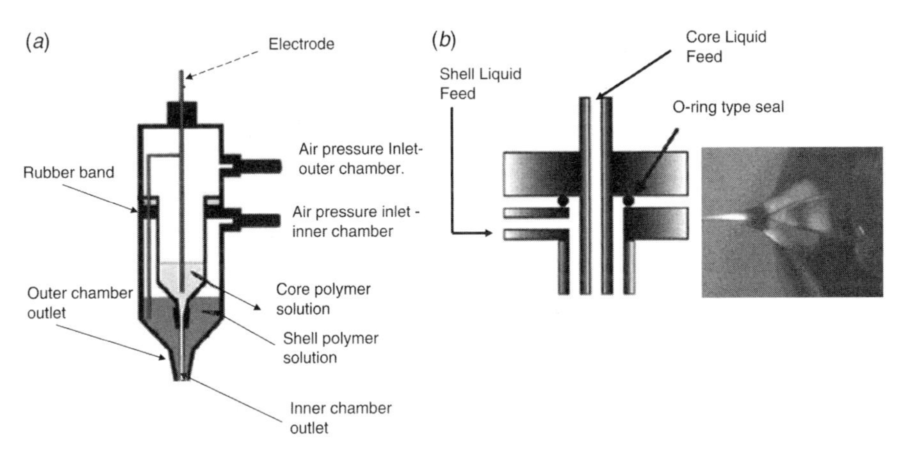

Figure 9.5 (*a*) The coaxial double-capillary tip arrangement used in electrospinning core–shell nanofibers. Reprinted with permission from Sun et al. (2003). Copyright 2003. John Wiley & Sons. (*b*) A coaxial tip design used by Larsen et al. (2003) for olive oil–water (liquid–liquid) coaxial jets. The inset shows the double-layered Taylor's cone. Reprinted with permission from Larsen et al. (2003). Copyright 2003. American Chemical Society.

as PEO–polysulfone or PEO–poly(dodecylthiophene) were fabricated in early experiments using this approach (Sun et al. 2003). Other designs of coaxial double capillaries for electrospinning core–shell nanofibers have been described in the literature (Zhang et al. 2004; McCann et al. 2005), but differ little from the one depicted in the diagram. Basically, two independent polymer solutions are pumped into the concentric capillaries or needles, with the shell polymer being guided to the annular space between the tubes; the liquids contact each other only at the tip of the coaxial capillary.

The core–shell fiber morphology is initiated from the two-layered droplet and the associated structured Taylor's cone, produced using the coaxial capillary tip. A liquid jet emerges simultaneously from the vertices of the outer meniscus as well as from the inner core meniscus to yield a compound jet of co-flowing solutions (Zhang et al. 2004). However, it is not necessary for both liquids to form stable jets for the process to work successfully, but at least one of these polymers needs to be electrospinnable. The ratio of core vs shell content in the compound fiber is controlled by changing the concentration of the polymer solutions or the relative feed rates used in spinning. Figure 9.6 shows the composition of a compound droplet formed from a coaxial tip, with a shell of PAN/DMF (12 wt% solution) and a core of PMMA/acetone : DMF (60 : 40) (15 wt% solution) (Zussman et al. 2006). The PMMA core can be pyrolized or oxidized away to yield hollow

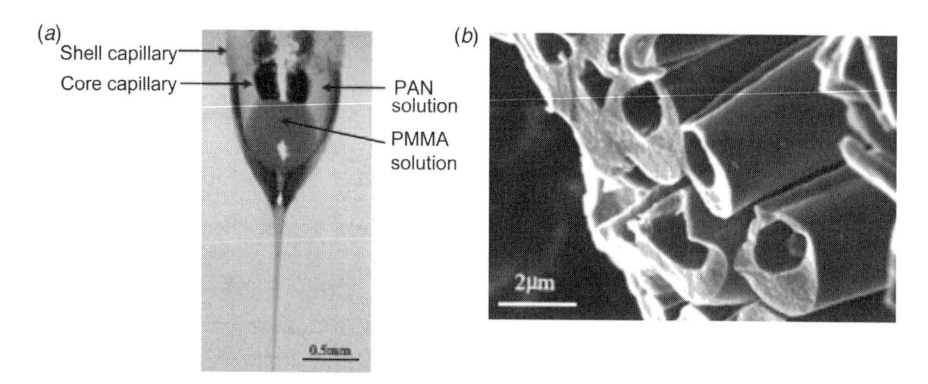

Figure 9.6 (*a*) Composition of the compound Taylor's Cone from a core–shell spinneret. (*b*) Hollow nanofibers formed by pyrolysis of the core–shell nanofibers. Reprinted with permission from Zussman et al. (2006). Copyright 2006. Wiley-VCH.

nanofibers of PAN. The characteristic time associated with mutual diffusion at the liquid–liquid boundary is much slower than the millisecond timescales associated with bending instability and drying of the jet. For instance, in coaxial electrospinning of silk/PEO from the common aqueous solvent (LiBr solutions) in which both polymers are miscible, clear core–shell morphology was obtained in the nanofibers (Wang, H., et al. 2005). However, this need not be true of all systems. For example, McCann and colleagues reported that for the system of core (PS in DMF/THF) and shell (PVP/titanium isopropoxide), the coaxially spun nanofibers lacked the characteristic core–shell morphology (McCann et al. 2005).

With a pair of incompatible polymers such as PCL and GT (both dissolved in 2,2,2-trifluoroethanol) (TFE) (Zhang et al. 2004), mutual diffusion during jet formation can be essentially ruled out. When core and shell polymers are dissolved in different solvents for coaxial spinning, however, these must be selected to avoid any possibility of precipitation at the liquid–liquid interface (the solubility of the individual polymers in both solvents does not necessarily guarantee this). The use of immiscible solvents in the core and shell generally contributes towards the stability of the jet. Critical physical properties of solutions that invariably determine the stability of coaxial jets are not well understood. Loscertales et al. (2004) found pairs of liquids that satisfied the condition $\sigma_c > \sigma_s$ to form stable structured cones (where σ is the liquid/dielectric atmosphere surface tension and subscripts c and s refer to core and shell liquids, respectively). Any instability of the jet may lead to the core being asymmetrically placed within the nanofiber (Jiang et al. 2005). A bimodal distribution or a mix of thick and thin fibers is sometimes obtained in coaxial spinning (Huang et al. 2006), with both types of fibers still showing the core/shell morphology. This suggests multiple jets from a two-layered cone rather than the splitting or branching of the main jet during electrospinning.

Interestingly, in the case of GT/PCL core/shell nanofibers, mechanical properties of the mat were found to be better than that of mats of individual components. In nanofibers that showed the best tensile properties (electrospun with 7.5 w/v% GT solution and 10 w/v % PCL), the tensile strength and extensibility were 8.4 MPa and 178%, respectively. Comparable values for GT fibers were 5 MPa and 8%, while for PCL they were 1.3 MPa and 52%. A good explanation of this synergy is not available (Huang et al. 2006).

A particularly interesting application of nanofibers with core–shell geometry is in tissue scaffolding, where they perform the dual role of mechanical support and the delivery of bioactive agents concurrently. Nanofibers with the water-soluble polymer PEG used as the core and biodegradable PCL used as the shell polymer have been reported (Zhang et al. 2006c). A model agent (BSA conjugated with fluorescein isothiocyanate, ftc–BSA) was blended with the core polymer and electrospun using a coaxial capillary tip. A composite blend of (ftc–BSA/PEG/PCL), blended in the same ratio of the constituents was also electrospun into blend nanofibers. Unlike with the core–shell fiber mat, the composite blend nanofibers were beaded and were of poor quality. Also, comparative BSA release studies showed the burst release phase (see Chapter 7) to be suppressed only in core–shell nanofibers and that sustainability of the release was also superior for the same geometry relative to that in blend fibers.

The coaxial electrospinning technique can be adapted to prepare hollow tubular nanofibers (Loscertales et al. 2004) by using an easily extractible low molecular weight liquid as the core material. However, a shell polymer sheath, stiff enough to avoid the collapse of the fiber after the removal of the core liquid by solvent extraction, needs to be used. Using a sol–gel precursor material as the shell polymer is particularly useful in developing a stronger inorganic sheath material. The hollow nanofibers made with PVP blended with titanium isopropoxide as the shell polymer and an oil as the core (Li and Xia 2004; McCann et al. 2005) and post-processed by calcinations illustrate this. Figure 9.7 shows TEM images of hollow nanofibers prepared in this manner. Core–shell nanofibers where the core material can undergo slow hydrolysis or biodegradation in a buffer can also yield hollow nanofibers. But in this case the core material is removed at a much slower rate that is determined by the thickness of the sheath. As with the case of nanofibers where the sheath layer is made of PCL and the core is BSA-loaded PEG, the organic polymer shell (without

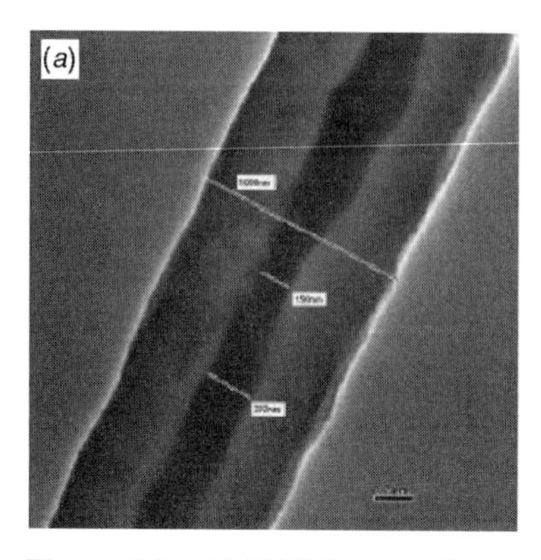

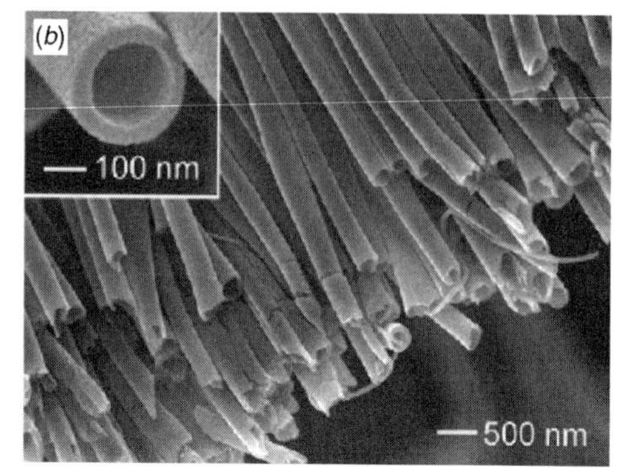

Figure 9.7 (*a*) TEM image of a coaxially spun nanofiber with a shell layer of PEO and a core of poly(dodecylthiophene) (PDT). Both polymers were electrospun from $CHCl_3$ (1–2% solutions). Reprinted with permission from Sun et al. (2003). Copyright 2003. John Wiley & Sons. (*b*) Hollow nanofibers of titania obtained by calcination of PVP (containing titanium isopropoxide) coaxially electrospun with an oil core. Reprinted with permission from Li and Xia (2004). Copyright 2004. American Chemical Society.

reinforcement) tends to collapse into a flat fiber on removal of the core protein material (Jiang et al. 2005).

In electrospinning regular nanofibers it is the concentration of polymer and the applied electric field that primarily determines average fiber diameters. The same must hold true for at least the shell polymer in electrospinning a core–shell nanofiber. An additional factor, however, is a "die–swell" effect where the viscoelastic core fiber swells against the wet shell material, increasing the fiber dimensions. This was observed to occur with the core composition of fitc-BSA/PCL when spun with a PEG shell polymer. As the feed rate of the core polymer solution is increased, the average fiber diameters increased as well (Fig. 9.8). As the core material is not electrospinnable by itself, the swelling mechanism is likely to be responsible for this increase. Under the same experimental conditions, varying the feed rate of a non-viscoelastic core material (hexane) did not alter the fiber diameters (Song, T., et al. 2005).

Coaxial spinning can also be used to form polymer–particle composite nanofibers as discussed in Chapter 6. Li et al. (2005a), in their research on decorating the interior of hollow nanofibers with nanoparticles, used this technique very effectively. The core fluid used was a ferrofluid of magnetic iron oxide particles and the shell layer was a mixture of poly(vinyl pyrollidone)/titanium isopropoxide. Extraction of the core phase of the core–shell nanofibers with octane yielded hollow, magnetically susceptible nanofibers, with their interiors decorated with oxide nanoparticles.

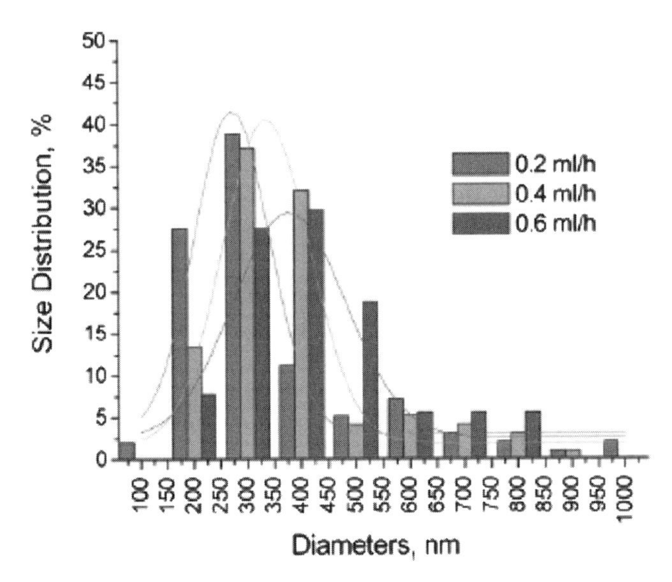

Figure 9.8 The effect of varying the feed rate of the core stream on the average fiber diameter of the core/shell nanofiber with a fitc–BSA/polycaprolactone core and a PEG shell. Reprinted with permission from Zhang et al. (2006c). Copyright 2006. American Chemical Society.

9.2.2 Core–Shell Geometry by Post-Treatment of Nanofibers

Core–shell nanofibers can also be made by coating the exterior of an electrospun nanofiber with a polymer solution or other material. This is carried out either by gas-phase or solution deposition (dip coating) on the exterior of electrospun nanofibers.

The TUFT (tubes by fiber template) process (first described by Bognitzky et al. 2000) is an example of a gas-phase technique where a wall material or a conformal coating is deposited on the nanofiber surface and the core fiber is subsequently removed to yield a tubular structure (Hou et al. 2002; Liu 2004). Mild coating conditions allows thin (as thin as 50 nm) conformal defect-free coatings to be applied in this manner without causing any thermo-oxidative degradation of nanofibers in the process (Hou et al. 2002). Protein delivery characteristics of PVA nanofibers were shown to be modified by applying a thin coating of poly(*p*-xylylene) (PPX), as already pointed out in Chapter 7 (Zeng et al. 2005a). Bognitzki et al. (2000) electrospun PLA nanofibers ($d = 300–3500$ nm) from CH_2Cl_2 solutions and deposited a layer of PPX on the nanofiber surface by chemical vapor deposition (CVD). This was converted into a hollow fiber by pyrolytic removal of the PLA core material at 250°C. Hollow tubular nanofibers with a thin metal coating (of Al or Au) either on the outside or the inner walls of fiber were obtained by combining this approach with physical vapor deposition. The TUFT approach was used by Ochanda and Jones (2005) to create metal shell nanofibers. A thin metallic shell of Au, Cu, or Ni was then deposited on the fiber by electroless plating with solutions of the metal salt and a reducing agent. On pyrolysis, the shell of the core–shell nanofibers was converted into metal nanotubes ($d = 450$–730 nm) with a wall thickness of 50–150 nm. The metal was in polycrystalline form with a particle size in the range of 5–25 nm. Other related coating methods such as physical vapor deposition (PVD; Wei et al. 2006a; Liu et al. 2002), plasma-enhanced PVD (PEPVD; Buldum et al. 2005), initiated CVD (aluminum coating on poly(*m*-phenylene isophthalimide) nanofibers; Liu et al. 2002) have been used to deposit thin coating layers of metal on the nanofibers. Initiated CVD has been used to render PCL nanofibers more hydrophobic by coating them with a polymerized perfluoroalkyl ethyl methacrylate layer (Ma, M. L., et al. 2005b). To ensure that coatings adhere well, nanofiber surfaces may be plasma treated prior to coating as in the case of P(LLA-CA 70 : 30) copolymer nanofibers coated by chitosan (He et al. 2005).

Simple dip coating has also been used with nanofibers. For example, electrospun PLA nanofiber mats were coated with a 1–4% solution of polyamic acid followed by heat treatment at 150–285°C to convert the coating into a polyimide (Bognitzki et al. 2000). Dip coating has also successfully been used to convert an organic nanofiber into a hollow inorganic nanofiber. PLLA nanofibers electrospun from CH_2Cl_2 solutions were soaked first in a metal oxide precursor solution of titanium isopropoxide and then in isopropanol/water solutions to convert the alkoxide into titania. The core PLLA was removed by calcination to obtain hollow titania nanofibers (Caruso et al. 2001). Others have combined physical coating from solution with electrostatic layer-by-layer (LBL) deposition (Ding et al. 2005b) to construct multilayer core–shell fibers of poly (acrylic acid) (PAA)/titania layers for sensor applications. The layer-by-layer technique was also used to deposit a layer of fluorescent probe molecules onto the surface of electrospun cellulose acetate nanofibers. These nanofibers were shown to be effective in detecting ppb levels of methyl viologen and cytochrome C in aqueous solution (Wang, X. Y., et al. 2004). Coating ionically functionalized nanofibers in liquid phase is particularly facile because of the electrostatic interactions (Drew et al. 2005).

Dip coating while simple and cost effective does not always lead to a continuous uniform shell layer as obtained in coaxial electrospinning. A comparison of the relative effectiveness of collagen/PCL core-shell nanofiber mats and collagen-coated PCL nanofiber mats as scaffolding for human dermal fibroblasts (HDF) was carried out by Y. Z. Zhang et al. (2005c).

The results showed significantly reduced density of cells attachment on the dip-coated nanofiber mat relative to that on core-shell nanofibers.

9.3 HIGHLY ALIGNED NANOFIBER MATS

Chaotic movements of the electrospinning jet typically result in a fiber mat with sections of nanofibers randomly placed within it. However, highly aligned nanofibers and layered mats where different layers have different fiber alignment are desirable in a variety of applications. These include nanofiber-reinforced polymer composites, nanofibrous scaffolding where tissue growth is guided by the directionality of the fibers, and in some sensor applications where a high degree of fiber alignment enhances performance. Layered constructs of fiber mats with each layer showing a different angular fiber alignment are used in the design of fiber-reinforced composites (Lee, C. H., et al. 2005). With human ligament fibroblast (HLF) growing on polyurethane nanofiber scaffolds, for instance, not only did cells growing on the aligned mats produce relatively more collagen, but the proliferating cells were found to arrange themselves in the direction of the alignment of nanofibers (Lee, C. H., et al. 2005). Alignment of nanofibers in mats has also been shown to guide cell alignment of meniscal fibroblasts and organization of actin filaments in mesenchymal stem cells growing on PCL nanofibers (Li et al. 2007). The same was shown to be true of human coronary artery smooth muscle cells (SMCs) growing on P(LLA-CL) nanofibers (Xu et al. 2004a) and for neural stem cells (NSCs) growing on PLLA nanofibers (Yang, F., et al. 2005). Recently, a scaffolding study investigated if alignment in PCL nanofiber mats would better augment matrix content and organization, leading to improved mechanical integrity of the scaffold (Baker and Mauck, 2007). A significant increase in the modulus of these constructs (by a factor of $\sim$7) was observed for meniscal fibrochondrocytes (MFCs) or mesenchymal stem cells (MSCs) growth. Also, these constructs yielded a relatively greater amount of extracellular matrix (on a par with that formed by native MFCs).

Mechanical properties of aligned nanofiber mats tend to be much higher in comparison to those for random mats. With electrospun polyurethane mats the Young's modulus of comparable aligned and random mats were reported to be 3550 kPa and 1630 kPa, respectively; the corresponding values for tensile strength were 3520 kPa and 1130 kP, respectively when tested at a strain rate of 1.25%/s (Lee, C. H., et al. 2005). Similar results were reported with collagen nanofibers (electrospun from HFP solutions; Matthews et al. 2002); the modulus of mats measured in the direction of alignment was 52.3 ± 5.2 MPa, while that in the cross direction was 26.1 ± 4.0 MPa.

Several techniques for obtaining highly aligned nanofibers have been described in the literature, but not all of them are readily amenable to scale-up.

9.3.1 Parallel Electrode Collector

A simple means of aligning nanofibers is to collect them on a grounded pair of electrodes on a nonconducting surface.[4] For example, a pair of conducting silicon strips or gold electrodes patterned on a quartz surface might be used as the collector. Li and colleagues showed that PVP nanofibers of diameter exceeding $\sim$150 nm could be deposited as a highly aligned mat in the void space between a pair of such electrodes (Li, D., et al. 2003b). The electrodes were placed on an insulating substrate having a bulk resistivity[5] of 1×10^{22} Ω.cm. In the vicinity of the collector, the electric field is split into two and directed at each of the electrodes. Depending on the instantaneous position of the whipping, undulating jet surface, it is attracted first to the electrode closest to the jet segment and is then stretched rapidly to the adjacent electrode, placing it in perpendicular alignment with it (Li, D., et al. 2003b). As polymer nanofibers have poor electrical conductivity, the sections of the fiber in the gap between the electrodes lose their residual charge very slowly (Liu and Hsieh 2002). The fibers therefore tend to repel each other and their tautness as well as their parallel placement between the collector electrodes is believed to be due, at least in part, to this surface charge (Li, D., et al. 2003b). This is consistent with the observation that the fraction of aligned nanofibers in the mats collected on parallel electrodes increases with the collection time (or the duration of electrospinning). The residual charge on the nanofibers in the vicinity of the electrodes inducing an opposite charge on the electrode surface and contributing to the attraction of the fiber towards them has been suggested (Li, D., et al. 2004c). Thinner nanofibers (less than $d \sim 150$ nm in the case of PVP) were too fragile to be oriented in this manner. The sequence of events is schematically depicted in Fig. 9.9. The fibers tend to orient along a direction that minimizes the net torque of electrostatic forces in the segment of the fiber (Li, D., et al. 2005a).

[4]With point-plate electrospinning equipment oriented vertically, the collector plate is horizontal and the grounded parallel electrodes of interest are placed on the collector itself. However, it is also possible to collect aligned nanofibers on a parallel pair of stainless steel wires placed vertically within the gap (Chuangchote and Supaphol 2006).

[5]The resistivity of the substrate appears to be important to obtain good alignment; using a glass substrate of resistivity 1×10^{12} Ω/cm in this experiment yielded a random isotropic mat!

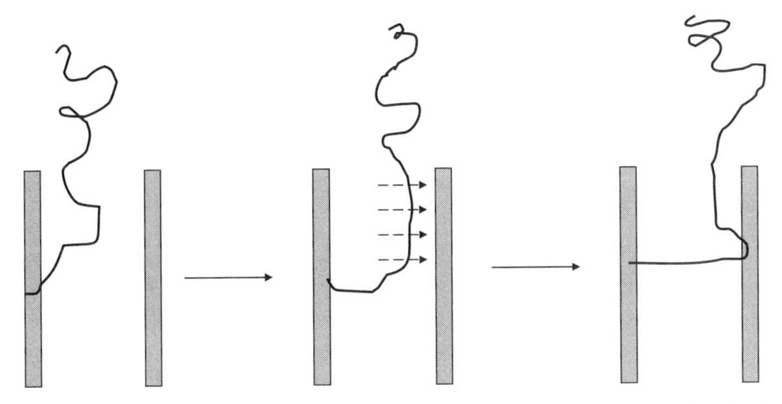

Figure 9.9 Stretching of a charged jet segment across a pair of parallel electrodes. Redrawn after Jalili et al. (2006). Copyright 2006. John Wiley & Sons.

As demonstrated by D. Li et al. (2003b, 2004c), the approach is easily extended to a set of four or six electrodes, and alternate pairs of these might be grounded at different times during electrospinning to change the direction of fiber alignment. Hierarchical structures consisting of multilayered mats with separate nanofiber layers aligned in different directions were collected in the void space between grounded electrodes (Fig. 9.10). Such mats can in principal serve as very effective reinforcement in composite materials, but this particular technique is not well suited for making mats that are more than 1–2 cm^2 in area. A variant of this approach involves the use of a static (or rotating) grounded metal frame as the collector. With a rectangular frame several centimeters in width, small samples of highly aligned nanofibers might be conveniently electrospun (Fig. 9.10). Also, when the fibers are of PAN, the aligned fiber assemblies might be thermally converted into carbon fiber mats that retain the same alignment (Li, D., et al. 2003b).

Teo and colleagues found knife edges to be more effective than parallel electrodes in collecting highly aligned nanofibers (Teo and Ramakrishna 2005) and demonstrated a marked difference in nanofiber deposition on a grounded knife-edge as opposed to one carrying a negative bias. In the latter case the fibers tended to aggregate relatively densely in the middle of the collection field (i.e., along the axis of the knife edges).

9.3.2 Rotating Cylinder Collectors

Rotating drum or mandrel collectors, described by Formhals back in 1934, are perhaps the most popular arrangements for generating aligned fiber mats in

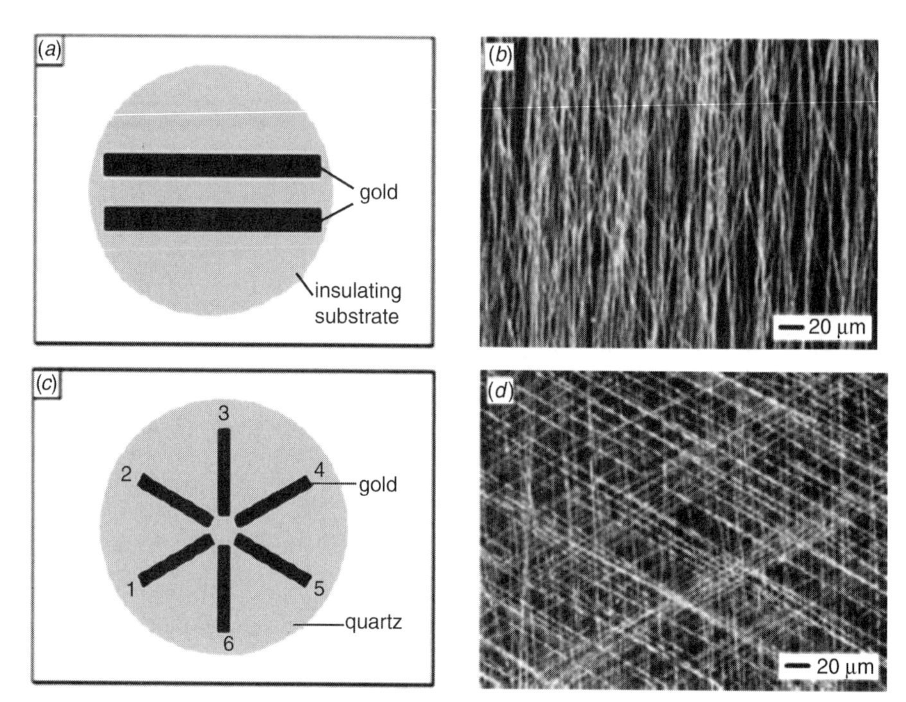

Figure 9.10 (*a*, *b*) and The alignment of PVP nanofibers collected between a pair of gold electrodes on an insulating surface. (*c*, *d*) A six-electrode assembly and a three-layered construct of aligned nanofibers collected at their center by sequentially grounding pairs of the electrodes. Reprinted with permission from Li, D., et al. (2004c). Copyright 2004. John Wiley & Sons.

the laboratory[6] (Boland et al. 2001; Li, W.-J., et al. 2007; Lee, C. H., et al. 2005; Matthews et al. 2002). A segment of the whipping charged jet contacting the rotating drum surface, attaches to it electrostatically. This allows the trailing section of the jet to straighten out, straightening its spiraling trajectory and facilitating fiber alignment as well as extension by drawing. Courtney and colleagues, using a rotating mandrel 4.5 inches in diameter, found electrospun poly(ester urethane) urea to yield isotropic nanofiber mats at low tangential velocities (0.3–1.5 m/s) of the drum, but aligned anisotropic mats at velocities >3.0 m/s (Courtney et al. 2006). The fiber alignment distribution $R(\theta)$ function[7] has an impact not only on the strain energy of the mat but also the crystallinity of nanofibers and therefore on its mechanical properties. Figure 9.11 shows the increase in alignment as a function of the mandrel

[6] A disk collector rotating at 400–500 rpm was used to collect tangentially oriented chitosan/PEO nanofibers by Subramanian et al. (2005). Chitosan being a difficult polymer to spin even in a blend, a mat of thick oriented fibers ($d \sim 3$ μm) bridged by numerous fine unoriented fibers, was obtained.

[7] $R(\theta)$ is normalized so that $\int R(\theta)\,d(\theta)$ (from $-\pi/2$ to $\pi/2$) = 1.

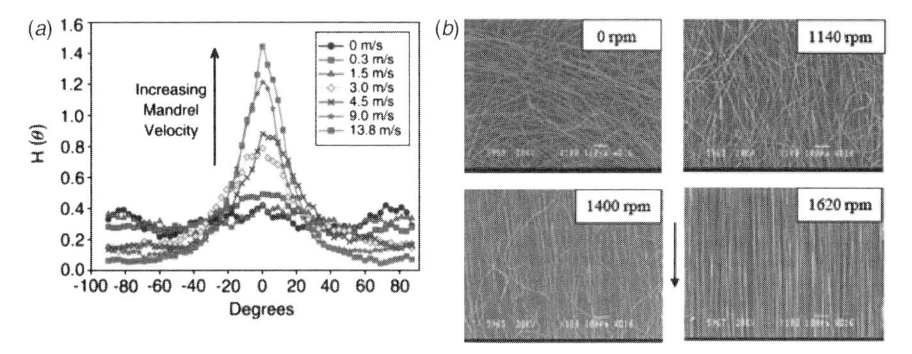

Figure 9.11 (*a*) Increase in fiber alignment with mandrel velocity in electrospinning poly(ester urethane) urea (PEUU) from 5 wt% HFP solution. Reprinted with permission from Courtney et al. (2006). Copyright 2006. Elsevier. (*b*) Increasing alignment of PS ($M_w = 25{,}000$ g/mol) nanofibers as the rotation speed of the mandrel is increased (the linear speed of the rotating drum was in the range of 2.5–3 m/s). (B. Sundaray, PhD Thesis, Department of Physics, Indian Institute of Technology, Chennai, India, 2006).

velocity. When the velocity of the drum was increased from 0 to 13.8 m/s, the alignment of nanofibers collected on it increased from 37% to nearly 100%. Also, the crystallinity of the nanofibers increased at the higher degrees of alignment. At very high rpm, however, fibers may tend to break due to excessive elongational stresses imposed on them (Zussman et al. 2003). Electrospinning PAN fibers from a 10% DMF solution, Fennessey and Farris (2004) also reported an increasing degree of fiber alignment with the speed of rotation. At a rotational velocity of 9.84 m/s they obtained nanofibers with a maximum orientation parameter of 0.23 (as estimated by FTIR spectroscopy). An "open" frame rotating drum as opposed to a solid one may also be used as the collector, with the nanofibers aligned in the gaps in the wire frame of the drum as with nylon electrospun from formic acid (Katta et al. 2004). As the removal of residual charges is less efficient with an open drum compared to a solid one, mat thicknesses likely increases (but mat density decreases) due to the accumulation of charge on the nanofibers.

Tubular nanofiber constructs have been made with the rotating mandrel technique for several different polymers (Teo et al. 2005). These include scaffolding P(LLA-CL) (75 : 25) (Inoguchi et al. 2006; Mo and Weber 2004), P(LL-G) for *in vivo* nerve regeneration (Bini et al. 2004), PLA/collagen for fibroblast growth (Kitazono et al. 2004), and segmented polyurethane and multipolymer layered constructs (Kidoaki et al. 2005).

In developing electrospun artificial blood vessel scaffolds (Boland et al. 2004a; Telemeco et al. 2005; Xu et al. 2004a), obtained tubular three-dimensional structures with a circumferential orientation of nanofibers.

This helped guide the proliferation of seeded cells and also improved the structural integrity of the scaffold. Electrospinning collagen solutions in HFP onto rotating mandrels, Matthews et al. (2002) found random alignment at low rpm ($\sim$500), but preferred alignment of fibers at 4500 rpm (surface velocity 1.4 m/s), that was reflected in the mechanical properties of the mats. Placing an auxiliary knife-edge electrode below the rotating mandrel, Mo and Weber (2004) obtained nanofibers with a high degree of circumferential alignment, even at the lower speed of rotation of 830 rpm. A two-layered tubular scaffold of PCL/PLA nanofibers has been fabricated using a rotating mandrel collector (6 mm outer diameter and 6 cm in length) (Vaz et al. 2005). A randomly aligned PCL nanofiber mat was first spun (from a 12.5% solution in $CHCl_3$) onto the mandrel, followed by a second layer of highly aligned PLA nanofibers (spun from a 14% solution in $CHCl_3$: DMF (15 : 3)). The PLA layer was spun at a higher speed of rotation to effect circumferential orientation of the nanofibers. The Ramakrishna group used an aluminum grid electrode (maintained at −8 kV and placed 8 cm below a rotating Teflon mandrel) to assist in the alignment of fibers (Huang et al. 2003). A comparison of nanofiber mats collected with and without voltage on the grid electrode showed substantial alignment only in the former case. The finding is also consistent with the patent issued to Bornat et al. (U.S. patent # 4,689,186, 1987). Placing a set of auxiliary knife-edge electrodes carrying a charge opposite to that of the capillary tip below the mandrel also helps alignment.

High degrees of fiber alignment have also been obtained by collecting electrospun nanofibers on a liquid surface, and the resulting fiber bundle drawn off as a yarn by a motorized take-up roller operated at 0.05 m/s (Smit et al. 2005). The drawing process results in substantial fiber alignment.

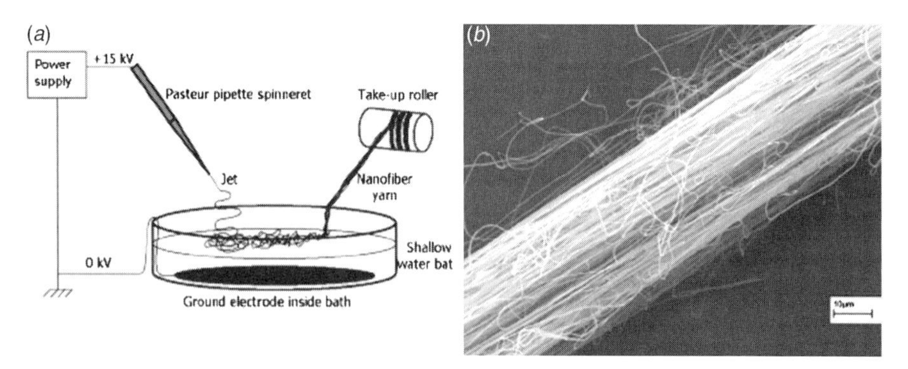

Figure 9.12 (*a*) Electrospinning continuous yarn of nanofibers by collection in a water bath with an immersed grounded collector followed by drawing out the nanofiber mat as a yarn. (*b*) An image of the resulting highly aligned yarn of polyacrylonitrile (PAN) nanofibers electrospun from 5% DMF solution. Reprinted with permission from Smit et al. (2005). Copyright 2005. Elsevier.

All three polymers processed in this manner (PVDF, PVAc, and PAN) yielded highly aligned nanfiber yarn (see example in Fig. 9.12). About 3700 nanofibers per cross-section of yarn and a single capillary tip throughput of 180 m yarn per hour was estimated for this process. Although some SEM images of the yarn show high levels of beads, the nanofibers within them are still highly aligned for the most part. Nanofiber yarn was also produced in a recent study using two grounded rotating disc collectors placed perpendicular to each other (in X-Y and X-Z planes). Nylon nanofibers (with 1 wt% MWCNT) was electrospun from 98% formic acid solution (20 wt%) on to the collectors where the first rotating disc twisted the fibers into a yarn that was collected on the second disk.

9.3.3 Chain Orientation During Fiber Alignment

An electrospun random mat of polymer nanofibers that is isotropic with respect to fiber alignment may still show a high degree of molecular-level orientation within the individual fibers. Chain orientation is purely a geometric descriptor of all or some of the polymer chains in the fiber. Oriented polymer chains display a preferred direction that forms an angle with the fiber axis. Orientation is a consequence of the rapid extension of the fiber that results in chain ordering (as with "cold drawing" in conventional fiber drawing processes). Cold drawing, however, also results in the formation of voids in the "neck" area of the material (often rendering the neck opaque during fiber necking) while drawing of the viscous polymer solution in electrospinning does not result in comparable void formation. Also, the whipping instability results in draw ratios that are several orders of magnitude higher than the maximum values encountered in drawing of fibers in conventional processing. Orientation in cold drawing also involves the deformation of crystalline morphologies, with the crystalline lamella breaking up to reform as longer thinner lamellae. Drawing viscous solutions does not involve these complex processes, as crystalline morphology is absent in the spinning jet.

Chain orientation invariably results in stronger covalent interactions between polymer chains in the draw direction along the chain axis. The resulting anisotropy in fiber properties can be easily demonstrated using polarized infrared spectroscopy and optical birefringence measurements and is also apparent from modulus measurements.[8] Infrared dichroism (Fennessey and Farris 2004) and X-ray diffraction (Fang et al. 2004; Fong and Reneker 1999; Gu et al. 2005c) have been used extensively to study chain orientation

[8] At moderate strains the modulus of an isotropic fiber mat of pellathene (polyurethane) nanofibers electrospun from 7% DMF was higher than that of the bulk polymer (Pedicini and Farris 2003). Similar data were reported for PAN nanofibers (Fennessey et al. 2006).

in highly aligned electrospun nanofibers. Aligning nanofibers using any of the described techniques invariably contributes to even greater stretching of nanofibers over and above that obtained in conventional electrospinning. This additional macroscopic strain in the fibers contributes to additional macromolecular orientation.

Therefore, fiber alignment is generally accompanied by changes in the crystalline morphology of the fiber. Electropsun nylon-6 (from 7.5 wt% HFP solutions) nanofibers can be aligned using an oscillating grounded metal frame as the collector (Fong et al. 2002). X-ray diffraction patterns of the aligned fiber mats showed the presence of γ-phase crystallites in the nanofibers; the crystallinity of the fibers was in the range 0.24–0.28, somewhat lower than for the comparable film samples where a value of 0.36 has been reported (Hong, K. H., et al. 2005). Given the different rates of processing or drying in films and fibers, significant differences in crystallinity are to be expected.

9.3.4 Infrared Dichroism

Pedicini and Farris (2003) used the dichroic ratio of the —NH stretching band ($3320\ \mathrm{cm}^{-1}$) in polyurethane nanofibers electrospun from 7% DMF to estimate the orientation function f for nanofibers. Herman's function, f, varies from $f = 0$ (isotropic) to $f = 1$ (perfect chain orientation).[9] The value of f is related to the average angle θ made by the oriented chain axis to that of the fiber (Fig. 9.13):

$$f = (3\langle\cos^2\theta - 1\rangle)/2. \tag{9.1}$$

The dichroic ratio D is the ratio of the absorbance of polarized light measured with the electric vector direction of the polarizer oriented parallel to the fiber draw direction, $A_{\parallel}$, and that corresponding to the perpendicular orientation, $A_{\perp}$. (i.e., $D = A_{\parallel}/A_{\perp}$). Then, f can be expressed in terms of D (Pedicini and Farris 2003) as follows:

$$f = (3\langle\cos^2\theta - 1\rangle)/2 = (D - 1)(D_0 + 2)/(D_0 - 1)(D + 2), \tag{9.2}$$

where θ is the angle of the chain segment relative to the fiber axis and D_0 is the dichroic ratio of the perfectly oriented fiber. This is a versatile technique in that dichroism can even be assigned to different groups of repeat units or even specific side chains in the polymer. D can also be measured

[9]Herman's function is a mathematical description of the degree of orientation of the axis of polymer chains within the fiber relative to some other axis of interest (in this case the fiber axis).

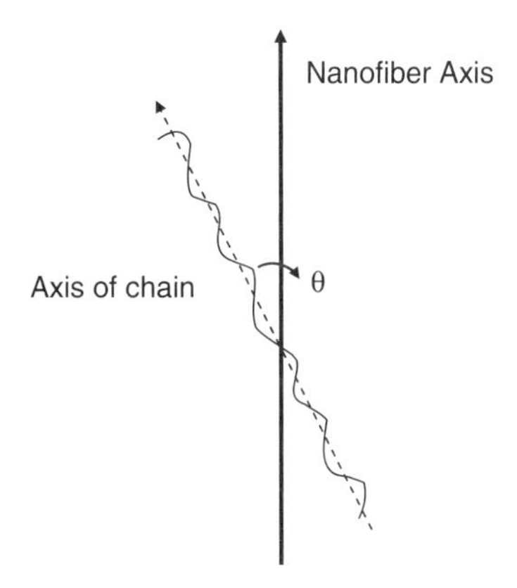

Figure 9.13 Definition of angle θ.

using ultraviolet or visible light and can be useful in studying polymers with the appropriate chromophoric groups. X-ray diffraction techniques are particularly useful in studying the crystalline morphology of polymers and have been extensively used in nanofiber characterization. The application of these technique to nanofibers was discussed in Chapter 5 on characterization. Wide angle X-ray diffraction (WAXD) studies have been particularly useful in establishing the lack of significant crystalline content in electrospun nanofibers of even the semi-crystalline polymers (Deitzel et al. 2001a; Liu et al. 2000).

Both techniques (infra-red dichroism and WAXD) have been used to characterize electrospun aligned polyacrylonitrile (PAN) nanofibers. With PAN nanofibers, a high degree of chain orientation is often desirable as the mechanical integrity of carbon nanofibers derived from these (via pyrolysis) can be influenced by the orientation. Isotropic random mats of PAN electrospun from 15 wt% in DMF solution showed no optical birefringence. However, aligned fibers collected using a rotating drum (surface velocity 2.5–12.3 m/s) were birefringent under cross-polars and the dichroic ratio D associated with the —CN stretching vibration also decreased, indicating increased orientation. The maximum value of f calculated was 0.2 (Fennessey and Farris 2004) for the fibers collected on a drum surface moving at 8.1–9.8 m/s. As conformational orientation is essentially a consequence of fiber drawing, as orientation increases so the fiber diameter decreases. Liquid-crystalline poly(hexyl isocyanate) nanofibers electrospun from $CHCl_3$ solutions (5–25 wt%) showed a banded structure reflecting local organization of the chains. The X-ray diffraction data indicated two equatorial peaks; a strong peak at $2\theta = 16^\circ$ corresponding to 1010 reflection (a d-spacing of 5.3 A°), in particular, indicated the high degree

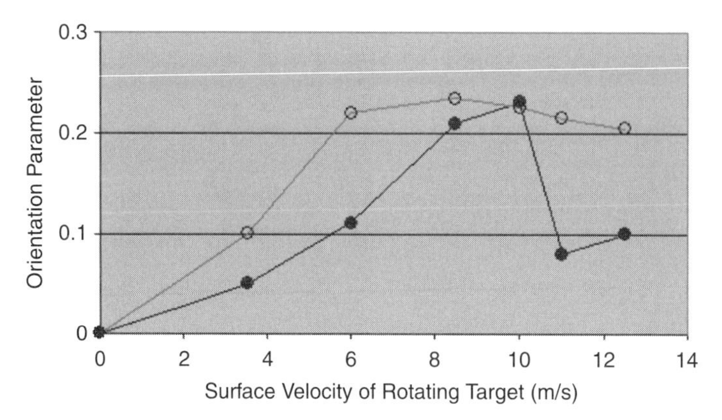

Figure 9.14 Change in chain orientation parameter with the speed of rotation of the cylindrical target. The chain orientation parameter from dichroism measurements (filled symbols) and from the 5.3 Å equatorial peak in WAXD measurements (open symbols) are shown in the figure. Redrawn from Fennessey and Farris (2004).

of orientation in the nanofibers. As expected, the higher take-up speed of the rotating collector resulted in a higher degree of alignment as well as chain orientation.

Correlation between the fiber diameter and the development of chain orientation for PHIC nanofibers was also established by Lin and Martin (2006). The relationship between the velocity of the rotating surface and the chain orientation parameter for PAN nanofibers is shown in Fig. 9.14 (Fennessy and Farris 2004). As the alignment varies with drum velocity, this qualitatively illustrates the relationship between chain orientation and fiber alignment. Others (Inai et al. 2005b) have shown that changing the velocity of the rotating collector drum from 63 m/min to 630 m/min results in a qualitative increase in the arc of diffraction in WAXD data, again suggesting a moderate increase in the orientation in PLLA nanofibers.

9.4 MIXED POLYMER NANOFIBERS AND NANOFIBER MATS

In designing nanofiber mat constructs for specific applications, the engineering as well as surface-chemistry requirements cannot sometimes be satisfied by a single material. For instance, an application may need good nanofibers of magnetic susceptibility as well as enzyme activity via surface-grafted bioactive moieties. In such instances composite mats made of either layered or mixed nanofibers of two or more different polymers might be employed.

Sequential electrospinning of several nanofiber mats on top of each other on a single grounded target is the simplest (layer by layer) approach to fabricating a multicomponent mat. Sequential deposition of different layers on a rotating mandrel has also been attempted to obtain tubular constructs where the wall composition (from inside to the outside of the tube) was PEO/collagen I/SPU polymer (Kidoaki et al. 2005). However, the success of the process was limited due to adhesion problems between the PEO and collagen mats. A hierarchical layered structure with layers differing in polymer types, average fiber diameters, as well as the degree of alignment may be constructed using this approach. Kidoaki et al. (2005) sequentially electrospun three polymers—segmented polyurethane (SPU) from a 12.5% solution in THF, styrenated gelatin (S-GT) from a 10% solution in HFP, and Type I collagen 5% solution in HFP—to obtain a layered composite mat. Although the flow rates used with all solutions were about the same, both the concentration and the spinning voltage were varied, resulting in different layer thicknesses in the construct. Dyes dissolved in the different polymers allowed laser confocal microscopy to be used to visualize the layers of fiber and observed a transition zone between them. The SPU layer was intended as the structural component, the photocrosslinked GT provided a good drug delivery matrix, and the collagen layer afforded improved cell adhesion.

Electrospinning of two or more polymers simultaneously from two different capillary tips onto the same collector (mixed electrospinning) yields a mat of intimately mixed nanofibers. As fibers from adjacent tips will otherwise be collected on adjacent regions on a rotating mandrel, the mandrel collector is rotated and transversely moved to obtain the mixing of fibers into a single mat. For instance, SPU and PEO have been electrospun (Kidoaki et al. 2005) into a mixed nanofiber mat by concurrent electrospinning from a pair of capillary tips. As the nanofibers spun from both tips have the same surface charge, they do not intermix particularly well in the mat. The same technique (with the modification that capillary tips were moved transversely instead of the collector) was used to make mixed nanofiber mats of cellulose acetate (CA) (in 10% solution in acetone/DMAC, 2/1) mixed with poly(vinyl alcohol) (PVA) (in 10% solution in water) (Ding et al. 2004c). The same approach was also used to fabricate mixed mats of regenerated chitin with PLGA (Min et al. 2004d). In this instance, four capillary tips arranged in a row were used and the approximate compositional ratio of CA : PVA in the mats was altered by selecting the number of tips assigned to each polymer solution (for example, two for PVA and two for CA yielded a weight ratio of 51 : 49 for PVA : CA in the mat). Mixed mats of chitin (formic acid solution) with PLGA copolymer (LA : GA 50 : 50) in HFP were also produced by simultaneous electrospinning from two tips maintained at the same polarity. Chitin (3–17 wt% of regenerated chitin in formic acid) did not electrospin as a continuous fiber, but yielded nanosized particles embedded in the PLGA fiber mat (Min et al. 2004d).

Electrospinning two separate polymer solutions from a single tip was also shown to be feasible by Gupta and Wilkes (2003), who used a compartmentalized syringe holding two polymer solutions, each connected to a separate Teflon needle. The pair of needles was adhered together and to allow a single droplet to form at their tip. Nanofibers of PVC/PVFD blends and PVC/SPU were prepared using this technique. With this arrangement, using short gap lengths (of <9 cm when using 14 kV and 3 mL/h flow rate with 20–25% solutions of polymer) resulted in two separate Taylor's cones (each corresponding to a single polymer) emanating from the droplet and spraying a pair of adjacent fiber mats on the collector. However, at larger gap lengths only a single cone was observed and some amount of mixing of components into single nanofibers was noticeable from energy-dispersive spectroscopic measurements. This tip geometry differs considerably from that used in core–shell spinning discussed earlier in that the liquid columns are not concentric and are charged using separate electrodes. A silicone microfluidic capillary tip assembly with three capillary channels ($d \sim 630$ μm internal diameter) that similarly accommodates two adjacent streams of polymer solutions has also been described (Lin et al. 2005b) (Fig. 9.15). The two polymer solutions exhibited laminar flow in the outlet channel with little or no dispersion. Self-crimping bicomponent nanofibers of PAN and elastomeric polyurethane (PU) have been prepared using this technique. Existence of the side-by-side bicomponent structure in the fiber was demonstrated by the removal of one polymer by solvent extraction, revealing a U-shaped cross-section of the extracted nanofiber.

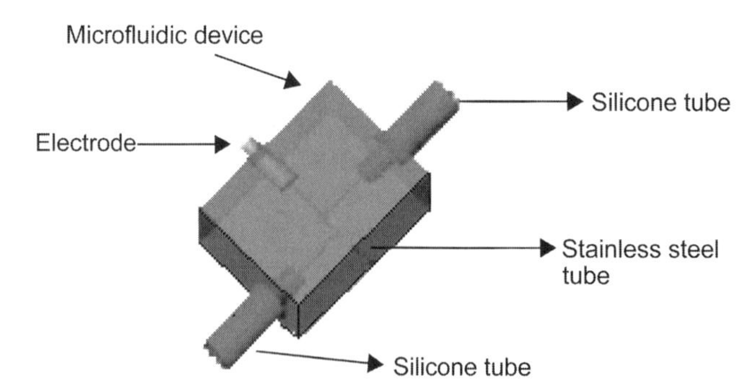

Figure 9.15 Schematic diagram of the microfluidics spinneret used by Lin et al. (2005b). Redrawn from Lin et al. (2005b).

9.5 CROSSLINKED NANOFIBERS

A crosslink covalently links polymer chains and is therefore a site from which three or more chains emanate.[10] With enough of the polymer chains in a sample linked to each other in this manner, the loose assembly of chains is converted into a network structure or a gel and is said to be "crosslinked." When crosslinked to an extent referred to as the "gel point," all the polymer chains in the sample are interlinked and the polymer is in effect a single giant macromolecule. At levels of crosslinking below the gel point the system is essentially a highly branched polymer. Polymer networks are usually characterized in terms of their crosslink density (number of crosslinks per unit volume) and the average functionality at the crosslink sites. (Other measures such as the cycle rank of networks and the average molecular weight between crosslinks are also used.) In nanofibers, crosslinking is important for at least two reasons:

1. Crosslinking renders the nanofibers insoluble in all solvents. When a crosslinked fiber mat is placed in a good solvent, it absorbs the solvent and swells rather than dissolving in it. In filters, sensors, and biological applications such insolubility can be desirable.
2. The strength and modulus of nanofiber mats increase significantly on crosslinking. With the mechanical integrity of mats being generally poor (nanofibers being somewhat fragile), crosslinking can often provide a means of strengthening the material. Also, this leads to better dimensional stability for the fiber, with the fibrillar morphology being better retained during its service life.

Random crosslinking is conveniently carried out by either heating or exposing to UV radiation a polymer that contains a small amount of a crosslinking agent (or initiator). Free-radical initiators may be used as crosslinking agents. The resulting crosslink density (crosslinks per unit volume) is determined by the weight fraction of the initiator or the agent used. Peroxides or azo compounds are commonly used as radical initiators in crosslinking bulk polymers and the same approach has been used with nanofibers (e.g., PET nanofibers crosslinked using azides; Baker and Brown 2005).

9.5.1 Photocrosslinked Nanofibers

Methacrylate-modified elastin-mimetic proteins were crosslinked by visible light irradiation of the preformed nanofibers to yield water-insoluble

[10]Some authors consider four chains (as opposed to three chains) to emanate from a crosslink, regarding a trifunctional crosslink as merely a junction point.

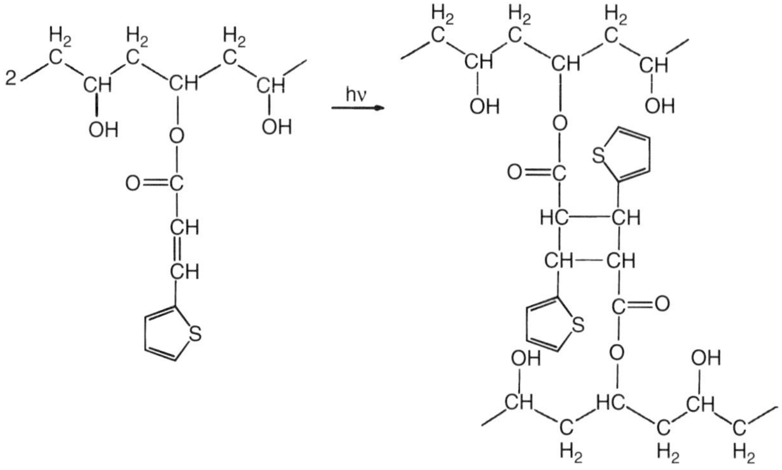

Scheme 9.1 Crosslinking reaction via unsaturation in the side chain.

mats of greatly improved mechanical integrity (Nagapudi et al. 2002). The crosslinking reaction in this case was confirmed using solid-state C-13 NMR spectroscopic analysis by observing the decreasing signal intensity for the unsaturated carbons in the methacrylate moieties in the crosslinked nanofibers. PVA modified by reaction with thienyl acryloyl chloride can also be similarly crosslinked by UV radiation ($\lambda \approx 310$ nm) to obtain a nanofiber mat that is insoluble in water or steam (Zeng et al. 2005b). The photocrosslinking reaction involved is shown in Scheme 9.1.

Poly(vinyl cinnamate) (PVCi) nanofibers (and those of PVCi/PHBV blends) electrospun from chloroform solutions also undergo facile crosslinking

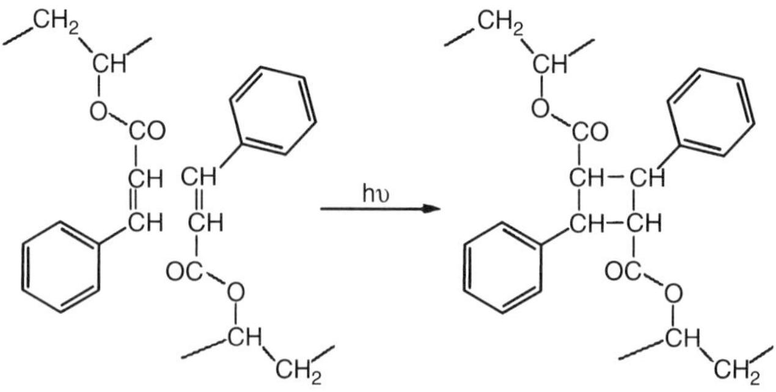

Scheme 9.2 Photocrosslinking of the pendent vinyl cinnamate moiety on copolymers.

on exposure to UV-B radiation as shown in Scheme 9.2 (Lyoo et al. 2005). In this case, the crosslinking was carried out in a post-processing step with the preformed nanofibers irradiated by UV radiation. However, this reaction is rapid enough to allow crosslinking of the jet during the short duration of exposure (estimated to be only about $\sim$0.2 s) on its passage from the capillary tip to the grounded collector. Gupta et al. (2004) demonstrated concurrent crosslinking during electrospinning for the first time using this chemistry. Poly(methyl methacrylate-*co*-2-hydroxyethyl acrylate) (85/15) [P(MMA-*co*-HEA)] copolymers were functionalized with the cinnamate functionality by esterifying the pendent —OH of the HEA repeat units with cinnamoyl chloride. Electrospinning was carried out using 20 wt% solutions of the copolymers in DMF, with an applied voltage of 15 kV and a gap distance of 20 cm, using a Teflon tip as the spinneret. A UV-B lamp (0.135 W/sq.cm intensity) was used to irradiate the electrospinning jets of several of these copolymers containing at least 4% cinnamate functionalization to obtain crosslinked, gel-containing nanofibers. Crosslinking was confirmed using IR spectroscopy showing the decreased intensity of the vinylene stretching band.

With particularly fast polymerization reactions, electrospun oligomeric solutions can be designed to polymerize (not merely crosslink) during passage of the jet from the capillary tip to the collector; polymerization during electrospinning is also a form of reactive electrospinning. Poly(2-hydroxyethyl methacrylate) (PHEMA), for instance, can be reactively electrospun in this manner, as demonstrated by H. S. Kim et al. (2005). A mixture of HEMA and ethylene glycol dimethacrylate (EGDMA), 2,2′-azobis(isobutyronitrile) (AIBN), and a photo-initiator constituted the electrospinning solution (a low viscous liquid with no polymer and hence not electrospinnable). The mix was heated prior to spinning to initiate polymerization to obtain a viscous prepolymer and the solution cooled to quench the process when the required viscosity was reached. This prepolymer solution was then electrospun and the resulting jet irradiated with a 200 W Hg vapor lamp during spinning. Insoluble nanofibers ($d \approx 100$–500 nm) that displayed enhanced mechanical properties were obtained.

9.5.2 Crosslinking Agents

A crosslinking agent can be electrospun along with the polymer and subsequently activated by heat or UV–visible radiation, or alternatively the mats can be immersed in a crosslinking agent to effect curing. For example 2–8 wt% of glyoxal (OCH—CHO) was added to PVA solution (7–15 wt%

in water[11]) and phosphoric acid was added to lower the pH value to 2–3. Nanofibers electrospun from this solution were heated at 120°C for a period of 2–30 min to crosslink and insolubilize the nanofibers (Ding et al. 2002a). Alternatively, the electrospun PVA nanofiber mats could be crosslinked by immersing them for 24 h in a solution of gluteraldehyde (GA) in 0.01 N HCl in acetone (Wang, X. F., et al. 2005).

Gluteraldehyde (GA) is a particularly versatile crosslinking agent used to crosslink a number of different types of polymer nanofibers. Reaction of PVA mats with GA could also be conveniently carried out in the vapor phase (Wu, L. L., et al. 2005) by exposing the mats to saturated GA vapor. After exposure, the mats were rinsed with a suitable solution (e.g., glycine) to remove any unreacted GA. This is particularly important in scaffolding studies where even low levels of GA residues can affect cell proliferation rates (Zhang et al. 2006b). Vapor phase reactions have been used to crosslink PVA (Wu, L. L., et al. 2005), gelatin (Zhang et al. 2006b), blends of collagen with glycosaminoglycan (GAG) (Zhong et al. 2005), type I collagen (Rho et al. 2006), and poly(styrene-*co*-maleic anhydride) (Kim, B. C., et al. 2005) nanofibers. The extent of crosslinking obtained in each case depends on the levels of crosslinking agent used and the extent of reaction. Above an optimum level of crosslinking, however, polymers tend to be too brittle for use.

Li and Hsieh (2005a) reported the use of β-cyclodextrin (CD) as a crosslinking agent for poly(acrylic acid) PAA nanofibers. The —COOH groups in PAA are believed to condense into six-membered cyclic anhydride rings that react with the —OH functionalities in CD to yield water-insoluble ester groups. Others have shown an inorganic polyoxometalate compound, $H_3PW_{12}O_{40}$, at 20 wt%, that primarily acts via hydrogen-bonded interactions, to be an effective crosslinking agent for PVA (Gong et al. 2004).

Polyelectrolyte hydrogel nanofibers are of interest in biomedical and sensor applications because their degree of swelling is particularly sensitive to factors such as pH, temperature, ionic strength, and solvent polarity (Li and Hsieh 2005a; Seki and Okahata 1984). The kinetics of these responses are diffusion-controlled and sensitive to the available specific surface area of the material (Fei et al. 2002). Among the most widely studied are the polyelectrolyte gel formed when PAA is mixed with PVA in aqueous solution (Li and Hsieh 2005a, 2005b). The anionic polymer PAA undergoes copious hydrogen bonding between its —COOH repeat units and the —OH units of PVA, and the complex can be thermally esterified into a covalently bonded network. Several investigators have electrospun the PVA/PAA

[11]PVA is best dissolved in water by stirring in the powder at a temperature of 90°C for 6 h or more. Incomplete dissolution leads to invisible microgel particles in the solution.

system. The electrospun mats of the polyelectrolyte, heated at 120–140°C, resulted in crosslinking via esterification. Both the duration of heating (Jin and Hsieh 2005b) as well as the curing temperature (Li and Hsieh 2005a) have been used to control the crosslink density. In nanofiber mats, most of the crosslinking is intrafiber, as might be expected, but a significant amount of interfiber crosslinking has also been observed (Jin and Hsieh 2005b) at points where the nanofibers overlap.

APPENDIX I

ELECTROSPUN POLYMERS USED IN TISSUE ENGINEERING AND BIOMEDICAL APPLICATIONS

APPENDIX I: Electrospun Polymers Used in Tissue Engineering and Biomedical Applications

Polymer	Molecular Weight (M_w)	Solvent (Concentration)	d (nm)	Cell Type	Reference
Polyurethane (Pellethane 2102–75 A) Dow	—	DMF (20%)	657 ± 183	Human ligament fibroblasts.	Lee, C. H., et al. (2005)
Polyurethane nanofibers on PCL microfiber mesh	—	DMAC (20%)	—	Human umbilical vein endothelial cells.	Williamson et al. (2006)
Poly(ester urethane) urea PCL and BDI	230,000	HFP (5–12%)	<1000	Vascular smooth muscle cells.	Stankus et al. (2006)
	87,600	HFP (~5%)	941	Showed anisotropy in mats. No cell-culture studies.	Courtney et al. (2006)
Poly(ethylene terephthalate) surface grafted with gelatin	$\eta = 0.82 \pm 0.02$	TFA (0.2 g/ml)	200–600	Human coronary artery endothelial cells.	Ma, Z. W., et al. (2005d)
Poly(L-lactide-*co*-ε-caprolactone)	$M_n = 3.9 \times 10^4$	HFP (3%)	~700	No cell culture studies.	Inoguchi et al. (2006)
Poly(lactic acid)	—	HFP (8%)	246 ± 79	MC3T3-E1 osteo-progenitor cells.	Badami et al. (2006)
Poly(ethylene glycol) Poly(lactic acid) diblock copolymers	—	HFP (8–26%)	179–2100	No cell culture study.	Xu, X. L., et al. (2005)
PGA	—	HFP (67–143 mg/mL)	220–880	Rat cardiac fibroblasts, rat intra-muscular implantations.	Boland et al. (2004b)
PCL	85,000	$CHCl_3$: DMF	196–1297	No cell culture studies.	Bölgen et al. (2005)

PCL	80,000	MeOH : $CHCl_3$ (1 : 1)	250	Cardiomyocytes from neonatal rats.	Shin et al. (2004a)
PCL	80,000	THF : DMF (1 : 1)	500–900	Human bone marrow derived mesenchymal stem cells.	Li, W. J., et al. (2005c)
PCL	80,000	$CHCl_3$ (10%)	20–5000	Rat and human bone marrow derived mesenchymal stem cells.	Yoshimoto et al. (2003)
Poly(-lactide-*co*-glycolide) LA/GA ratio (10 : 90)	7.5×10^4	HFP (10%)	900–1000	Primary cardiomyocytes.	Zong et al. (2005)
PCL/$CaCO_3$	80,000	MeOH : $CHCl_3$ (1 : 3)	600 ± 230	Human osteoblasts.	Fujihara et al. (2005)
Poly(3-hydroxybutyrate-*co*-3-hydroxyvalerate) (PHBV)	—	TFE (2%)	~185	Chondrocytes derived from rabbit ear.	Lee, I. S., et al. (2004)
Poly(3-hydroxybutyrate-*co*-3-hydroxyvalerate) (PHBV)	680,000	TFE (2%)		Kidney cells COS-7.	Ito et al. (2005)
Block copolymer poly(3-hydroxybutyric acid-*co*-(ε-caprolactone-*co*-glycolide)	—	$CHCl_3$ (30%)	~10 microns	C2C12 murine myoblast cell line. L2 rat myoblast cell line. Human satellite cells.	Riboldi et al. (2005)

(*Continued*)

APPENDIX I *Continued*

Polymer	Molecular Weight (M_w)	Solvent (Concentration)	d (nm)	Cell Type	Reference
Gelatin	—	HFP (2–8%)	485 ± 187	Human embryonic palata mesenchymal cells.	Li, M. Y., et al. (2005)
Collagen	—	HFP (>5%)	100–4600		
α-Elastin, recombinant human tropoelastin		HFP (10–20%)			
Polyaniline (PANi) + gelatin	100,000 for polyaniline	HFP (3–5%) (0–5% PANi)	61–803	Cardiac rat myoblast cells.	Li, M. Y., et al. (2006)
Polystyrene	230,000	$CHCl_3$ (13–15%)	290–1900	Smooth muscle cells.	Baker et al. (2006)
Polydiaxanone (PDO) PDO/collagen blends	—	HFP	100–600	Human dermal fibroblasts.	Boland et al. (2005)
Collagen type II	—	HFP	110–1800	Chondrocytes.	Matthews et al. (2002)
Polycaprolactone and collagen	80,000	MeOH : $CHCl_3$ (1 : 3)	300–700	Human coronary artery smooth muscle cells.	Venugopal et al. (2005a)
Poly(L-lactic acid)-*co*-poly(ε-caprolactone) (70 : 30)	150,000	DCM : DMF (70 : 30) (10%)	470 ± 130	Human coronary artery endothelial cells.	He, W., et al. (2005a)
Collagen and elastin with PEO. Crosslinked	8×10^6 (PEO)	Water	220–600	Smooth muscle cells.	Buttafoco et al. (2006)
Collagen type I. Crosslinked collagen type I and III	—	HFP (8%)	100–1200	Human keratinocyte.	Rho et al. (2006)
Silk/chitosan blends. Up to 30% chitosan	—	Formic acid	130–780	No cell culture studies.	Park et al. (2004)
Chitin	910,000	HFP (3–6%)	100–200	No cell culture studies.	Min et al. (2004c)

Chitosan (with PEO)	~100,000	—	Submicron	Canine chondrocytes.	Subramanian et al. (2005)
Chitosan[a] (with PEO[b])	190,000[a] 900,000[b]	0.5 M acetic acid	40 microns	Chondrocytes (HTB-94) and osteoblasts (MG-63).	Bhattarai et al. (2005)
Chitin[a]/poly(glycolic acid)[b] blend	91,000[a] $20–40 \times 10^{3}$[b]	HFP (5–8%)	200–600	Normal human epidermal fibroblasts.	Park et al. (2006)
Chitin	91,000	HFP (3–6%)	163	Normal human oral keratinocytes, human gingival fibroblasts, and human epidermal keratinocytes.	Noh et al. (2006)
Poly(L-lactide-*co*-ε-caprolactone)/collagen coated	$M_n = 150,000$	$CHCl_2$: DMF (70 : 30) (10%)	406 ± 126; 470 ± 80	Human coronary artery endothelial cells.	He, W., et al. (2006)
Poly(L-lactide-*co*-ε-caprolactone)	$M_n = 1.8–4.5 \times 10^5$	DCM (4–11%)	200–1160	Human umbilical vein endothelial cells.	Kwon et al. (2005)
Poly(ε-caprolactone-*co*-ethyl ethylene phosphate)	70,760	Acetone (22%)	760	Rat hepatocytes.	Chua et al. (2005)
Poly(L-lactic acid) (PLLA)	300,000	DCM : DMF (70 : 30)	150–500	Neural stem cells.	Yang, F., et al. (2005)
Poly(D,L-lactide-*co*-glycolide) PLGA (85 : 15)	—	DMF : THF (50 : 50) (5%)	500–800	Fibroblasts and bonemarrow-derived mesenchymal stem cells (MSCs).	Li, W.-J., et al. (2002)

(Continued)

Polymer	Molecular Weight (M_w)	Solvent (Concentration)	d (nm)	Cell Type	Reference
Poly(D,L-lactide-*co*-glycolide) PLGA (50 : 50)	250,000	HFP (15%)	310	Human oral keratinocytes derived from gingival tissue and primary normal human gingival fibroblasts.	Min et al. (2004d)
Poly(L-lactide-*co*-ε-caprolactone) (75 : 25)		Acetone (3–9%)	500–1500	Endothelial cells and smooth muscle cells.	Mo et al. (2004)
Poly(L-lactide-*co*-ε-caprolactone) (75 : 25)		Acetone (5%)	200–800	Human coronary artery smooth muscle cells.	Xu et al. (2004)
Silk/PEO blends (80/20 w/w)	900,000 PEO	Water (7.5%)	700 ± 50	Human bone marrow stromal cells.	Jin, H.-J., et al. (2004)
Silk*		Formic acid (3–15%)	30–120	Normal human keratinocytes and human fibroblasts.	Min et al. (2004a, 2004b)
Silk		HFP (0.2–1.2%)	7–200	No cell culture studies.	Zarkoob et al. (2004)
Silk		Formic acid (5–20%)	<100	No cell culture studies.	Sukigara et al. (2003, 2004)
Silk		Formic acid (3%)	30–120	Normal human oral keratinocyte.	Min et al. (2004a)

Silk (genetically engineered spider drag-line silk)		HFP (15%)	100–500	No cell culture studies.	Stephens et al. (2005)
Silk		Formic acid (9%)	8–223	No cell culture studies.	Ayutsede et al. (2005)
Silk/PEO blends (82 : 18)		Water (7.8%)	590 ± 60	Human bone marrow derived mesenchymal stem cells.	Li, C. M., et al. (2006)
Silk/chitosan blends (up to 30% chitosan)	$M_v = 2.2 \times 10^5$ (for chitosan)	Formic acid	130–450	No cell culture studies.	Park, W. H., et al. (2004)
Recombinant hybrid silk		HFA (2–10%)	100–1000	No cell culture studies.	Ohgo et al. (2003)

TFA: Trifluoroacetic acid; TFE: Trifluoroethanol; HA: Hydroxyapatite filler; PEO: Poly(ethylene oxide); PANi: Polyaniline (emeraldine base); BDI: Diisocyanatobutane; DCM: Dichloromethane; DMAC: Dimethylacetamide; THF: Tetrahydrofuran; HFA: Hexafluoroacetone; HFP: 1,1,1,3,3,3-Hexafluoro-2-Propanol; D: Fiber diameter.

*The silk is isolated from cocoons of silkworm *B. mori* and degummed. The silk derived has an average molecular weight of around 9×10^5 (g/mol).

APPENDIX II

SUMMARY TABLE OF ELECTROSPUN POLYMER NANOFIBERS

Science and Technology of Polymer Nanofibers. By Anthony L. Andrady

	Polymer	Molecular Weight (Mw)	Solvent (Concentration)	Application Area	Reference
	Aramid (Kevlar 49)		Sulfuric acid (95–98%) (2–3 wt%)	Nanofibers having a circular cross-section, birefringent and stable at temperatures >400°C electrospun from solution.	Srinivasan and Reneker (1995)
	P4VP/PMMA + (1 : 1) blend	160,000/350,000	DMF	Decoration of nanofiber surface with nanoparticles. Synthesis and assembly of metal nanoparticles on electrospun poly(4-vinylpyridine) fibers and blend fibers.	Dong, H., et al. (2006)
	PES	55,000	DMSO (20 wt%)	Enhanced adhesion and expansion of human umbilical cord blood hematopoietic stem/progenitor cells via surface-aminated electrospun nanofibers reported.	Chua et al. (2006)
Cellulosics	α-Cellulose	DP = 700 and 800	N-methylmorpholine-N-oxide/water	Electrospinning and characterization of cellulose fibers. The effect of electrospinning process conditions in obtaining cellulose fibers, a nonwoven fiber network, and a cellulose membrane reported.	Kulpinski (2005)

Cellulose	DP = 1070 and 1140	LiCl/DMAc (1–3 wt%)	Effects of the temperature of the collector, type of metal used as the collector, and postspinning treatment is reported.	Kim, C. W., et al. (2005)
Cellulose acetate (CA)	30,000	DMAc/Acetone (2 w/w) (~15 wt%)	Enzyme immobilization on nanofiber surface studied. Cellulose fiber surfaces reacted with polyethylene glycol (PEG) diacylchloride to simultaneously attach amphiphilic spacers and reactive end groups for coupling with a lipase enzyme.	Wang and Hsieh (2004)
CA (acetyl content ~40%)	30,000	Acetone/Water (0/100 to 80/200) (9–21 wt%)	Electrospinning of antimicrobial ultrafine fibers with silver nanoparticles by direct electrospinning of a cellulose acetate (CA) solution with small amounts of silver salts discussed.	Son et al. (2004a, 2004b, 2004d)
CA (acetyl content ~50%)		MC and MC/EtOH mixtures (5 wt%)	Ultrafine porous triacetate fibers electrospun from MC having isolated circular shaped pores with a narrow size distribution electrospun.	Han et al. (2005)

(*Continued*)

Polymer	Molecular Weight (Mw)	Solvent (Concentration)	Application Area	Reference
CA (acetyl content ~50%)	$M_n = 30,000$	Acetone, acetic acid, DMF blends	Use of mixed solvents in electrospinning polymers. Where single solvents did not allow electrospinning of continuous fibers, two-component mixtures of these produced suitable solvent systems.	Liu and Hsieh (2002)
CA (acetyl content ~50%)	$M_n = 29,000$ (40% acetyl)	Acetone/DMF/TFE (3 : 1 : 1) (~16 w/v%)	Water filtration using electrospun regenerated cellulose nanofiber membranes studied and compared with commercial micro-filtration membranes.	Ma et al. (2005c)
Cellulose acetate	$M_n = 30,000$	Acetone/DMAc (2 : 1) (15 wt%)	Ultrafine cellulose nanofiber surfaces grafted with polyacrylic acid (PAA) and lipase adsorbed on to the fiber to obtain a functionalized mat.	Chen and Hsieh (2004)
Cellulose acetate	$M_n = 40,000$	Acetone/DMAc (2 : 1) (10 wt%)	Electrospinning of poly(vinyl alcohol) (PVA) and cellulose acetate (CA) via multi-jet electrospinning reported.	Ding et al. (2004c)

	Carboxymethyl cellulose	700,000	Methanol/water (50/50) (0.01 wt%)	Nanofibers spun into mats using both DC and AC driving potentials, with the AC potential resulting in a significant reduction in fiber 'whipping' during spinning with resulting mats exhibiting a higher degree of fiber alignment.	Kessick et al. (2004)
	Ethyl cellulose	160,000	THF/DMF (various) (13 wt%)	The influence of the composition of multicomponent solvent on the surface morphology and diameter distribution of fibers produced by electrospinning investigated.	Wu, X. H., et al. (2005)
	Ethyl-cyanoethyl cellulose	$M_n = 97,000$	THF (17 wt%)	Microcavities or surface porosity of fibers attributed to the volatilization of the solvent. (The crystallinity of the fibers varied with the voltage of the electrostatic field.)	Zhao et al. (2003, 2004)
Chitin/chitosan	Dibutyrylchitin (natural)	—	DMAc	A new method of manufacturing nonwoven products made from dibutyrylchitin, which can be used as wound dressing materials, discussed.	Błasińska et al. (2004)

(*Continued*)

APPENDIX II *Continued*

Polymer	Molecular Weight (Mw)	Solvent (Concentration)	Application Area	Reference
Chitosan/PEO	Range of PEO polymers	Water (4–6 wt%)	In electrospinning chitosan/PEO blend solutions spectroscopic and thermal analysis techniques showed heterogeneity in the mat with larger diameter fibers being entirely made of PE.	Duan et al. (2004)
Chitosan/PEO	106,000	90% aqueous acetic acid (7 wt%)	Chitosan nanofibers electrospun from aqueous acetic acid solution. Only chitosan of a M_w = 106,000 g/mol shown to produce bead-free chitosan nanofibers. (Neither the low- or high-M_w chitosans of 30,000 and 398,000 g/mol did.)	Geng et al. (2005)
Chitosan/PEO (ratio 0.05 : 1)	PEO = 800,000	Water (0.05–5 wt%)	The first successful preparation of chitosan-containing nanofibers achieved by electrospinning chitosan/poly(ethylene oxide) (PEO) blend solutions.	Spasova et al. (2004)
Chitosan/PVA (blends)	Chitosan: M_v = 200,000	TFA (7–8 wt%)	The effect of the type of solvent and chitosan concentration on the morphology of the resulting nanofibers studied.	Ohkawa et al. (2004a)

Other bio-polymers	Casein and blends with PEO or PVA	PEO: $M_v = 60{,}000$	5 wt% aqueous ethanolamine (10–30 wt%)	Casein nanofibers electrospun as blends with water-soluble polymers to obtain blend nanofiber and subsequently crosslinked with isocyanate.	Xie and Hsieh (2003)
	Collagen type I, St-gelatin		Collagen - HFP (1–7%), St-GT (7–10%)	Two new electrospinning techniques proposed: multilayer electrospinning and mixed mat electrospinning. A trilayered mat, in which individual fiber meshes (type I collagen, ST-gelatin, and SPU) deposited layer by layer by sequential electrospinning.	Kidoaki et al. (2005)
	Copolymer (TMC-CL)		DMF/MC (50/50)	The effects of solvent composition, concentration, voltage, and tip-collector distance on the morphology of the electrospun fibers investigated by scanning electron microscopy.	Jia et al. (2006)
	Dextran	64,000 to 76,000	DMSO/DMF, DMSO/water, water	Uniform nanofibrous dextran membranes electrospun from water and solvent mixtures by adjusting the processing conditions.	Jiang, H. L., et al. (2004a)
	Gelatin (GT)		TFE (2.5–15 wt%)	Electrospinning of the biopolymer, gelatin, and the mass concentration-mechanical property relationship of the resulting nanofiber mats investigated.	Huang et al. (2004)

(*Continued*)

APPENDIX II *Continued*

Polymer	Molecular Weight (Mw)	Solvent (Concentration)	Application Area	Reference
Gelatin (GT)		98% formic acid (7–12 wt%)	The parameters, electric field, gap distance, and concentration studied for their effects on electrospinnability and fiber morphology.	Ki et al. (2005)
GT		TFE (10 wt/v%)	Vapor phase crosslinking of nanofibers discussed.	Zhang, Y. Z., et al. (2006b)
GT and GT + PCL	PCL: $M_n = 80{,}000$	TFE (2.5 to 12.5 wt%)	Scaffolding applications and electrospinning of core-shell type nanofibers, discussed.	Zhang Y. Z., et al. (2005b); Zhang et al. (2004)
DNA		Water 0.3–1.5%	Electrospinning of calf thymus Na-DNA reported. However, bead-like structures were observed on many of the fibers.	Fang and Reneker (1997)
Gluten (wheat)			Wheat gluten from two sources electrospun. The highest molecular weight glutenin polymer chains in the wheat protein appeared to be responsible for the lower threshold concentration for fiber formation.	Woerdeman et al. (2005)
Hyaluronic acid (HA)	45,000 and 3,500,000	Water (HCl) (1.3–2 wt/v%)	Electroblowing of natural polymers to obtain nanofibers. The temperature of air-blowing was the most effective parameter in ensuring HA nanofiber formation.	Um et al. (2004); Wang, X., et al. (2005)

	Hyaluronic acid/ Gelatin	HA = 2,000,000 $M_n = 80,000$	DMF/Water	Electrospinning of hyaluronic acid (HA) and hyaluronic acid/gelatin (HA-GT) blends in N,N-dimethylformamide (DMF)/water-mixed solvents investigated.	Li, J. X., et al. (2006)
	PANi + gelatin (bovine skin)	PANi: $M_w = 100,000$	HFP (3 wt% for PANi) + (8 wt% GT)	Results suggesting that PANi-gelatin blend nanofibers might provide a novel conductive material that is also a biocompatible scaffold for tissue engineering, reported.	Li, M. Y., et al. (2006)
	Ethylene-vinyl alcohol copolymer		Isopropanol : water (70 : 30) (2.5–20 wt%)	Electrospun EVOH mats shown to support the culture of smooth muscle cells and fibroblasts.	Kenawy et al. (2003)
	EVA, PLA, EVA/PLA	P(EVA): $M_w = 60,400$	Chloroform (14% w/v)	Electrospun fiber mats explored as drug delivery vehicles using tetracycline hydrochloride as the model drug.	Kenawy et al. (2002)
Polyamides	Nylon-4,6		99% formic acid (pyridine additive)	Electrospinning ultrafine fibers and process parameters. A small amount of pyridine added to the electrospinning solution avoided formation of beaded nanofibers at low concentrations.	Huang, C. B., et al. (2006a)

(Continued)

APPENDIX II *Continued*

Polymer	Molecular Weight (Mw)	Solvent (Concentration)	Application Area	Reference
Nylon-6	22,910	MC/TCA (50/50 wt/wt) (16 wt%)	The morphological properties of electrospun fiber webs investigated and compared with those of spunbond nonwoven fabrics.	Hong, K. H., et al. (2006)
Nylon-6	43,000	HFP (15 wt%)	Fundamental study on the crystallinity of nanofibers. Change in the chain conformation on electrospinning studied using Raman spectroscopy.	Stephens et al. (2004)
Nylon-6	17,000, 20,000, 32,000	85% formic acid (10–46 wt%)	Effect of polarity of electrodes and other key process variables on nanofiber quality studied.	Supaphol et al. (2005a)
Nylon-12	32,000	HFP (15 wt%)	Coformational changes accompanying electrospinning using Raman spectroscopy, studied.	Stephens et al. (2004)
Polyamide-6	20,000	85% formic acid (10–46 wt%)	Effect of process variables including solvent quality on the quality of electrospun nanofibers, investigated.	Mit-uppatham et al. (2004a, 2004b)
Polyacrylamide	9,000,000	Water (0.3–1.5 wt%)	The effect of process variables on the morphology of fibers and beads, studied.	Zhao, Y. Y., et al. (2005)

Polyacid	Poly(acrylic acid) (PAA)	250,000	Water/EtOH (1 : 0; 5 : 5; 0 : 1) (~6 wt%)	Gas sensor applications (for ammonia) using nanofiber mats on quartz crystal microbalance.	Ding et al. (2005c)
	PAA	450,000	Water or NaCl solution (5 wt%)	The effect of ionic strength and other variables on nanofiber quality investigated.	Kim, B., et al. (2005)
	PAA	250,000, 450,000	Ethanol/water (40/60) (5–6 wt%)	Study of process parameters on the volume and surface charge density in the polymer jets investigated.	Theron et al. (2004)
	Poly(2-acrylamido-2-methyl-1-propane sulfonic acid) (PAMPS)	2,000,000	Ethanol/water (2–10 wt%)	The effect of electric current on nanofiber morphology investigated using a field emission scanning electron microscopy (FE-SEM). Charge-to-mass ratio of a highly conducting liquid on nanofiber uniformity also studied.	Kim, S. J., et al. (2005)
	PAMPS		Water (6 wt%)	The dependence of morphology of electrospun nanofiber on the applied voltage and the influence of added ionic salts on nanofiber uniformity, studied.	Lee, C. K., et al. (2005)

(*Continued*)

APPENDIX II *Continued*

	Polymer	Molecular Weight (Mw)	Solvent (Concentration)	Application Area	Reference
Polyacrylonitrile	PAN	75,000	DMF (1 wt%) with 1–4 wt% LiCl	Theoretical analysis showing the stable length of electrospinning jet to decrease at higher levels of LiCl in the dope, reported.	Qin et al. (2004, 2005)
	PAN	86,000	DMF (7 wt%)	Electrospinning of composite nanofibers with MWCNTs and their spectroscopic characterization discussed.	Hou et al. (2005)
	PAN	86,200	DMF (7.5 wt%)	Wet process for coating electrospun nanofibers with metal oxides reported.	Drew et al. (2003a, 2005)
	PAN	150,000	DMF (7 wt%)	A single-step method of preparing ultrafine fibers with nanoparticles reported.	Lee, H. K., et al. (2005)
	PAN	150,000	DMF (7 wt%) with 1–4 wt% graphite	Electrospinning composite nanofiber with graphite filler reported. Thin graphite nanoplatelets synthesized by an intercalation/exfoliation process incorporated into nanofibers on electrospinning.	Mack et al. (2005)
	PAN	150,000	DMF	Electrospinning composite nanofiber with MWCNTs or SWCNTs and a study of their failure modes, reported.	Ye (2006)

PAN	150,000	DMF (8 wt%)	Mechanical characterization of individual carbonized nanofibers reported.	Zussman et al. (2005)
PAN	210,000	DMF (5 wt%)	A technique consisting of spinning onto a water reservoir collector and drawing the resulting non-woven web of fibers in the form of a continuous yarn demonstrated.	Smit et al. (2005)
PAN	114,000	DMF 9.9 and 14.8 wt%	Alignment and orientation study on nanofibers reported.	Fennessey and Farris (2004)
PAN		DMF (10 wt%)	Graphitization of electrospun nanofibers and their characterization studied. Hydrogen storage in nanofibers discussed.	Kim, C., and Yang (2003); Kim, C., et al. (2004d, 2004f)
PAN		DMF (15–17 wt%)	The influence of electrospinning parameters on the thermal stability of nanofibers on pyrolysis discussed.	Klata et al. (2005)
PAN		DMF (0.5 to 10 wt%)	Anisotropic electrical conductivity in carbon nanotube-containing electrospun nanofibers investigated.	Ra et al. (2005)
PAN (with dye/TiO_2)	80,000	DMF (2 wt%)	Photovoltaic cell application of nanofibers addressed.	Drew et al. (2005)

(*Continued*)

APPENDIX II *Continued*

	Polymer	Molecular Weight (Mw)	Solvent (Concentration)	Application Area	Reference
	PAN		DMF	The effect of electrospinning variables on fiber quality discussed.	Buer et al. (2001)
	PANi + PEO	PEO = 900,000	Chloroform (~2.5 wt%)	EPR investigation of conducting polymer blend nanofibers.	Kahol and Pinto (2002, 2004)
	PANi/PEO	PEO = 900,000	PANi (chloroform 1% and PEO ~2%)	Electronic applications of conducting polymer nanofibers in FETs reported.	Pinto et al. (2003)
	PANi/PS (2 : 7.5)	PANi = 65,000, PS = 200,000	Chloroform	Nanofiber based sensors for hydrogen peroxide and humidity demonstrated.	Aussawasathien et al. (2005)
Polycaprolactone	PCL		Chloroform/MeOH (75 : 25) (9 wt%)	Melt and solution electrospinning using ring-shaped collectors and the conversion of fibers into yarn reported.	Dalton et al. (2005, 2006)
	PCL	40,000	Chloroform (4–9 wt%)	Porous nanofiber and bead formation discussed, including the use of additives in spinning solution to improve fiber quality.	Hsu and Shivkumar (2004a)
	PCL	M_n = 80,000	MC/DMF (various) (8–10 wt%)	Scaffolding effectiveness for MCF-7 carcinoma cells.	Khil et al. (2005)
	PCL	80,000	MF/MC (25/75) (8 wt%)	Electrospinning of porous nanofibers discussed.	Kim, H. Y., et al. (2003)
	PCL		Chloroform and THF/DMF (1 : 1 v/v)	Morphological features of electrospun fibers studied as a function of the solvent and the processing voltage.	Krishnappa et al. (2003)

PCL	$M_n = 80{,}000$	MC/DMF & MC/toluene (10–15 wt%)	Effect of nanofiber quality as a function of different spinning solvents. Characterization of nanofiber mats carried out.	Lee, K. H., et al. (2003b)
PCL	80,000	Chloroform/MeOH (3 : 1 w/w) (2–12 wt%)	Superhydrophobic nanofiber mats by iCVD coating of surface. Improvement of tissue scaffolding by surface grafting with gelatin, demonstrated.	Ma, Z. W., et al. (2005a)
PCL	$M_n = 80{,}000$	MC/DMF (85 : 15 v/v) (7, 8, 9 wt%)	Biodegradation of nanofiber mats by soil fungi determined.	Ohkawa et al. (2004b)
PCL	80,000 and 120,000	Acetone (14–18 wt%)	The formation of garland-like nanofibers reported.	Reneker et al. (2002)
PCL	27,000	THF/DMF (various) (13, 14, 15 wt%)	The effects of processing, parameters, particularly the choice of solvent, on nanofiber quality, reported.	Shawon and Sung (2004)
PCL	80,000	Chloroform/MeOH (1 : 1) (10 wt%)	Bone formation from mesenchymal stem cells (MSCs) on novel nanofibrous scaffolding materials studied. *In vitro* growth of cardiomyocytes on nanofiber scaffolding studied.	Shin et al. (2004a)
PCL	$M_n = 80{,}000$	Chloroform/MeOH (90 : 10) (~12 wt%)	The use of steel blades to control the electric field to electrospin highly aligned nanofibers reported.	Teo et al. (2005)

(*Continued*)

APPENDIX II *Continued*

	Polymer	Molecular Weight (Mw)	Solvent (Concentration)	Application Area	Reference
	PCL	80,000	MC/DMF (75/25 and 40/60) (10–14 wt%)	A study of the influence of process parameters on the volume and surface charge density in the polymer jet studied.	Theron et al. (2004)
	PCL	80,000	Chloroform/MeOH (3 : 1) (7.5 wt%)	Biocomposite nanofiber scaffolding for coronary artery smooth muscle cells investigated. The use of coated nanofibers to improve scaffolding performance also reported.	Venugopal and Ramakrishna (2005); Venugopal et al. (2005b)
	PCL	280,000	Chloroform/acetone (1 : 1 v/v) (5 wt%)	Parameters influencing the spinning process and fiber morphology examined. The influence of cationic and anionic surfactants on enzymatic degradation rates investigated.	Zeng, J. X., et al. (2004a)
	PCL		Chloroform/DMF	The effects of concentration, solvent composition, applied voltage and tip-collector distance on fiber diameter and fiber morphology investigated.	Bölgen et al. (2005)
	PDLA	109,000	DMF (20–35 wt%)	Effect of concentration on electrospinnability studied using several polymer/solvent systems.	Shenoy et al. (2005a)
Polyoxides	Polyethylene oxide (PEO)	400,000	EtOH/water (0.7 : 0.3)	Nanofiber based sensors for hydrogen peroxide and humidity reported.	Aussawasathien et al. (2005)

PEO	100,000 and 2,000,000	Water or chloroform	Electrospinning of nanofibers, measurement of birefringence and fiber dimensions in the mat, discussed.	Buer et al. (2001)
PEO	400,000	Water (4–10 wt%)	Effect of process variables on nanofiber quality studied.	Dalton et al. (2005)
PEO	400,000	Water (4–10 wt%)	Real-time observations of electrospinning process using high-speed, high-magnification imaging techniques, reported.	Deitzel et al. (2001c)
PEO	300,000	Water/EtOH (2 : 7)	Effect of solution viscosity on nanofiber morphology studied in polymer/titania systems.	Drew et al. (2003b)
PEO	900,000	Water and water/EtOH mix	The bead formation during electrospinning of polymer solutions studied.	Fong et al. (1999)
PEO	100,000	EtOH/water (0.5 : 0.5) (5–30 wt%)	Formation of oriented polymeric nanofibers using electrospinning from an integrated microfluidic device investigated.	Kameoka and Craighead (2003)
PEO	2,000,000	Water (2 wt%) (with 5 wt% carbon)	Color change on heating mats of nanofibers containing carbon nanotubes, reported.	Pedicini and Farris (2004)
PEO	400,000	Water (4–10 wt%)	Effects of concentration of polymer on electrospinnability investigated.	Shenoy et al. (2005a)
PEO	2,000,000	Water (10 wt%)	Effect of spinning parameters on nanofiber quality studied.	Shin et al. (2001a)

(*Continued*)

APPENDIX II *Continued*

	Polymer	Molecular Weight (Mw)	Solvent (Concentration)	Application Area	Reference
	PEO	300,000	DMF, EtOH, water, chloroform (2–8 wt%)	Effect of solvent parameters such as the dielectric constant on fiber quality, explored.	Son et al. (2004d)
	PEO	600,000	Water (3 wt%)	Observation of multiple jets in electrospinning reported.	Theron et al. (2005)
	PEO	600,000, 1×10^6, 4×10^6	Ethanol/water (40 : 60), water, (1–6 wt%)	Study of process parameters and their effect on fiber quality in electrospinning.	Theron et al. (2004)
	PEO/PPy (4 : 1)	400,000/62,300	DMF and water	Electrospinning of blends containing conductive polymers investigated.	Chronakis et al. (2006a)
Polyesters	Poly(ethylenetere-phthalate) (PET)		DCM/TFA (1 : 1) (~20 wt%)	Molecular imprinting of nanofibers reported.	Chronakis et al. (2006b)
	PET		TFA (~20% w/v)	Application as a blood vessel engineering scaffold studied.	Ma et al. (2005d)
	PET (crystalline and amorphous)		TFA/MC (50 : 50 v/v)	The effects of blend composition and heat treatment on nanofiber mat characteristics.	Kim, K. W., et al. (2005)
	PET-*co*-PEI	11,700–106,000	Chloroform/DMF (70 : 30 w/w) (8–20 wt%)	Effect of normalized concentration (C/Ce) on electrospinnability, particularly the dependence of fiber diameter on the zero shear rate viscosity, determined.	McKee et al. (2004b)
	PHBV (~5% 3-HV)	680,000	TFE (2 wt%)	Biomimicking composite with hydroxyapatite. Also, nanofibrous mats that enhanced cell adhesion over that of the thin film, reported.	Ito et al. (2005)

	PHBV (~6% 3-HV units)	680,000	Chloroform (3–23 wt%)	Effect of additives in controlling fiber morphology studied. Conductivity of the solution was found to be a major parameter affecting fiber morphology.	Choi, J. S., et al. (2004)
	Poly(butylene succinate) (PBS)	$M_n = 75{,}000$	DMF, CE, 3-CP, MC (13–15 wt%)	Electrospinning of ultrafine fibers and their crystalline morphology studied. Lamellar stack morphology containing crystalline and amorphous layers reported.	Jeong et al. (2005)
	Poly(trimethylene terephthalate)	Inherent viscosity 0.92	TFA/MC (50 : 50) (5–16 wt%)	Effect of processing parameters on nanofiber quality investigated. Periodic feature of surface roughness, such as diamond-shaped structure reported.	Khil et al. (2004)
Polyacids	PLGA (L : G 50 : 50)	56,000	THF/DMF (75 : 25) (~10–30%)	Nanofibers mats that could serve as a biodegradable gauze, and as an antibiotic delivery system discussed.	Katti et al. (2004)
	PLGA	90,000–126,000	DMF (45 wt%)	Biologically active functionalized electrospun matrix that allows immobilization and long-term delivery of growth factors is described.	Casper et al. (2005)

(Continued)

APPENDIX II *Continued*

Polymer	Molecular Weight (Mw)	Solvent (Concentration)	Application Area	Reference
PLGA (LA/GA 50 : 50)	25,000	Chloroform, HFP (15 wt%)	Effect of copolymer composition on degradation rates of biodegradable polymer nanofibers. Concluded that conductivity or dielectric constant of solution determined fiber morphology.	You et al. (2005a, 2006a, 2006b)
PLGA (LA : GA 75 : 25 and 50 : 50)		DMF (30–50 wt%)	Concentration effects in electrospinning of triblock copolymer and polylactide on cell proliferation and hydrophilicity of the electrospun mats, investigated.	Kim, K. S., et al. (2003)
PLGA (LA : GA 75 : 25) and blends	75,000	DMF (33 wt%)	Drug release behavior from electrospun scaffolds and antimicrobial effects of the released drugs investigated.	Kim, K. S., et al. (2004)
PLA		MC (10 wt%)	The effect of gaseous environments on the morphology of electrospun nanofibers reported.	Larsen et al. (2004a)
PLLA	100,000 and 300,000	MC/MeOH & MC/Pyr (7.5–12.5 wt%)	Take-up velocity found as a dominant parameter inducing a highly ordered structure in nanofibers compared to other parameters (such as solution conductivity and polymer concentration).	Inai et al. (2005b)
PLLA	670,000	MC (5 wt%)	Effect of materials variables on the quality of fibers reported.	Jun et al. (2003)

	PLLA	670,000	DCM (1–4 wt%)	Effect of solution concentration on electrospinnability investigated.	Shenoy et al. (2005b)
	PLLA	100,000 and 300,000	MC, MC/DMF, MC/Py (1–4 wt%)	A processing map summarizing effects of solutions properties and processing conditions on the nanofiber morphology developed.	Tan, S.-H., et al. (2005); Tan, S. T., et al. (2005)
	PLLA	300,000	MC/DMF (70 : 30) (1–5 wt%)	Good interaction between neural stem cells and the nanofibrous scaffolds due to the greatly improved surface roughness of electrospun nanofibrous scaffold reported.	Yang, F., et al. (2004, 2005)
	PLLA	225,000	Chloroform/acetone (2 : 1 v/v) (3.5 wt%)	Drug delivery by nanofiber matrix involving slow bioerosion of the polymer, demonstrated.	Zeng et al. (2004a)
Acrylic polymers	Poly(methyl methacrylate) (PMMA)	495,000	Anisole (4–5.5 wt%)	Novel electrospinning tip made of the small apex of a microfabricated source used to make nanofibers.	Czaplewski et al. (2003)
	PMMA	12,470–365,700	DMF	The effects of solution concentration, molecular weight, molecular weight distribution and viscosity on fiber formation and morphological features, studied.	Gupta et al. (2005)

(*Continued*)

APPENDIX II *Continued*

Polymer	Molecular Weight (Mw)	Solvent (Concentration)	Application Area	Reference
PMMA	95,000–150,000	Various solvents	Concentration dependence of fiber morphology and unique bead shapes obtained by electroprocessing within a lower concentration range, reported.	Liu and Kumar (2005)
PMMA	120,000	THF/DMF (15 wt%) (with 1 wt% carbon)	Color change on heating nanofiber mats containing carbon filler materials, observed.	Pedicini and Farris (2004)
PMMA	120,000	Chloroform (15, 20 and 25 wt%)	Aligned polymer fibers, several cm in length, with separation between in the range of 5–100 μm fabricated.	Sundaray et al. (2004)
PMMA	120,000	Acetone (12.5 wt%)	Luminescent polymer fibers with "beads on a string" morphology prepared by electrospinning and their photonic characteristics, reported.	Tomczak et al. (2005)
Poly[bis(2,2,2-trifluoroethoxy) phosphazene]		MEK, acetone (10 wt%), THF (0.5–10 wt%)	Electrospun nonwoven mats with enhanced surface hydrophobicity compared to spun cast films. Variation of hydrophobicity with fiber diameter and surface morphology, reported.	Singh et al. (2005)

	Poly(3-hexylthiophene)	87,000	Chloroform (~10 w/v%) (~2 wt%)	Single nanofiber FETs from electrospun regioregular poly(3-hexylthiophene) reported.	Liu, H. Q., et al. (2005); González and Pinto (2005)
Other Polymers	Polysulfone (PSU)	$M_n = 26{,}000$	Pyridine (25 wt%)	Surface modification via carboxyl groups introduced onto the fiber surface through grafting co-polymerization.	Ma et al. (2006)
	PSU		DMAc/acetone (9 : 1) (20 w/v%)	Effects of processing parameters, including the applied voltage, flow rate, and gap distance as well as their importance in controlling fiber morphology, investigated.	Yuan et al. (2004)
	Poly(etherimide) (PEI)		TFE	The effect of adding carbon black on the average diameter of fibers under the same electrospinning conditions to that of the polymer solution without filler, compared.	Lee, S.-G., et al. (2002); Han et al. (2004)
	PEI (ULTEM 1000)	12,000–30,000*	TCE (14 wt%)	Interfiber bonding in the electrospun nanofiber web by thermal treatment above its T_g shown to improve its physical properties.	Choi et al. (2004a)

(*Continued*)

APPENDIX II *Continued*

Polymer	Molecular Weight (Mw)	Solvent (Concentration)	Application Area	Reference
Poly(ferrocenyl dimethylsilane)	87,000	THF/DMF (9 : 1) (30 wt%)	Novel organometallic polymer electrospun into nanofibers and the crystal structure of nanofibers studied using electron diffraction.	Chen, Z. H., et al. (2001)
Poly(meta-phenylene isophthalamide)	90,000	DMAc (16 wt%) with LiCl	Nanofibers used as substrates for creating nanoscale carbon-based materials and metals. Nanofibers coated with carbon, copper, and aluminum fabricated using plasma enhanced chemical vapor deposition and physical vapor deposition.	Liu, W. X., et al. (2002)
Poly(vinylphenol)	20,000 and 100,000	THF (20 w/v% and w/v 60%)	Antimicrobial polymer nanofiber systems studied. Polymer morphology and molecular weight shown to affect on biocidal effectiveness against test microorganisms.	Kenawy and Abdel–Fattah (2002)
P(LA-EG)	$M_n = 92{,}100$ and 84,800	Chloroform (5.5–6 wt%)	Water-soluble drugs successfully electrospun from water-in-oil (W/O) emulsions, with an aqueous phase carrying an water-soluble drugs and a solution of an amphiphilic diblock copolymer.	Xu et al. (2005)

PEI + PHBV (~6% 3-HV) blends	PEI = 680,000	Chloroform (23% PEI) (21% PHBV)	Porous ultra-fine fibers obtained via selective thermal degradation of electrospun polymer blend nanofibers.	Han et al. (2004)
Polydiaxanone	Isolated from sutures	HFP	The ability to control fiber diameter as a function of concentrations as well as fiber orientation.	Boland et al. (2005)
Polyimide (Matrimide 5218)		DMAc (~20 wt%)	The effects of iron salts on carbonization and the resulting morphology of polymer nanofibers, reported.	Chung et al. (2005)
Polyphosphazene	1,909,000	Chloroform, DMF, THF/DMF (1 : 1) (1–3%)	Electrospinning of a class of inorganic polymers of high biocompatibility, high-temperature stability, and low-temperature flexibility, studied.	Nair et al. (2004)
Polypyrrole		Chloroform (5%)	Electrically conducting polymer nanofiber nonwoven web fabricated using the electrospinning technique and the electrical conductivity of the fibers measured.	Kang et al. (2005)
MA-based copolymers	135,000 and 150,000	Ethanol (10–35 wt%)	At a given applied electrical potential, increasing polymer concentration of spinning solutions shown to change nanofiber morphology and electrospinnability.	Pornsopone et al. (2005)

(*Continued*)

APPENDIX II *Continued*

Polymer	Molecular Weight (Mw)	Solvent (Concentration)	Application Area	Reference
MEH-PPV/PFO poly[2-methoxy-5-(2-ethylhexoxy)-1,4-phenylenevinylene] (MEH) poly(9,9-dioctylfluorene) (PFO)	MEH-PPV = 1,000,000/ PFO = 65,000	Chloroform	Phase-separated domains in these nanofibers (30–50 nm) shown to be much smaller compared to those of comparable blend thin films (100–150 nm).	Babel et al. (2005)
Polybenzimidazol (PBI)		DMAc (20 wt%)	Polymer nanofibers carbonized, activated by steam, and converted into activated carbon nanofibers. Their specific capacitance shown to depend on the activation temperature.	Kim, C., et al. (2004e)
PBI		DMAc + LiCl (~4 wt%) (~20 wt%)	A nonwoven fabric of polymer nanofibers electrospun, and thermally treated to improve its strength. Electrospun nanofibers shown to be birefringent.	Kim and Reneker (1999b)
P(LLA-CL) 25/75, 75/25, 50/50	195,000–440,000	Acetone (5–10 wt%)	The deformation behavior of electrospun membranes with randomly oriented nanofibers evaluated under uniaxial tensile loading.	Inai et al. (2005a)

	P(LLA-SA-1, 4BD)	210,000	MC/DMF	Proliferation of mouse fibroblasts on electrospun nanofiber mat surfaces compared with those on bulk polymer films.	Jin, H.-J., et al. (2005)
	Poly(3-hexylthiophene-2,5-diyl) (P3HT)		Chloroform (2 wt%)	Electrospun regio-regular fiber-based field effect transistor reported.	González and Pinto (2005)
	EVAc and modified EVAc	EVAc 86,000	DMAc (10 wt%)	Nanofibrous membranes evaluated as scaffolding for bone-marrow stromal cell (BMSC) culture. Cells not only attached but grew well on the surface and were also able to migrate inside the scaffold.	Zhang et al. (2005c)
Styrenics	Polystyrene (PS)	280,000	THF/DMF (50 : 50) (5–20 wt%)	Shear modulation force microscopy (SMFM) utilized to investigate the surface nanomechanical properties of electrospun fibers as a function of the fiber diameter and temperature.	Ji et al. (2006b)
	PS	$M_n = 140,000$	THF/DMF (7–13 wt%)	The surface tension of polymer solution was shown to have a linear correlation with the critical voltage and nanofiber throughput was shown to be dependent on solution conductivity.	Lee et al. (2003a)

(*Continued*)

APPENDIX II *Continued*

Polymer	Molecular Weight (Mw)	Solvent (Concentration)	Application Area	Reference
PS	100,000	THF/DMF (50/50) (5–15 wt%)	Polymer/surfactant interaction in electrospinning of nanofibers studied. Cationic surfactants additives reported to improve the solution conductivity, but with no effect on the viscosity.	Lin et al. (2004)
PS	212,400	THF (8 wt%)	Conducting electronic polymers, their blends and conventional polymers are fabricated into nanofibers by a non-mechanical, electrostatic dispersion method.	MacDiarmid et al. (2001)
PS	190,000	THF (18–35 wt%)	Effect of concentration on electrospinnability. Several polymer/solvent systems, studied.	Shenoy et al. (2005b)
PS	250,000	THF (15, 20 and 25 wt%)	Electrospinning of aligned polymer fibers, several centimeters in length, with separation between the fibers in the range of 5–100 μm, reported.	Sundaray et al. (2004)
PS (expanded)	160,000–260,000	*d*-Limonene (35 wt%) DMAc (10–20%)	Electrospinning styrofoam from *d*-limonene suggested as a recycling technique. Mats with fiber diameters 10–500 nm electrospun.	Shin and Chase (2005); Shin et al. (2005)

	PS and PSMA (7% MA)	PS = 860,000	THF (9–23 wt%)	Covalent attachment of enzymes onto styrenic nanofibers followed by glutaraldehyde (GA) treatment that cross-links additional enzyme molecules to obtain highly-active bound enzymes.	Kim, B. C., et al. (2005)
	Poly(styrene-b-dimethyl-siloxane)	$M_n = 114{,}000$	THF/DMF (3 : 1 w/w) (21 wt%)	Block copolymer fibers with submicrometer diameters in the range 150–400 nm electrospun to obtain hydrophobic non-woven mat surfaces.	Ma, M. L., et al. (2005a)
Polyurethane	PU block copolymer		THF/DMF (60 : 40) (30, 40 or 50 wt%)	Shape-memory polyurethane (PU) block copolymers used to prepare electrospun hard-segment concentrations of 40 and 50 wt%, found to have a shape recovery of more than 80%.	Cha et al. (2005)
	Polyurethane (medical grade)	110,000	DMF/THF (70 : 30 & 30 : 70)	Electrospun nanofibrous wound dressing membrane shown to have controlled evaporative water loss, excellent oxygen permeability and fluid drainage ability, still inhibited microorganisms.	Khil et al. (2003)

(Continued)

APPENDIX II *Continued*

	Polymer	Molecular Weight (Mw)	Solvent (Concentration)	Application Area	Reference
	Polyurethane (SPU), PEO		SPU-THF (15 wt%) PEO (4% $CHCl_3$)	A trilayered electrospun mesh, in which individual layers (collagen, ST-gelatin, and PU) deposited layer by layer, by sequential electrospinning, described.	Kidoaki et al. (2005, 2006)
	Poly(urethane urea) copolymer		DMF (~21 wt%)	Electrospinning behavior of elastomeric polyurethane urea copolymer in solution studied. The effects of electrical field, temperature, conductivity, and viscosity of the solution evaluated.	Demir et al. (2002)
	Polyurethane SPU		THF/DMF (DMF 0–30 v/v%) (10–17 wt%)	The effect of electrospinning parameters (e.g., concentration and solvent ratio) and operational parameters (e.g., applied voltage, air gap, and flow rate) on nanofiber morphology studied.	Kidoaki et al. (2006)
Polyalcohols	Poly(vinyl alcohol)	115,000	Water (6, 8, and 10 wt%)	Near the gelation threshold, the combination of thermo-reversible junctions and chain entanglements shown to stabilize the liquid jet and overcome capillary forces allowing the electrospinning.	Shenoy et al. (2005a)

PVA	10,000	Ethanol/water (50 : 50) (6 wt%)	Effect of process parameters on the volume and surface charge density in the polymer jet, investigated.	Theron et al. (2004)
PVA	$M_n = 72{,}000$	Water (15 wt%)	Concentric electrospinning head used to optimize efficiency and quality of the nanofibers. Throughput of 1 mg of dry PVA nanofibers from tip per minute demonstrated.	Tomaszewski and Szadkowski (2005)
PVA	2,000,000	Chloroform (0.5 wt%)	Luminescent polymer fibers with "beads on a string" morphology prepared by electrospinning and their photonic characteristics studied.	Tomczak et al. (2005)
PVA	86,000	Water (10 wt%)	Photochromic oxide/polymer nanofibers with fiber aggregates that changes from white to blue under ultraviolet irradiation obtained. (Photochromism is reversible under ambient conditions.)	Yang, G. C., et al. (2005)
PVA	$M_n = 115{,}000$	Water (10 wt%)	Treatment of the fiber mats with methanol shown to stabilize the mats against disintegration in contact with water. Also the mats showed increased mechanical strength following treatment with methanol.	Yao et al. (2003)

(Continued)

Polymer	Molecular Weight (Mw)	Solvent (Concentration)	Application Area	Reference
PVA	195,000	Water (10 wt%)	Poly(*p*-xylylene) coated polymer/BSA nanofibers prepared by chemical vapor deposition on electrospun fiber surface. The altered diffusion kinetics of bovine serum albumin from the nanofibers into buffer studied.	Zeng et al. (2005a)
PVA (PVA/CA blends)	66,000	Water (10 wt%)	Blend nanofibrous mats with good dispersibility electrospun using a multi-jet process. Mechanical properties of blend nanofiber mats were influenced by the weight ratio in blends.	Ding et al. (2004c)
PVA (80–99% hydrolyzed)	$M_n = \sim 57{,}000$	Water (6–8 wt%)	At higher degrees of hydrolysis of 98% gap distance observed to have no significant effect on the fiber morphology. The additions of sodium chloride and ethanol found to have significant effects on the fiber morphology.	Zhang, C. X., et al. (2005a)
PVA (96% hydrolyzed)	65,000*	Water	Nanofibers efficiently crosslinked using glyoxal. Phosphoric acid used as a catalyst activator to reduce strength losses during crosslinking.	Ding et al. (2002a)
PVA (98% hydrolyzed)	78,000	Water (0.2 to 1.2)	A novel high-flux filtration medium for oil/water emulsion separations, demonstrated.	Wang, X. F., et al. (2005)

PVA (98–99% hydrolyzed)	9,000–186,000	Water (9 to 31 wt%)	The effects of polymer MW on the fiber structure of electrospun PVA was studied. In general, fiber diameters increased with both molecular weight and concentration.	Koski et al. (2004)
PVA (99% hydrolyzed)	145,000	Water (7 wt%)	UV cross-linking of thienyl acrylate modified polymer nanofibers in the solid state, demonstrated. Water stability of these proven by steam test at 95°C.	Zeng et al. (2005a)
PVA (99.7% hydrolyzed)	78,000	Water (7 wt%)	The effect of pH on the morphology and diameter of nanofibers electrospun from aqueous media studied. Straighter and finer fibers at higher pH values.	Son et al. (2005)
PVA + PAA	124,000–186,000 & 450,000	Water (6 wt%) of blend	Interconnected fibrous membranes generated by electrospinning an aqueous mixture of polymers. The crosslinked nanofibers shown to act like anionic hydrogels.	Jin and Hsieh (2005a, 2005b)
PVA + PAA (75/25)	PVA = 145,000 PAA = 250,000	Water (7 wt%)	The water stability of electrospun nanofibers shown to improve significantly by annealing with poly(acrylic acid).	Zeng et al. (2004b)

(Continued)

APPENDIX II *Continued*

	Polymer	Molecular Weight (Mw)	Solvent (Concentration)	Application Area	Reference
	PVAC	500,000	DMF (18 wt%)	Spinning continuous uniaxial fiber bundle yarns from electrospun fibers, demonstrated.	Smit et al. (2005)
Vinyl polymer	Poly(vinyl chloride) (PVC)	DP = 800	THF/DMF (10–15 wt%)	The effects of solvent composition, concentration, applied electric field, and gap distance on the fiber morphology studied. Mechanical properties of nonwoven mats shown to depend on fiber orientation and linear velocity of the collector.	Lee, K. H., et al. (2002)
	PVC, PU, and blends		THF/DMF (60 : 40 v/v) (13 wt%)	The relationship between the morphology and mechanical behavior of blend fibers studied. The point-bonded structures in the nanofiber mats shown to increase with increasing PU content in blend.	Lee, K. H., et al. (2003c)
	PVC, PVDF, PU	135,900, 250,000	DMAc (~20–25 wt%)	Simultaneous electrospinning of two polymer solutions in a side-by-side fashion conducted. A new experimental device to electrospin bicomponent fibers described.	Gupta and Wilkes (2003)

Poly(vinylidene Fluoride) (PVDF)		Acetone/DMAc (7/3) (18 wt%)	Prototype cell (MCMB/PVdF-based fibrous electrolyte/$LiCoO_2$) with microporous membrane shown to have a very stable charge-discharge behavior with a slight capacity loss under constant current and voltage conditions.	Kim, J. R., et al. (2004, 2005)
PVDF	2,500	DMAc (15–20 wt%)	A technique consisting of spinning onto a water reservoir collector and drawing the resulting non-woven web in the form of a continuous yarn described.	Smit et al. (2005)
PVDF		DMF/acetone (7/3; 8/2; 9/1) (13–15 wt%)	Polymer electrospun into fibrous membranes and the effects of solvent, polymer concentration, and the gap distance on the morphology of the membranes studied.	Zhao, Z. Z., et al. (2005)
PVDF (Kynar 761)	55,000	Acetone/DMAc (3 : 7) (15 wt%)	A nanofibrous polymer electrolyte using the electrospun mat of fiber shown to have excellent mechanical as well as electrochemical properties.	Choi, S. W., et al. (2003)

(*Continued*)

	Polymer	Molecular Weight (Mw)	Solvent (Concentration)	Application Area	Reference
	PVDF (Solef 1010)	45,000*	DMAA (25 wt%)	Application of eletrospun nanofiber webs as an electrolyte binder or a separator for a battery studied.	Choi, S. S., et al. (2004b)
Polyamine	Poly(vinyl pyrrolidone (PVP)	55,000	DMF (47 wt%)	Silver nanoparticles introduced into nanofibers. Two methods to prepare PVP nanofibers containing Ag nanoparticles described.	Jin, W.-J., et al. (2005)
	PVP	1,300,000	Water/buffer (25 wt%)	Rod-shaped M13 viruses electrospun with polymer to fabricate one-dimensional micro- and nano-sized fibers.	Lee and Belcher (2004)
	PVP	1,130,000	EtOH (~10%)	Composite nanofiber with rod-like nanostructures of CdS and PbS incorporated into polymer fiber matrices developed.	Lu, X. F., et al. (2005a, 2005b)
	PVP	1,300,000	EtOH (3–9 wt%)	Effect of concentration on the electrospinnability studied using several polymer/solvent systems.	Shenoy et al. (2005b)
	PVP	1,300,000	MC, EtOH, DMF and blends (4 wt%)	The influence of solvents on micro-/nanofibers of PVP investigated. The DMF/ethanol ratio in mixed solvent investigated in particular.	Yang, Q. B., et al. (2004)

SBS copolymer	151,000	THF/DMF (3/1 w/w) (14 wt%)	Commercial styrene-butadiene-styrene triblock copolymer electrospun from solution to obtain nanofibers that were elastic, birefringent, and most had diameters around 100 nm.	Fong and Reneker (1999)
Silk + PEO	PEO = 900,000	HFIP (4.8–8.8 wt%)	Adhesion, spreading and proliferation of human bone marrow stromal cells (BMSCs) on silk matrices (blends with PEO) studied.	Jin et al. (2002)

*The molecular weight of the polymer is an estimate from literature. Also note where not specified, the average molecular weight is Mw.

REFERENCES

Abbas, M. A. and J. Latham (1967). "The instability of evaporating charged drops." *Journal of Fluid Mechanics* **30**(4):663–670.

Abidian, M. R., D.-H. Kim, and D. C. Martin (2006). "Conducting-polymer nanotubes for controlled drug release." *Advanced Materials* **18**(4):405–409.

Acatay, K., E. Simsek, C. Ow-Yang, and Y. Z. Menceloglu (2004). "Tunable, superhydrophobically stable polymeric surfaces by electrospinning." *Angewandte Chemie-International Edition* **43**(39):5210–5213.

Alberts, B., D. Bray, K. Hopkin, A. Johnson, J. Lewis, M. Raff, K. Roberts, and P. Walter (2003). *Essential Cell Biology*. Second Edition. Garland Science Publishing.

Arayanarakul, K., N. Choktaweesap, D. Ant-ong, C. Meechaisve, and P. Supaphol (2006). "Effects of poly(ethylene glycol), inorganic salt, sodium dodecyl sulfate, and solvent system on electrospinning of poly(ethylene oxide)." *Macromolecular Materials and Engineering* **291**(6):581–591.

Arichi, S. and S. Himuro (1989). "Solubility parameters of poly(4-acetoxystyrene) and poly(4-hydroxystyrene)." *Polymer* **30**(4):686–692.

Aryal, S., N. Dharmaraj, S. R. Bhattarai, M. S. Khil, and H. Y. Kim (2006). "Deposition of gold nanoparticles on electrospun $MgTiO_3$ ceramic nanofibers." *Journal of Nanoscience and Nanotechnology* **6**(2):510–513.

Aussawasathien, D., J.-H. Dong, and L. Dai (2005). "Electrospun polymer nanofiber sensors." *Synthetic Metals* **154**(1–3):37–40.

Aussawasathien, D., P. He, and L. Dai (2006). "Polymer nanofibers and polymer sheathed carbon nanotubes for sensors." In: *Polymeric Nanofibers*. ACS Symposium Series 918. Edited by D. H. Reneker and H. Fong. Oxford University Press (USA), p. 246.

Ayutsede, J., M. Gandhi, S. Sukigara, M. Micklus, H.-E. Chen, and F. Ko (2005). "Regeneration of *Bombyx mori* silk by electrospinning. Part 3: Characterization of electrospun nonwoven mat." *Polymer* **46**(5):1625–1634.

Ayutsede, J., M. Gandhi, S. Sukigara, H. H. Ye, C. M. Hsu, Y. Gogotsi, and F. Ko (2006). "Carbon nanotube reinforced *Bombyx mori* silk nanofibers by the electrospinning process." *Biomacromolecules* **7**(1):208–214.

Babel, A., D. Li, Y. Xia, and S. A. Jenekhe (2005). "Electrospun nanofibers of blends of conjugated polymers: morphology, optical properties, and field-effect transistors." *Macromolecules* **38**(11):4705–4711.

Badami, A. S., M. R. Kreke, M. S. Thompson, J. S. Riffle, and A. S. Goldstein (2006). "Effect of fiber diameter on spreading, proliferation, and differentiation of osteoblastic cells on electrospun poly(lactic acid) substrates." *Biomaterials* **27**(4):596–606.

Baker, B. M. and R. L. Mauck (2007). "The effect of nanofiber alignment on the maturation of engineered meniscus constructs." *Biomaterials* **28**(11):1967–1977.

Baker, D. A. and P. J. Brown (2005). "Crosslinked electrospun PET webs." *AATCC Review* **5**(7):28–33.

Baker, S. C., N. Atkin, P. A. Gunning, N. Granville, K. Wilson, D. Wilson, and J. Southgate (2006). "Characterisation of electrospun polystyrene scaffolds for three-dimensional in vitro biological studies." *Biomaterials* **27**(16):3136–3146.

Balzer, F., V. G. Bordo, A. C. Simonsen, and H.-G. Rubahn (2003). "Isolated hexaphenyl nanofibers as optical waveguides." *Applied Physics Letters* **82**(1):10–12.

Batteas, J. D., C. A. Michaels, and G. C. Walker (Eds.) (2005). *Applications of Scanned Probe Microscopy to Polymers*. ACS Symposium Series 897. Oxford University Press (USA).

Baughman, R. H., A. A. Zakhidov, and W. A. de Heer (2007). "Carbon nanotubes — the route toward applications." *Science* **297**(5582):787–792.

Baumgarten, P. K. (1971). "Electrostatic spinning of acrylic microfibers." *Journal of Colloid and Interface Science* **36**(1):71–79.

Bellan, L. M., J. Kameoka, and H. G. Craighead (2005). "Measurement of the Young's moduli of individual polyethylene oxide and glass nanofibres." *Nanotechnology* **16**(8):1095–1099.

Bellan, L. M., G. W. Coates, and H. G. Craighead (2006). "Poly(dicyclopentadiene) submicron fibers produced by electrospinning." *Macromolecular Rapid Communications* **27**(7):511–515.

Bergshoef, M. M. and G. J. Vancso (1999). "Transparent nanocomposites with ultrathin, electrospun nylon-4,6 fiber reinforcement." *Advanced Materials* **11**(16):1362–1365.

Benoit, J. M., J. P. Buisson, O. Chauvet, C. Godon, and S. Lefrant (2002). "Low-frequency Raman studies of multiwalled carbon nanotubes: experiments and theory." *Physical Review B* **66**(7):073417.

Berkland, C., D. W. Pack, and K. K. Kim (2004). "Controlling surface nano-structure using flow-limited field-injection electrostatic spraying (FFESS) of poly(D,L-lactide-*co*-glycolide)." *Biomaterials* **25**(25):5649–5658.

Bhattarai, N., D. Il Cha, S. R. Bhattarai, M. S. Khil, and H. Y. Kim (2003). "Biodegradable electrospun mat: novel block copolymer of poly(*p*-dioxanone-*co*-L-lactide)-*block*-poly(ethylene glycol)." *Journal of Polymer Science Part B: Polymer Physics* **41**(16):1955–1964.

Bhattarai, N., D. Edmondson, O. Veiseh, F. A. Matsen, and M. Zhang (2005). "Electrospun chitosan-based nanofibers and their cellular compatibility." *Biomaterials* **26**(31):6176–6184.

Bhowmick, S., A. Fowler, S. B. Warner, T. Meressi, and P. Gibson (2007). "Transport in 3-D nanofab geometries." National Textile Center Annual Reports, NTC Project F06-MD04.

Bini, T. B., S. J. Gao, T. C. Tan, S. Wang, A. Lim, L. B. Hai, and S. Ramakrishna (2004). "Electrospun poly(L-lactide-*co*-glycolide) biodegradable polymer nanofibre tubes for peripheral nerve regeneration." *Nanotechnology* **15**(11): 1459–1464.

Bini, T. B., S. J. Gao, S. Wang, and S. Ramakrishna (2006). "Poly(L-lactide-*co*-glycolide) biodegradable microfibers and electrospun nanofibers for nerve tissue engineering: an *in vitro* study." *Journal of Materials Science* **41**(19):6453–6459.

Birdi, K. S. S. (2003). *Scanning Probe Microscopes: Applications in Science and Technology*. CRC Press, Boca Raton, FL.

Bishop, A. and P. Gouma (2005). "Leuco-emeraldine based polyaniline — polyvinyl-pyrrolidone electrospun composites and bio-composites: a preliminary study of sensing behavior." *Reviews on Advanced Materials Science* **10**(3):209–214.

Bishop-Haynes, A. and P. Gouma (2007). "Electrospun polyaniline composites for NO_2 detection." *Materials and Manufacturing Processes* **22**(6):764–767.

Błasińska, A., I. Krucińska, and M. Chrzanowski (2004). "Dibutyrylchitin nonwoven biomaterials manufactured using electrospinning method." *Fibres and Textiles in Eastern Europe* **12**(48):51–55.

Bognitzki, M., H. Hou, M. Ishaque, T. Frese, M. Hellwig, C. Schwarte, A. Schaper, J. H. Wendorff, and A. Greiner (2000). "Polymer, metal, and hybrid nano- and mesotubes by coating degradable polymer template fibers (TUFT process)." *Advanced Materials* **12**(9):637–640.

Bognitzki, M., W. Czado, T. Frese, A. Schaper, M. Hellwig, M. Steinhart, A. Greiner, and J. H. Wendorff (2001a). "Nanostructured fibers via electrospinning." *Advanced Materials* **13**(1):70–72.

Bognitzki, M., T. Frese, M. Steinhart, A. Greiner, J. H. Wendorff, A. Schaper, and M. Hellwig (2001b). "Preparation of fibers with nanoscaled morphologies: electrospinning of polymer blends." *Polymer Engineering and Science* **41**(6):982–989.

Boland, E. D., G. E. Wnek, D. G. Simpson, K. J. Pawlowski, and G. L. Bowlin (2001). "Tailoring tissue engineering scaffolds using electrostatic processing techniques: a study of poly(glycolic acid) electrospinning." *Journal of Macromolecular Science Part A: Pure and Applied Chemistry* **38**(12):1231–1243.

Boland, E. D., J. A. Matthews, K. J. Pawlowski, D. G. Simpson, G. E. Wnek, and G. L. Bowlin (2004a). "Electrospinning collagen and elastin: preliminary vascular tissue engineering." *Frontiers in Bioscience* **9**(2):1422–1432.

Boland, E. D., T. A. Telemeco, D. G. Simpson, G. E. Wnek, and G. L. Bowlin (2004b). "Utilizing acid pretreatment and electrospinning to improve biocompatibility of poly(glycolic acid) for tissue engineering." *Journal of Biomedical Materials Research Part B: Applied Biomaterials* **71B**(1):144–152.

Boland, E. D., B. D. Coleman, C. P. Barnes, D. G. Simpson, G. E. Wnek, and G. L. Bowlin (2005). "Electrospinning polydioxanone for biomedical applications." *Acta Biomaterialia* **1**(1):115–123.

Boland, E. D., K. J. Pawlowski, C. P. Barnes, D. G. Simpson, G. E. Wnek, and G. L. Bowlin (2006). "Electrospinning of bioresorbable polymers for tissue engineering scaffolds." In: *Polymeric Nanofibers*. ACS Symposium Series 918. Edited by D. H. Reneker and H. Fong. Oxford University Press (USA). pp. 188–204.

Bölgen, N., Y. Z. Menceloğlu, K. Acatay, I. Vargel, and E. Pişkin (2005). "*In vitro* and *in vivo* degradation of non-woven materials made of poly(ε-caprolactone) nanofibers prepared by electrospinning under different conditions." *Journal of Biomaterials Science, Polymer Edition* **16**(12):1537–1555.

Bordi, F., R. H. Colby, C. Cametti, L. De Lorenzo, and T. Gili (2002). "Electrical conductivity of polyelectrolyte solutions in the semidilute and concentrated regime: the role of counterion condensation." *Journal of Physical Chemistry Part B* **106**(27):6887–6893.

Boudriot, U., B. Goetz, R. Dersch, A. Greiner, and J. H. Wendorff (2005). "Role of electrospun nanofibers in stem cell technologies and tissue engineering." *Macromolecular Symposia* **225**(1):9–16.

Brewster, M. E., G. Verreck, I. Chun, J. Rosenblatt, J. Mensch, A. Van Dijck, M. Noppe, A. Ariën, M. Bruining, and J. Peeters (2004). "The use of polymer-based electrospun nanofibers containing amorphous drug dispersions for the delivery of poorly water-soluble pharmaceuticals." *Pharmazie* **59**(5):387.

Briot, P. and M. Primet (1991). "Catalytic oxidation of methane over palladium supported on alumina. Effect of aging under reactants." *Applied Catalysis* **68**(1):301–314.

Buchko, C. J., L. C. Chen, Y. Shen, and D. C. Martin (1999). "Processing and microstructural characterization of porous biocompatible protein polymer thin films." *Polymer* **40**(26):7397–7407.

Buer, A., S. C. Ugbolue, and S. B. Warner (2001). "Electrospinning and properties of some nanofibers." *Textile Research Journal* **71**(4):323–328.

Bugarski, B., B. Amsden, R. J. Neufeld, D. Poncelet, and M. F. A. Goosen (1994). "Effect of electrode geometry and charge on the production of polymer microbeads by electrostatics." *The Canadian Journal of Chemical Engineering* **72**(June):517–521.

Buldum, A., C. B. Clemons, L. H. Dill, K. L. Kreider, G. W. Young, X. Zheng, E. A. Evans, G. Zhang, and S. I. Hariharan (2005). "Multiscale modeling, simulations, and experiments of coating growth on nanofibers. Part II. Deposition." *Journal of Applied Physics* **98**(4):044304.

Bunyan, N. N., J. Chen, I. Chen, and S. Farboodmanesh (2006). "Electrostatic effects of electrospun fiber deposition and alignment." In: *Polymeric Nanofibers*. ACS Symposium Series 918. Edited by D. H. Reneker and H. Fong. Oxford University Press (USA), p. 106.

Burger, C., B. S. Hsiao, and B. Chu (2006). "Nanofibrous materials and their applications." *Annual Review of Materials Research* **36**:333–368.

Burke, J. (1984). "Solubility parameters: Theory and application." In: the *AIC Book and Paper Group Annual,* Volume 3. Edited by Craig Jensen, p. 13–58.

Buttafoco, L., N. G. Kolkman, A. A. Poot, P. J. Dijkstra, I. Vermes, and J. Feijen (2005). "Poster abstracts: electrospinning collagen and elastin for tissue engineering small diameter blood vessels." *Journal of Controlled Release* **101**(1–3):322–324.

Buttafoco, L., N. G. Kolkman, P. Engbers-Buijtenhuijs, A. A. Poot, P. J. Dijkstra, I. Vermes, and J. Feijen (2006). "Electrospinning of collagen and elastin for tissue engineering applications." *Biomaterials* **27**(5):724–734.

Caruso, R. A., J. H. Schattka, and A. Greiner (2001). "Titanium dioxide tubes from sol-gel coating of electrospun polymer fibers." *Advanced Materials* **13**(20):1577–1579.

Casper, C. L., J. S. Stephens, N. G. Tassi, D. B. Chase, and J. F. Rabolt (2004). "Controlling surface morphology of electrospun polystyrene fibers: effect of humidity and molecular weight in the electrospinning process." *Macromolecules* **37**(2):573–578.

Casper, C. L., N. Yamaguchi, K. L. Kiick, and J. F. Rabolt (2005). "Functionalizing electrospun fibers with biologically relevant macromolecules." *Biomacromolecules* **6**(4):1998–2007.

Casper, C. L., W. Yang, M. C. Farach-Carson, and J. F. Rabolt (2006). "Understanding the effects of processing parameters on electrospun fibers and applications in tissue engineering." In: *Polymeric Nanofibers*. ACS Symposium Series 918. Edited by D. H. Reneker and H. Fong. Oxford University Press (USA), p. 205.

Cha, D. I., H. Y. Kim, K. H. Lee, Y. C. Jung, J. W. Cho, and B. C. Chun (2005). "Electrospun nonwovens of shape-memory polyurethane block copolymers." *Journal of Applied Polymer Science* **96**(2):460–465.

Chakrabarti, K., P. M. G. Nambissan, C. D. Mukherjee, K. K. Bardhan, C. Kim, and K. S. Yang (2006). "Positron annihilation spectroscopy of polyacrylonitrile-based carbon fibers embedded with multi-wall carbon nanotubes." *Carbon* **44**(5):948–953.

Chen, G.-X., Y. J. Li, and H. Shimizu (2007). "Ultrahigh-shear processing for the preparation of polymer/carbon nanotube composites." *Carbon* **45**(12): 2334–2340.

Chen, H. and Y.-L. Hsieh (2004). "Enzyme immobilization on ultrafine cellulose fibers via poly(acrylic acid) electrolyte grafts." *Biotechnology and Bioengineering* **90**(4):405–413.

Chen, X., C. Burger, D. Fang, I. Sics, X. Wang, W. He, R. H. Somani, K. Yoon, B. S. Hsiao, and B. Chu (2006). "In-situ x-ray deformation study of fluorinated multi-walled carbon nanotube and fluorinated ethylene-propylene nanocomposite fibers." *Macromolecules* **39**(16):5427–5437.

Chen, Z. H., M. D. Foster, W. S. Zhou, H. Fong, D. H. Reneker, R. Resendes, and I. Manners (2001). "Structure of poly(ferrocenyldimethylsilane) in electrospun nanofibers." *Macromolecules* **34**(18):6156–6158.

Chen, Z., X. Mo, and F. Qing (2007). "Electrospinning of collagen–chitosan complex." *Materials Letters* **61**(16):3490–3494.

Chew, S. Y., J. Wen, E. K. F. Yim, and K. W. Leong (2005). "Sustained release of proteins from electrospun biodegradable fibers." *Biomacromolecules* **6**(4): 2017–2024.

Cho, J. and I. M. Daniel (2008). "Reinforcement of carbon/epoxy composites with multi-wall carbon nanotubes and dispersion enhancing block copolymers." *Scripta Materialia* **58**(7):533–536.

Choi, J. S., S. W. Lee, L. Jeong, S. H. Bae, B. C. Min, J. H. Youk, and W. H. Park (2004). "Effect of organosoluble salts on the nanofibrous structure of electrospun poly(3-hydroxybutyrate-co-3-hydroxyvalerate)." *International Journal of Biological Macromolecules* **34**(4):249–256.

Choi, S. S., S. G. Lee, C. W. Joo, S. S. Im, and S. H. Kim (2004a). "Formation of interfiber bonding in electrospun poly(etherimide) nanofiber web." *Journal of Materials Science* **39**(4):1511–1513.

Choi, S. S., Y. S. Lee, C. W. Joo, S. G. Lee, J. K. Park, and K. S. Han (2004b). "Electrospun PVDF nanofiber web as polymer electrolyte or separator." *Electrochimica Acta* **50**(2–3):339–343.

Choi, S.-S., B. Y. Chu, D. S. Hwang, S. G. Lee, W. H. Park, and J. K. Park (2005). "Preparation and characterization of polyaniline nanofiber webs by template reaction with electrospun silica nanofibers." *Thin Solid Films* **477**(1–2):233–239.

Choi, S. W., S. M. Jo, W. S. Lee, and Y. R. Kim (2003). "An electrospun poly(vinylidene fluoride) nanofibrous membrane and its battery applications." *Advanced Materials* **15**(23):2027–2032.

Chronakis, I. S. (2005). "Novel nanocomposites and nanoceramics based on polymer nanofibers using electrospinning process—a review." *Journal of Materials Processing Technology* **167**(2–3):283–293.

Chronakis, I. S., S. Grapenson, and A. Jakob (2006a). "Conductive polypyrrole nanofibers via electrospinning: electrical and morphological properties." *Polymer* **47**(5):1597–1603.

Chronakis, I. S., B. Milosevic, A. Frenot, and L. Ye (2006b). "Generation of molecular recognition sites in electrospun polymer nanofibers via molecular imprinting." *Macromolecules* **39**(1):357–361.

Chua, K.-N., W.-S. Lim, P. C. Zhang, H. F. Lu, J. Wen, S. Ramakrishna, K. W. Leong, and H.-Q. Mao (2005). "Stable immobilization of rat hepatocyte spheroids on galactosylated nanofiber scaffold." *Biomaterials* **26**(15):2537–2547.

Chua, K.-N., C. Chou, P.-C. Lee, Y.-N. Tang, S. Ramakrishna, K. W. Leong, and H.-Q. Mao (2006). "Surface-aminated electrospun nanofibers enhance adhesion and expansion of human umbilical cord blood hematopoietic stem/progenitor cells." *Biomaterials* **27**(36):6043–6051.

Chuangchote, S. and P. Supaphol (2006). "Fabrication of aligned poly(vinyl alcohol) nanofibers by electrospinning." *Journal of Nanoscience and Nanotechnology* **6**(1):125–129.

Chung, G. S., S. M. Jo, and B. C. Kim (2005). "Properties of carbon nanofibers prepared from electrospun polyimide." *Journal of Applied Polymer Science* **97**(1):165–170.

Colby, R. H. and M. Rubinstein (1990). "Two-parameter scaling for polymers in θ solvents." *Macromolecules* **23**(10):2753–2757.

Colby, R. H., L. J. Fetters, W. G. Funk, and W. W. Graessley (1991). "Effects of concentration and thermodynamic interaction on the viscoelastic properties of polymer solutions." *Macromolecules* **24**(13):3873–3882.

Comyn, J. (1985). *Polymer Permeability*. Elsevier Applied Science Publishers, New York.

Courtney, T., M. S. Sacks, J. Stankus, J. J. Guan, and W. R. Wagner (2006). "Design and analysis of tissue engineering scaffolds that mimic soft tissue mechanical anisotropy." *Biomaterials* **27**(19):3631–3638.

Crank, J. (1980). *The Mathematics of Diffusion*. Oxford University Press.

Cukierman, E., R. Pankov, D. R. Stevens, and K. M. Yamada (2001). "Taking cell-matrix adhesions to the third dimension." *Science* **294**(5547):1708–1712.

Czaplewski, D., J. Kameoka, and H. G. Craighead (2003). "Nonlithographic approach to nanostructure fabrication using a scanned electrospinning source." *Journal of Vacuum Science and Technology B* **21**(6):2994–2997.

Da Silva, S., L. Grosjean, N. Ternan, P. Mailley, T. Livache, and S. Cosnier (2004). "Biotinylated polypyrrole films: an easy electrochemical approach for the reagentless immobilization of bacteria on electrode surfaces." *Bioelectrochemistry* **63**(1–2):297–301.

Dai, H. Q., J. Gong, H. Kim, and D. Lee (2002). "A novel method for preparing ultrafine alumina-borate oxide fibres via an electrospinning technique." *Nanotechnology* **13**(5):674–677.

Dai, X. S. and S. Shivkumar (2007). "Electrospinning of PVA-calcium phosphate sol precursors for the production of fibrous hydroxyapatite." *Journal of the American Ceramic Society* **90**(5):1412–1419.

Dalton, P. D., D. Klee, and M. Möller (2005). "Electrospinning with dual collection rings." *Polymer* **46**(3):611–614.

Dalton, P. D., K. Klinkhammer, J. Salber, D. Klee, and M. Möller (2006). "Direct in vitro electrospinning with polymer melts." *Biomacromolecules* **7**(3):686–690.

Davies, C. N. (1973). *Air Filtration*. Academic Press, London.

de Gennes, P. G. (1979). *Scaling Concepts in Polymer Physics*. Cornell University Press, Ithaca, NY.

Deb, P. C. and S. R. Palit (1973). "Solution properties of poly(methyl methacrylate) in mixtures of two nonsolvents (carbon tetrachloride and *n*-alcohols)." *Die Makromolekulare Chemie* **166**(1):227–234.

Deitzel, J. M., J. Kleinmeyer, D. Harris, and N. C. B. Tan (2001a). "The effect of processing variables on the morphology of electrospun nanofibers and textiles." *Polymer* **42**(1):261–272.

Deitzel, J. M., J. D. Kleinmeyer, J. K. Hirvonen, and N. C. B. Tan (2001b). "Controlled deposition and collection of electro-spun poly(ethylene oxide) fibers." Army Research Laboratory ARL-TR-2415, Aberdeen Proving Grounds, March 2001.

Deitzel, J. M., J. D. Kleinmeyer, J. K. Hirvonen, and N. C. B. Tan (2001c). "Controlled deposition of electrospun poly(ethylene oxide) fibers." *Polymer* **42**(19):8163–8170.

Deitzel, J. M., C. Krauthauser, D. Harris, C. Perganis, and J. Kleinmeyer (2006). "Key parameters influencing the onset and maintenance of electrospinning jet." In: *Polymeric Nanofibers*. ACS Symposium Series 918. Oxford University Press (USA), p. 56.

Delozier, D. M., K. A. Watson, J. G. Smith, T. C. Clancy, and J. W. Connell (2006). "Investigation of aromatic/aliphatic polyimides as dispersants for single wall carbon nanotubes." *Macromolecules* **39**(5):1731–1739.

Demczky, B. G., Y. M. Wang, J. Cumings, M. Hetman, W. Han, A. Zettl, and R. O. Richie (2002). "Direct mechanical measurement of the tensile strength and elastic modulus of multiwalled carbon nanotubes." *Materials Science and Engineering A* **334**(1–2):173–178.

Demir, M. M., I. Yilgor, E. Yilgor, and B. Erman (2002). "Electrospinning of polyurethane fibers." *Polymer* **43**(11):3303–3309.

Demir, M. M., M. A. Gulgun, Y. Z. Menceloglu, B. Erman, S. S. Abramchuk, E. E. Makhaeva, A. R. Khokhlov, V. G. Matveeva, and M. G. Sulman (2004). "Palladium nanoparticles by electrospinning from poly(acrylonitrile-*co*-acrylic acid)-$PdCl_2$ solutions. Relations between preparation conditions, particle size, and catalytic activity." *Macromolecules* **37**(5):1787–1792.

Dersch, R., T. Q. Liu, A. K. Schaper, A. Greiner, and J. H. Wendorff (2003). "Electrospun nanofibers: internal structure and intrinsic orientation." *Journal of Polymer Science Part A: Polymer Chemistry* **41**(4):545–553.

Dersch, R., M. Steinhart, U. Boudriot, A. Greiner, and J. H. Wendorff (2005). "Nanoprocessing of polymers: applications in medicine, sensors, catalysis, photonics." *Polymers for Advanced Technologies* **16**(2–3):276–282.

DeShon, W. E. and R. S. Carson (1968). "Electric field investigations and a model for electrical liquid spraying." *Journal of Colloid and Interface Science* **28**(1):161–166.

Devienne, M. (1958). *Frottement et Echanges Thermiques dans la Gaz Rarefiles*. Gauthier-Villars, Paris, France.

Dharmaraj, N., H. C. Park, C. K. Kim, H. Y. Kim, and D. R. Lee (2004a). "Nickel titanate nanofibers by electrospinning." *Materials Chemistry and Physics* **87**(1):5–9.

Dharmaraj, N., H. C. Park, B. M. Lee, P. Viswanathamurthi, H. Y. Kim, and D. R. Lee (2004b). "Preparation and morphology of magnesium titanate nanofibers via electrospinning." *Inorganic Chemistry Communications* **7**(3):431–433.

Dharmaraj, N., C. H. Kim, K. W. Kim, H. Y. Kim, and E. K. Suh (2006a). "Spectral studies of SnO_2 nanofibres prepared by electrospinning method." *Spectrochimica Acta Part A: Molecular and Biomolecular Spectroscopy* **64**(1):136–140.

Dharmaraj, N., H. C. Park, C. H. Kim, P. Viswanathamurthi, and H. Y. Kim (2006b). "Nanometer sized tantalum pentoxide fibers prepared by electrospinning." *Materials Research Bulletin* **41**(3):612–619.

Ding, B., H. Y. Kim, S. C. Lee, D. R. Lee, and K. J. Choi (2002a). "Preparation and characterization of nanoscaled poly(vinyl alcohol) fibers via electrospinning." *Fibers and Polymers* **3**(2):73–79.

Ding, B., H.-Y. Kim, S.-C. Lee, C.-L. Shao, D.-R. Lee, S.-J. Park, G.-B. Kwag, and K.-J. Choi (2002b). "Preparation and characterization of a nanoscale poly(vinyl alcohol) fiber aggregate produced by an electrospinning method." *Journal of Polymer Science Part B: Polymer Physics* **40**(13):1261–1268.

Ding, B., H. Kim, C. Kim, M. Khil, and S. Park (2003). "Morphology and crystalline phase study of electrospun TiO_2–SiO_2 nanofibres." *Nanotechnology* **14**(5):532–537.

Ding, B., C. K. Kim, H. Y. Kim, M. K. Seo, and S. J. Park (2004a). "Titanium dioxide nanofibers prepared by using electrospinning method." *Fibers and Polymers* **5**(2):105–109.

Ding, B., J. H. Kim, E. Kimura, and S. Shiratori (2004b). "Layer-by-layer structured films of TiO_2 nanoparticles and poly(acrylic acid) on electrospun nanofibres." *Nanotechnology* **15**(8):913–917.

Ding, B., J. H. Kim, Y. Miyazaki, and S. M. Shiratori (2004c). "Electrospun nanofibrous membranes coated quartz crystal microbalance as gas sensor for NH_3 detection." *Sensors and Actuators B: Chemical* **101**(3):373–380.

Ding, B., E. Kimura, T. Sato, S. Fujita, and S. Shiratori (2004d). "Fabrication of blend biodegradable nanofibrous nonwoven mats via multi-jet electrospinning." *Polymer* **45**(6):1895–1902.

Ding, B., K. Fujimoto, and S. Shiratori (2005a). "Preparation and characterization of self-assembled polyelectrolyte multilayered films on electrospun nanofibers." *Thin Solid Films* **491**(1–2):23–28.

Ding, B., J. Gong, J. Kim, and S. Shiratori (2005b). "Polyoxometalate nanotubes from layer-by-layer coating and thermal removal of electrospun nanofibres." *Nanotechnology* **16**(6):785–790.

Ding, B., M. Yamazaki, and S. Shiratori (2005c). "Electrospun fibrous polyacrylic acid membrane-based gas sensors." *Sensors and Actuators B: Chemical* **106**(1):477–483.

Dole, M., L. L. Mack, R. L. Hines, R. C. Mobley, L. D. Ferguson, and M. B. Alice (1968). "Molecular beams of macroions." *Journal of Chemical Physics* **49**(5):2240–2249.

Dong, H., V. Nyame, A. G. MacDiarmid, and W. E. Jones, Jr. (2004). "Polyaniline/poly(methyl methacrylate) coaxial fibers: the fabrication and effects of the solution properties on the morphology of electrospun core fibers." *Journal of Polymer Science Part B: Polymer Physics* **42**(21):3934–3942.

Dong, H., E. Fey, A. Gandelman, and W. E. Jones, Jr. (2006). "Synthesis and assembly of metal nanoparticles on electrospun poly(4-vinylpyridine) fibers and poly(4-vinylpyridine) composite fibers." *Chemistry of Materials* **18**(8):2008–2011.

Dong, W., T. Zhang, M. McDonald, C. Padilla, J. Epstein, and Z. R. Tian (2006). "Biocompatible nanofiber scaffolds on metal for controlled release and cell colonization." *Nanomedicine: Nanotechnology, Biology and Medicine* **2**(4):248–252.

Doshi, J. and D. H. Reneker (1995). "Electrospinning process and applications of electrospun fibers." *Journal of Electrostatics* **35**(2–3):151–160.

Dosunmu, O. O., G. G. Chase, W. Kataphinan, and D. H. Reneker (2006). "Electrospinning of polymer nanofibres from multiple jets on a porous tubular surface." *Nanotechnology* **17**(4):1123–1127.

Drew, C., X. Y. Wang, K. Senecal, H. Schreuder-Gibson, J. N. He, J. Kumar, and L. A. Samuelson (2002). "Electrospun photovoltaic cells." *Journal of Macromolecular Science—Pure and Applied Chemistry* **A39**(10):1085–1094.

Drew, C., X. Liu, D. Ziegler, X. Y. Wang, F. F. Bruno, J. Whitten, L. A. Samuelson, and J. Kumar (2003a). "Metal oxide-coated polymer nanofibers." *Nano Letters* **3**(2):143–147.

Drew, C., X. Y. Wang, L. A. Samuelson, and J. Kumar (2003b). "The effect of viscosity and filler on electrospun fiber morphology." *Journal of Macromolecular Science—Pure and Applied Chemistry* **A40**(12):1415–1422.

Drew, C., X. Y. Wang, F. F. Bruno, L. A. Samuelson, and J. Kumar (2005). "Electrospun polymer nanofibers coated with metal oxides by liquid phase deposition." *Composite Interfaces* **11**(8–9):711–724.

Dror, Y., W. Salalha, R. L. Khalfin, Y. Cohen, A. L. Yarin, and E. Zussman (2003). "Carbon nanotubes embedded in oriented polymer nanofibers by electrospinning." *Langmuir* **19**(17):7012–7020.

Drummy, L. F., J. Y. Yang, and D. C. Martin (2004). "Low-voltage electron microscopy of polymer and organic molecular thin films." *Ultramicroscopy* **99**(4):247–256.

Du, F., J. E. Fischer, and K. I. Winey (2003). "Coagulation method for preparing single-walled carbon nanotube/poly(methyl methacrylate) composites and their modulus, electrical conductivity, and thermal stability." *Journal of Polymer Science Part B: Polymer Physics* **41**(24):3333–3338.

Du, F., R. C. Scogna, W. Zhou, S. Brand, J. E. Fischer, and W. I. Winey (2004). "Nanotube networks in polymer nanocomposites: rheology and electrical conductivity." *Macromolecules* **37**(24):9048–9055.

Duan, B., C. H. Dong, X. Y. Yuan, and K. D. Yao (2004). "Electrospinning of chitosan solutions in acetic acid with poly(ethylene oxide)." *Journal of Biomaterials Science—Polymer Edition* **15**(6):797–811.

Duft, D., T. Achtzehn, R. Müller, B. A. Huber, and T. Leisner (2003). "Coulomb fission: Rayleigh jets from levitated microdroplets." *Nature* **421**(9):128.

Dunn, M. G., L. D. Bellincampi, A. J. Tria, Jr., and J. P. Zawadsky (1997). "Preliminary development of a collagen-PLA composite for ACL reconstruction." *Journal of Applied Polymer Science* **63**(11):1423–1428.

Dzenis, Y. A. (2004). "Spinning continuous fibers for nanotechnology." *Science* **304**(5679):1917–1919.

Ensor, D. A., H. Walls, and A. L. Andrady (2006). "A novel nanofiber filter for protection against chem-bio aerosols." Scientific Conference for Chemical and Biological Defense Research, US Army Edgewood Chemical Biological Center Hunt Valley, MD.

Ensor, D. S., K. K. Foarde, J. T. Hanley, and D. W. van Osdell (2003). "The effect of filtration in heating, ventilation, and air-conditioning systems." In: *Indoor Environment: Airborne Particles and Settled Dust*. Edited by Lidia Morawska and Tunga Salthammer, Wiley-VCH, Weinheim, Germany.

Entov, V. M. and L. E. Shmarayan (1997). "Numerical modeling of the capillary breakup of jets of polymeric liquids." *Fluid Dynamics* **32**(5):696–703.

Erickson, J. (2003). *Incorporating Carbon Nanotubes into Polypropylene Fibers*. MSc Thesis. Department of Textile and Apparel Technology Management. North Carolina State University, Raleigh, NC.

Fang, D., C. Chang, B. S. Hsiao, and B. Chu (2006). "Development of multiple-jet electrospinning technology." In: *Polymeric Nanofibers*. ACS Symposium Series 918. Edited by D. H. Reneker and H. Fong. Oxford University Press (USA), p. 91.

Fang, X. and D. H. Reneker (1997). "DNA fibers by electrospinning." *Journal of Macromolecular Science: Physics* **B36**(2):169–173.

Fang, Z. X., L. Zhang, T. Han, and P. Hu (2004). "Studies on the morphology and structure of electrospun PHBV fibers." *Acta Polymerica Sinica* (4):500–505.

Faury, G., G. M. Maher, D. Y. Li, M. T. Keating, R. P. Mecham, and W. A. Boyle (1999). "Relation between outer and luminal diameter in cannulated arteries." *American Journal of Physiology – Heart and Circulatory Physiology* **277**(5):H1745–H1753.

Fei, J., Z. Zhang, and L. Gu (2002). "Bending behaviour of electroresponsive poly(vinyl alcohol)/poly(acrylic acid) semi-interpenetrating network hydrogel fibres under an electric stimulus." *Polymer International* **51**(6):502–509.

Feng, J. J. (2002). "The stretching of an electrified non-Newtonian jet: a model for electrospinning." *Physics of Fluids* **14**(11):3912–3926.

Feng, J. J. (2003). "Stretching of a straight electrically charged viscoelastic jet." *Journal of Non-Newtonian Fluid Mechanics* **116**(1):55–70.

Fennessey, S. F. and R. J. Farris (2004). "Fabrication of aligned and molecularly oriented electrospun polyacrylonitrile nanofibers and the mechanical behavior of their twisted yarns." *Polymer* **45**(12):4217–4225.

Fennessey, S. F., A. Pedicini, and R. J. Farris (2006). "Mechanical behavior of nonwoven electrospun fabrics and yarns." In: *Polymeric Nanofibers*. ACS Symposium Series 918. Edited by D. H. Reneker and H. Fong. Oxford University Press (USA), p. 403.

Fetters, L. J., D. J. Lohse, and R. H. Colby (2007). "Chain dimensions and entanglement spacings." In: *Physical Properties of Polymers Handbook*. Second Edition. Edited by James E. Mark. Springer-Verlag, New York. Chapter 25, pp. 447–454.

Fong, H. and D. H. Reneker (1999). "Elastomeric nanofibers of styrene-butadiene-styrene triblock copolymer." *Journal of Polymer Science Part B: Polymer Physics* **37**(24):3488–3493.

Fong, H., I. Chun, and D. H. Reneker (1999). "Beaded nanofibers formed during electrospinning." *Polymer* **40**(16):4585–4592.

Fong, H., W. D. Liu, C.-S. Wang, and R. A. Vaia (2002). "Generation of electrospun fibers of nylon 6 and nylon 6-montmorillonite nanocomposite." *Polymer* **43**(3):775–780.

Fornes, T. D., J. W. Baur, Y. Sabba, and E. L. Thomas (2006). "Morphology and properties of melt-spun polycarbonate fibers containing single- and multi-wall carbon nanotubes." *Polymer* **47**(5):1704–1714.

Fridrikh, S. V., J. H. Yu, M. P. Brenner, and G. C. Rutledge (2003). "Controlling the fiber diameter during electrospinning." *Physical Review Letters* **90**(14):144502.

Fridrikh, S. V., J. H. Yu, M. P. Brenner, and G. C. Rutledge (2006). "Nonlinear whipping behavior of electrified fluid jets." In: *Polymeric Nanofibers*. ACS Symposium Series 918. Edited by D. H. Reneker and H. Fong. Oxford University Press (USA), p. 36.

Frisch, H. L. and R. Simha (1956). In: *Rheology: Theory and Applications, Vol. I*. Edited by F. R. Eirich. Academic Press, New York, pp. xiv+716.

Fujihara, K., M. Kotaki, and S. Ramakrishna (2005). "Guided bone regeneration membrane made of polycaprolactone/calcium carbonate composite nanofibers." *Biomaterials* **26**(19):4139–4147.

Fultz, B. and J. M. Howe (2005). *Transmission Electron Microscopy and Diffractometry of Materials*. Third Edition. Springer-Verlag, Berlin pp. 704.

Gao, J., L. Niklason, and R. Langer (1998). "Surface hydrolysis of poly(glycolic acid) meshes increases the seeding density of vascular smooth muscle cells." *Journal of Biomedical Materials Research Part A* **42**(3):417–424.

Ge, J. J., H. Q. Hou, Q. Li, M. J. Graham, A. Greiner, D. H. Reneker, F. W. Harris, and S. Z. D. Cheng (2004). "Assembly of well-aligned multiwalled carbon nanotubes in confined polyacrylonitrile environments: electrospun composite nanofiber sheets." *Journal of the American Chemical Society* **126**(48):15754–15761.

Ge, S., Y. Pu, W. Zhang, M. Rafailovich, J. Sokolov, C. Buenviaje, R. Buckmaster, and R. M. Overney (2000). "Shear modulation force microscopy study of near surface glass transition temperatures." *Physical Review Letters* **85**:2340–2343.

Geng, X., O.-H. Kwon, and J. Jang (2005). "Electrospinning of chitosan dissolved in concentrated acetic acid solution." *Biomaterials* **26**(27):5427–5432.

Gibson, P., D. Rivin, C. Kendrick, and H. Schreuder-Gibson (1999). "Humidity-dependent air permeability of textile materials." *Textile Research Journal* **69**(5):311–317.

Gibson, P. and H. Schreuder-Gibson (2002). "Use of electrospun nanofibers for aerosol filtration in textile structures." US Army Soldier Systems Center, AMSSB-RSS-MS(N), Natick 01760–5020.

Gibson, P. and H. Schreuder-Gibson (2006). "Applications of electrospun nanofibers in current and future materials." In: *Polymeric Nanofibers*. ACS Symposium Series 918. Edited by D. H. Reneker and H. Fong. Oxford University Press (USA), p. 121.

Gong, J., X.-D. Li, B. Ding, D.-R. Lee, and H.-Y. Kim (2003). "Preparation and characterization of $H_4SiMo_{12}O_{40}$/poly(vinyl alcohol) fiber mats produced by an electrospinning method." *Journal of Applied Polymer Science* **89**(6):1573–1578.

Gong, J., C. L. Shao, Y. Pan, F. M. Gao, and L. Y. Qu (2004). "Preparation, characterization and swelling behavior of $H_3PW_{12}O_{40}$/poly(vinyl alcohol) fiber aggregates produced by an electrospinning method." *Materials Chemistry and Physics* **86**(1):156–160.

González, R. and N. J. Pinto (2005). "Electrospun poly(3-hexylthiophene-2,5-diyl) fiber field effect transistor." *Synthetic Metals* **151**(3):275–278.

Gonzalez, R. C. and R. E. Woods (2001). *Digital Image Processing*. Second Edition. Prentice-Hall, Upper Saddle River, New Jersey.

Gordeyev, S. A., J. A. Ferreira, C. A. Bernardo, and I. M. Ward (2001). "A promising conductive material: highly oriented polypropylene filled with shot vapour-grown carbon fibres." *Material Letters* **51**(1):32–36.

Graessley, W. W., R. L. Hazelton, and L. R. Lindeman (1967). "The shear-rate dependence of viscosity in concentrated solutions of narrow-distribution polystyrene." *Journal of Rheology* **11**(3):267–285.

Graessley, W. W. (1980). "Polymer chain dimensions and the dependence of viscoelastic properties on concentration, molecular weight and solvent power." *Polymer* **21**(3):258–262.

Graham, K., M. Ouyang, T. Raether, T. Grafe, B. McDonald, and P. Knauf (2002). "Polymeric nanofibers in air filter applications." Presented at the Fifteenth Annual Technical Conference & Expo of the American Filtration & Separations Society, Galveston, Texas, April 9–12, 2002.

Grant, P. V., C. M. Vaz, P. E. Tomlins, L. Mikhalovska, S. Mikhalovsky, S. James, and Vadgama, P. (2006). "Physical characterization of a polycaprolactone tissue scaffold." In: *Surface Chemistry in Biomedical and Environmental Science*.

Edited by J. P. Blitz and V. M. Gun'ko. NATO Science Series II: Mathematics, Physics, and Chemistry, Vol. 228. Springer, pp. 215–228.

Gu, S. Y. and J. Ren (2005). "Process optimization and empirical modeling for electrospun poly(D,L-lactide) fibers using response surface methodology." *Macromolecular Materials and Engineering* **290**(11):1097–1105.

Gu, S. Y., J. Ren, and Q. L. Wu (2005a). "Preparation and structures of electrospun PAN nanofibers as a precursor of carbon nanofibers." *Synthetic Metals* **155**(1):157–161.

Gu, S. Y., J. Ren, and G. J. Vancso (2005b). "Process optimization and empirical modeling for electrospun polyacrylonitrile (PAN) nanofiber precursor of carbon nanofibers." *European Polymer Journal* **41**(11):2559–2568.

Gu, S. Y., Q.-L. Wu, J. Ren, and G. J. Vancso (2005c). "Mechanical properties of a single electrospun fiber and its structures." *Macromolecular Rapid Communications* **26**(9):716–720.

Guan, G.-H., C.-C. Li, and D. Zhang (2005). "Spinning and properties of poly(ethylene terephthalate)/organomontmorillonite nanocomposite fibers." *Journal of Applied Polymer Science* **95**(6):1443–1447.

Guan, H. Y., C. L. Shao, S. B. Wen, B. Chen, J. Gong, and X. H. Yang (2003a). "A novel method for preparing Co_3O_4 nanofibers by using electrospun PVA/cobalt acetate composite fibers as precursor." *Materials Chemistry and Physics* **82**(3):1002–1006.

Guan, H. Y., C. L. Shao, S. B. Wen, B. Chen, J. Gong, and X. H. Yang (2003b). "Preparation and characterization of NiO nanofibres via an electrospinning technique." *Inorganic Chemistry Communications* **6**(10):1302–1303.

Guan, H. Y., C. L. Shao, Y. C. Liu, D. X. Han, X. H. Yang, and N. Yu (2004). "Fabrication of ZrO_2 nanofibers by electrospinning." *Chemical Journal of Chinese Universities (Chinese)* **25**(8):1413.

Guo, Q. P. (2003). "A DSC study on miscible blends containing two crystalline components: poly(ε-caprolactone)/poly[3,3-bis(chloromethyl) oxetane." *Die Makromolekulare Chemie* **191**(11):2639–2645.

Gupta, P. and G. L. Wilkes (2003). "Some investigations on the fiber formation by utilizing a side-by-side bicomponent electrospinning approach." *Polymer* **44**(20):6353–6359.

Gupta, P., S. R. Trenor, T. E. Long, and G. L. Wilkes (2004). "In situ photo-cross-linking of cinnamate functionalized poly(methyl methacrylate-*co*-2-hydroxyethyl acrylate) fibers during electrospinning." *Macromolecules* **37**(24):9211–9218.

Gupta, P., C. Elkins, T. E. Long, and G. L. Wilkes (2005). "Electrospinning of linear homopolymers of poly(methyl methacrylate): exploring relationships between fiber formation, viscosity, molecular weight and concentration in a good solvent." *Polymer* **46**(13):4799–4810.

Halpin, J. C. (1969). "Stiffness and expansion estimates for oriented short fiber composites." *Journal of Composite Materials* **3**(4):732–734.

Han, D.-H. (2002). "Correlations between workplace protection factors and fit factors for filtering facepieces in the welding workplace." *Industrial Health* **40**(2):328–334.

Han, L. and A. L. Andrady (2005). "Microscopic studies on electrospun polymer nanofibers." 230th American Chemical Society National Meeting, Washington, DC, August 28, 2005.

Han, S. O., W. K. Son, D. W. Cho, J. H. Youk, and W. H. Park (2004). "Preparation of porous ultra-fine fibres via selective thermal degradation of electrospun polyetherimide/poly(3-hydroxybutyrate-*co*-3-hydroxyvalerate) fibres." *Polymer Degradation and Stability* **86**(2):257–262.

Han, S. O., W. K. Son, J. H. Youk, T. S. Lee, and W. H. Park (2005). "Ultrafine porous fibers electrospun from cellulose triacetate." *Materials Letters* **59**(24–25):2998–3001.

Hansen, C. M. (1967a). "The three dimensional solubility parameter—key to paint component affinities. Part I: Solvents, plasticizers, polymers, and resins." *Journal of Paint Technology* **39**(505):104–117.

Hansen, C. M. and K. Skaarup (1967b). "The three dimensional solubility parameter—key to paint component affinities. Part III: Independent calculation of the parameter components." *Journal of Paint Technology* **39**(511):511–514.

Hawkes, P. W. and J. C. H. Spence (Eds.) (2008). *Science of Microscopy*. Springer-Verlag, Berlin, Germany.

Hayati, I., A. I. Bailey, and Th. F. Tadros (1987). "Investigations into the mechanisms of electrohydrodynamic spraying of liquids: I. Effect of electric field and the environment on pendant drops and factors affecting the formation of stable jets and atomization." *Journal of Colloid Interface Science* **117**(1):205–221.

He, C. H. and J. Gong (2003). "The preparation of PVA–Pt/TiO_2 composite nanofiber aggregate and the photocatalytic degradation of solid-phase polyvinyl alcohol." *Polymer Degradation and Stability* **81**(1):117–124.

He, J.-H., Y.-Q. Wan, and J.-Y. Yu (2004). "Application of vibration technology to polymer electrospinning." *International Journal of Nonlinear Sciences and Numerical Simulation* **5**(3):253–262.

He, J.-H., Y. Wu, and N. Pang (2005a). "A mathematical model for preparation by AC-electrospinning process." *International Journal of Nonlinear Sciences and Numerical Simulation* **6**(3):243–248.

He, J.-H., Y. Wu, and W.-W. Zuo (2005b). "Critical length of straight jet in electrospinning." *Polymer* **46**(26):12637–12640.

He, J.-H., Y.-Q. Wan, and J.-Y. Yu (2005c). "Scaling law in electrospinning: relationship between electric current and solution flow rate." *Polymer* **46**(8):2799–2801.

He, J.-H., Y. Liu, L. Xu, and J.-Y. Yu (2007a). "Micro sphere with nanoporosity by electrospinning." *Chaos, Solitons & Fractals* **32**(3):1096–1100.

He, J.-H., Y.-Q. Wan, and L. Xu (2007b). "Nano-effects, quantum-like properties in electrospun nanofibers." *Chaos, Solitons & Fractals* **33**(1):26–37.

He, J.-H., L. Xu, Y. Wu, and Y. Liu (2007c). "Mathematical models for continuous electrospun nanofibers and electrospun nanoporous microspheres." *Polymer International* **56**(11):1323–1329.

He, W., Z. W. Ma, T. Yong, W. E. Teo, and S. Ramakrishna (2005a). "Fabrication of collagen-coated biodegradable polymer nanofiber mesh and its potential for endothelial cells growth." *Biomaterials* **26**(36):7606–7615.

He, W., T. Yong, W. E. Teo, Z. Ma, and S. Ramakrishna (2005b). "Fabrication and endothelialization of collagen-blended biodegradable polymer nanofibers: potential vascular graft for blood vessel tissue engineering." *Tissue Engineering* **11**(9–10):1574–1588.

He, W., T. Yong, Z. W. Ma, R. Inai, W. E. Teo, and S. Ramakrishna (2006). "Biodegradable polymer nanofiber mesh to maintain functions of endothelial cells." *Tissue Engineering* **12**(9):2457–2466.

Hendricks, C. D., Jr., R. S. Carson, J. J. Hogan, and J. M. Schneider (1964). "Photomicrography of electrically sprayed heavy particles." *AIAA Journal* **2**(4):733–737.

Hertel, T., R. Martel, and P. Avouris (1998). "Manipulation of individual carbon nanotubes and their interaction with surfaces." *Journal of Physical Chemistry B* **102**(6):910–915.

Higgins, S. P., A. K. Solan, and L. E. Niklason (2003). "Effects of polyglycolic acid on porcine smooth muscle cell growth and differentiation." *Journal of Biomedical Materials Research Part A* **67A**(1):295–302.

Higuchi, T. (1961). "Rate of release of medicaments from ointment bases containing drugs in suspension." *Journal of Pharmaceutical Sciences* **50**(10):874–875.

Hinman, M. B. and R. V. Lewis (1992). "Isolation of a clone encoding a second dragline silk fibroin. *Nephila clavipes* dragline silk is a two-protein fiber." *Journal of Biological Chemistry* **267**(27):19320–19324.

Hirano, Y. and D. J. Mooney (2004). "Peptide and protein presenting materials for tissue engineering." *Advanced Materials* **16**(1):17–25.

Ho, B.-C., W.-K. Chin, and Y.-D. Lee (1991). "Solubility parameters of poly methacrylonitrile, poly(methacrylic acid) and methacrylonitrile/methacrylic acid copolymer." *Journal of Applied Polymer Science* **42**(1):99–106.

Hohman, M. M., M. Shin, G. Rutledge, and M. P. Brenner (2001a). "Electrospinning and electrically forced jets. I. Stability theory." *Physics of Fluids* **13**(8):2201–2220.

Hohman, M. M., M. Shin, G. Rutledge, and M. P. Brenner (2001b). "Electrospinning and electrically forced jets. II. Applications." *Physics of Fluids* **13**(8):2221–2236.

Hong, J. H., E. H. Jeong, H. S. Lee, D. H. Baik, S. W. Seo, and J. H. Youk (2005). "Electrospinning of polyurethane/organically modified montmorillonite nanocomposites." *Journal of Polymer Science Part B: Polymer Physics* **43**(22):3171–3177.

Hong, K. H., K. W. Oh, and T. J. Kang (2005). "Preparation of conducting nylon-6 electrospun fiber webs by the *in situ* polymerization of polyaniline." *Journal of Applied Polymer Science* **96**(4):983–991.

Hong, K. H., J. L. Park, I. H. Sul, J. H. Youk, and T. J. Kang (2006). "Preparation of antimicrobial poly(vinyl alcohol) nanofibers containing silver nanoparticles." *Journal of Polymer Science Part B: Polymer Physics* **44**(17):2468–2474.

Hong, K. H. (2007). "Preparation and properties of electrospun poly(vinyl alcohol)/silver fiber web as wound dressings." *Polymer Engineering and Science* **47**(1):43–49.

Hong, Y. L., T. C. Shang, Y. W. Jin, F. Yang, and C. Wang (2005). "Silica–polymer coaxial nanofibers." *Chemical Journal of Chinese Universities (Chinese)* **26**(5):985–987.

Hong, Y., T. Shang, Y. Li, L. Wang, C. Wang, X. Chen, and X. Jing (2006). "Synthesis using electrospinning and stabilization of single layer macroporous films and fibrous networks of poly(vinyl alcohol)." *Journal of Membrane Science* **276**(1–2):1–7.

Hou, H. Q., Z. Jun, A. Reuning, A. Schaper, J. H. Wendorff, and A. Greiner (2002). "Poly(*p*-xylylene) nanotubes by coating and removal of ultrathin polymer template fibers." *Macromolecules* **35**(7):2429–2431.

Hou, H. Q. and D. H. Reneker (2004). "Carbon nanotubes on carbon nanofibers: a novel structure based on electrospun polymer nanofibers." *Advanced Materials* **16**(1):69–73.

Hou, H. Q., J. J. Ge, J. Zeng, Q. Li, D. H. Reneker, A. Greiner, and S. Z. D. Cheng (2005). "Electrospun polyacrylonitrile nanofibers containing a high concentration of well-aligned multiwall carbon nanotubes." *Chemistry of Materials* **17**(5):967–973.

Hsu, C. M. and S. Shivkumar (2004a). "Nano-sized beads and porous fiber constructs of poly(ε-caprolactone) produced by electrospinning." *Journal of Materials Science* **39**(9):3003–3013.

Hsu, C. M. and S. Shivkumar (2004b). "*N,N*-dimethylformamide additions to the solution for the electrospinning of poly(ε-caprolactone) nanofibers." *Macromolecular Materials and Engineering* **289**(4):334–340.

Huang, C. B., S. L. Chen, C. L. Lai, D. H. Reneker, H. Y. Qiu, Y. Ye, and H. Q. Hou (2006a). "Electrospun polymer nanofibres with small diameters." *Nanotechnology* **17**(6):1558–1563.

Huang, C. B., S. L. Chen, D. H. Reneker, C. L. Lai, and H. Hou (2006b). "High-strength mats from electrospun poly(*p*-phenylene biphenyltetracarboximide) nanofibers." *Advanced Materials* **18**(5):668–671.

Huang, L., R. A. McMillan, R. P. Apkarian, B. Pourdeyhimi, V. P. Conticello, and E. L. Chaikof (2000). "Generation of synthetic elastin-mimetic small diameter fibers and fiber networks." *Macromolecules* **33**(8):2989–2997.

Huang, L., R. P. Apkarian, and E. L. Chaikof (2001a). "High-resolution analysis of engineered type I collagen nanofibers by electron microscopy." *Scanning* **23**(6):372–375.

Huang, L., K. Nagapudi, R. P. Apkarian, and E. L. Chaikof (2001b). "Engineered collagen-PEO nanofibers and fabrics." *Journal of Biomaterials Science: Polymer Edition* **12**(9):979–993.

Huang, Z. M., Y. Z. Zhang, M. Kotaki, and S. Ramakrishna (2003). "A review on polymer nanofibers by electrospinning and their applications in nanocomposites." *Composites Science and Technology* **63**(15):2223–2253.

Huang, Z. M., Y. Z. Zhang, S. Ramakrishna, and C. T. Lim (2004). "Electrospinning and mechanical characterization of gelatin nanofibers." *Polymer* **45**(15): 5361–5368.

Huang, Z. M. and Y. Z. Zhang (2005). "Micro-structures and mechanical performance of co-axial nanofibers with drug and protein cores and polycaprolactone shells." *Chemical Journal of Chinese Universities (Chinese)* **26**(5):968–972.

Huang, Z. M., Y. Z. Zhang, and S. Ramakrishna (2005). "Double-layered composite nanofibers and their mechanical performance." *Journal of Polymer Science Part B: Polymer Physics* **43**(20):2852–2861.

Huang, Z. M., C. L. He, A. Z. Yang, Y. Z. Zhang, X. J. Han, J. L. Yin, and Q. S. Wu (2006). "Encapsulating drugs in biodegradable ultrafine fibers through co-axial electrospinning." *Journal of Biomedical Materials Research Part A* **77A**(1):169–179.

Huang, Z. M. and A. H. Yang (2006). "Encapsulation of pure drugs into the central part of polycaprolactone ultrafine fibers." *Acta Polymerica Sinica* **1**:48–52.

Hugel, T., N. B. Holland, A. Cattani, L. Moroder, M. Seitz, and H. E. Gaub (2002). "Single-molecule optomechanical cycle." *Science* **296**(5570):1103–1106.

Hussain, M. M., N. Guven, and S. S. Ramkumar (2005). "Self detoxifying nanofiber webs." *Proceedings of the American Institute of Chemical Engineers (AIChE).* Annual Meeting and Fall Showcase Conference Proceedings. Session 365.

Ignatova, M., N. Manolova, and I. Rashkov (2007). "Novel antibacterial fibers of quaternized chitosan and poly(vinyl pyrrolidone) prepared by electrospinning". *European Polymer Journal* **43**(4):1112–1122.

Inai, R., M. Kotaki, and S. Ramakrishna (2005a). "Deformation behavior of electrospun poly(L-lactide-*co*-ε-caprolactone) nonwoven membranes under uniaxial tensile loading." *Journal of Polymer Science Part B: Polymer Physics* **43**(22):3205–3212.

Inai, R., M. Kotaki, and S. Ramakrishna (2005b). "Structure and properties of electrospun PLLA single nanofibres." *Nanotechnology* **16**(2):208–213.

Inoguchi, H., I. K. Kwon, E. Inoue, K. Takamizawa, Y. Maehara, and T. Matsuda (2006). "Mechanical responses of a compliant electrospun poly(L-lactide-*co*-ε-caprolactone) small-diameter vascular graft." *Biomaterials* **27**(8):1470–1478.

Ioannidis, M. A. and I. Chatzis (1993). "A mixed-percolation model of capillary hysteresis and entrapment in mercury porosimetry." *Journal of Colloid and Interface Science* **161**(2):278–291.

Iribarne, J. V. and B. A. Thompson (1976). "On the evaporation of small ions from charged droplets." *Journal of Chemical Physics* **64**(6):2287–2294.

Ito, Y., H. Hasuda, M. Kamitakahara, C. Ohtsuki, M. Tanihara, I.-K. Kang, and O. H. Kwon (2005). "A composite of hydroxyapatite with electrospun biodegradable nanofibers as a tissue engineering material." *Journal of Bioscience and Bioengineering* **100**(1):43–49.

Jaeger, C. R., H. Schönherr, and G. J. Vancso (1996). "Chain packing in electro-spun poly(ethylene oxide) visualized by atomic force microscopy." *Macromolecules* **29**(23):7634–7636.

Jaeger, C. R., M. M. Bergshoef, C. M. Batlle, H. Schönherr, and G. J. Vancso (1998). "Electrospinning of ultra-thin polymer fibers." *Macromolecular Symposia* **127**:141–150.

Jalili, R., S. A. Hosseini, and M. Morshed (2005). "The effects of operating parameters on the morphology of electrospun polyacrilonitrile nanofibres." *Iranian Polymer Journal* **14**(12):1074–1081.

Jalili, R., M. Morshed, S. Abdolkarim, and H. Ravandi (2006). "Fundamental parameters affecting electrospinning of PAN nanofibers as uniaxially aligned fibers." *Journal of Applied Polymer Science* **101**(6):4350–4357.

Janssen, L. L., M. D. Luinenburg, H. E. Mullins, and T. J. Nelson (2002). "Comparison of three commercially available fit-test methods." *Journal of Occupational and Environmental Hygiene* **63**(6):762–767.

Jarusuwannapoom, T., W. Hongrojjanawiwat, S. Jitjaicham, L. Wannatong, M. Nithitanakul, C. Pattamaprom, P. Koombhongse, R. Rangkupan, and P. Supaphol (2005). "Effect of solvents on electro-spinnability of polystyrene solutions and morphological appearance of resulting electrospun polystyrene fibers." *European Polymer Journal* **41**(3):409–421.

Jaworek, A. and A. Krupa (1999). "Classification of the modes of EHD spraying?" *Journal of Aerosol Science* **30**(7):873–893.

Jena, A. K. and K. M. Gupta (1999). "In-plane compression porometry of battery separators." *Journal of Power Sources* **80**(1–2):46–52.

Jeong, E. H., S. S. Im, and J. H. Youk (2005). "Electrospinning and structural characterization of ultrafine poly(butylene succinate) fibers." *Polymer* **46**(23): 9538–9543.

Jeong, J. S., S. Y. Jeon, T. Y. Lee, J. H. Park, J. H. Shin, P. S. Alegaonkar, A. S. Berdinsky, and J. B. Yoo (2006). "Fabrication of MWNTs/nylon conductive composite nanofibers by electrospinning." *Diamond and Related Materials* **15**(11–12):1839–1843.

Jeong, J. S., J. S. Moon, S. Y. Jeon, J. H. Park, P. S. Alegaonkar, and J. B. Yoo (2007). "Mechanical properties of electrospun PVA/MWNTs composite nanofibers." *Thin Solid Films* **515**(12):5136–5141.

Jeong, L., K. Y. Lee, J. W. Liu, and W. H. Park (2006). "Time-resolved structural investigation of regenerated silk fibroin nanofibers treated with solvent vapor." *International Journal of Biological Macromolecules* **38**(2):140–144.

Jeun, J. P., Y. M. Lim, and Y. C. Nho (2005). "Study on morphology of electrospun poly(caprolactone) nanofiber." *Journal of Industrial and Engineering Chemistry* **11**(4):573–578.

Jeun, J. P., Y. H. Kim, Y. M. Lim, J. H. Choi, C. H. Jung, P. H. Kang, and Y. C. Nho (2007). "Electrospinning of poly(L-lactide-*co*-D,L-lactide)." *Journal of Industrial and Engineering Chemistry* **13**(4):592–596.

Ji, Y., K. Ghosh, X. Z. Shu, B. Q. Li, J. C. Sokolov, G. D. Prestwich, R. A. F. Clark, and M. H. Rafailovich (2006a). "Electrospun three-dimensional hyaluronic acid nanofibrous scaffolds." *Biomaterials* **27**(20):3782–3792.

Ji, Y., B. Q. Li, S. R. Ge, J. C. Sokolov, and M. H. Rafailovich (2006b). "Structure and nanomechanical characterization of electrospun PS/clay nanocomposite fibers." *Langmuir* **22**(3):1321–1328.

Jia, H. F., G. Y. Zhu, B. Vugrinovich, W. Kataphinan, D. H. Reneker, and P. Wang (2002). "Enzyme-carrying polymeric nanofibers prepared via electrospinning for use as unique biocatalysts." *Biotechnology Progress* **18**(5):1027–1032.

Jia, Y.-T., H.-Y. Kim, J. Gong, and D.-R. Lee (2006). "Electrospun nanofibers of block copolymer of trimethylene carbonate and ε-caprolactone." *Journal of Applied Polymer Science* **99**(4):1462–1470.

Jiang, H. L., D. F. Fang, B. S. Hsiao, B. Chu, and W. Chen (2004a). "Optimization and characterization of dextran membranes prepared by electrospinning." *Biomacromolecules* **5**(2):326–333.

Jiang, H. L., D. F. Fang, B. Hsiao, B. Chu, and W. Chen (2004b). "Preparation and characterization of ibuprofen-loaded poly(lactide-*co*-glycolide)/poly(ethylene glycol)-*g*-chitosan electrospun membranes." *Journal of Biomaterials Science: Polymer Edition* **15**(3):279–296.

Jiang, H. L., Y. Q. Hu, Y. Li, P. C. Zhao, K. J. Zhu, and W. L. Chen (2005). "A facile technique to prepare biodegradable coaxial electrospun nanofibers for controlled release of bioactive agents." *Journal of Controlled Release* **108**(2–3):237–243.

Jiang, L., Y. Zhao, and J. Zhai (2004). "A lotus-leaf-like superhydrophobic surface: a porous microsphere/nanofiber composite film prepared by electrohydrodynamics." *Angewandte Chemie International Edition* **43**(33):4338–4341.

Jin, H.-J., S. V. Fridrikh, G. C. Rutledge, and D. L. Kaplan (2002). "Electrospinning *Bombyx mori* silk with poly(ethylene oxide)." *Biomacromolecules* **3**(6):1233–1239.

Jin, H.-J., J. S. Chen, V. Karageorgiou, G. H. Altman, and D. L. Kaplan (2004). "Human bone marrow stromal cell responses on electrospun silk fibroin mats." *Biomaterials* **25**(6):1039–1047.

Jin, H.-J., M.-O. Hwang, J. S. Yoon, K. H. Lee, I.-J. Chin, and M.-N. Kim (2005). "Preparation and characterization of electrospun poly(L-lactic acid-*co*-succinic acid-*co*-1,4-butane diol) fibrous membranes." *Macromolecular Research* **13**(1):73–79.

Jin, W.-J., H. K. Lee, E. H. Jeong, W. O. Park, and J. H. Youk (2005). "Preparation of polymer nanofibers containing silver nanoparticles by using poly (*N*-vinylpyrrolidone)." *Macromolecular Rapid Communications* **26**(24):1903–1907.

Jin, X. and Y.-L. Hsieh (2005a). "Anisotropic dimensional swelling of membranes of ultrafine hydrogel fibers." *Macromolecular Chemistry and Physics* **206**(17): 1745–1751.

Jin, X. and Y.-L. Hsieh (2005b). "pH-responsive swelling behavior of poly(vinyl alcohol)/poly(acrylic acid) bi-component fibrous hydrogel membranes." *Polymer* **46**(14):5149–5160.

Jose, M. V., B. W. Steinert, V. Thomas, D. R. Dean, M. A. Abdalla, G. Price, and G. M. Janowski (2007). "Morphology and mechanical properties of nylon 6/MWNT nanofibers." *Polymer* **48**(4):1096–1104.

Jun, Z., H. Q. Hou, A. Schaper, J. H. Wendorff, and A. Greiner (2003). "Poly-L-lactide nanofibers by electrospinning—influence of solution viscosity and electrical conductivity on fiber diameter and fiber morphology." *e-Polymers* 009[2003]:1–9.

Jun, Z., H. Q. Hou, J. H. Wendorff, and A. Greiner (2005). "Poly(vinyl alcohol) nanofibres by electrospinning: influence of molecular weight on fibre shape." *e-Polymers* 038[2005]:1–7.

Jung, Y. H., H. Y. Kim, D. R. Lee, S. Y. Park, and M. S. Khil (2005). "Characterization of PVOH nonwoven mats prepared from surfactant-polymer system via electrospinning." *Macromolecular Research* **13**(5):385–390.

Kahol, P. K. and N. J. Pinto (2002). "Electron paramagnetic resonance investigations of electrospun polyaniline fibers." *Solid State Communications* **124**(5–6): 195–197.

Kahol, P. K. and N. J. Pinto (2004). "An EPR investigation of electrospun polyaniline-polyethylene oxide blends." *Synthetic Metals* **140**(2–3):269–272.

Kalayci, V. E., P. K. Patra, A. Buer, S. C. Ugbolue, Y. K. Kim, and S. B. Warner (2004). "Fundamental investigations on electrospun fibers." *Journal of Advanced Materials* **36**(4):43–47.

Kalayci, V. E., P. K. Patra, Y. K. Kim, S. C. Ugbolue, and S. B. Warner (2005). "Charge consequences in electrospun polyacrylonitrile (PAN) nanofibers." *Polymer* **46**(18):7191–7200.

Kalayci, V., M. Ouyang, and K. Graham (2006). "Polymeric nanofibres in high efficiency filtration applications." *Filtration* **6**(4):286–293.

Kameoka, J. and H. G. Craighead (2003). "Fabrication of oriented polymeric nanofibers on planar surfaces by electrospinning." *Applied Physics Letters* **83**(2): 371–373.

Kameoka, J., R. Orth, Y. Yang, D. Czaplewski, R. Mathers, G. W. Coates, and H. G. Craighead (2003). "A scanning tip electrospinning source for deposition of oriented nanofibres." *Nanotechnology* **14**(10):1124–1129.

Kang, T. S., S. W. Lee, J. Joo, and J. Y. Lee (2005). "Electrically conducting polypyrrole fibers spun by electrospinning." *Synthetic Metals* **153**(1–3):61–64.

Kang, X. H., Y. B. Xie, H. M. Powell, L. J. Lee, M. A. Belury, J. J. Lannutti, and D. A. Kniss (2007). "Adipogenesis of murine embryonic stem cells in a three-dimensional culture system using electrospun polymer scaffolds." *Biomaterials* **28**(3):450–458.

Kang, Y.-S., H.-Y. Kim, Y.-J. Ryu, D.-R. Lee, and S.-J. Park (2002). "The effect of processing parameters on the diameter of electrospun polyacrylonitrile (PAN) nano fibers." *Polymer (Korea)* **26**(3):360.

Kannan, P., S. J. Eichhorn, and R. J. Young (2007). "Deformation of isolated single-wall carbon nanotubes in electrospun polymer nanofibres." *Nanotechnology* **18**(23):235707.

Karageorgiou, V. and D. Kaplan (2005). "Porosity of 3D biomaterial scaffolds and osteogenesis." *Biomaterials* **26**(27):5474–5491.

Kataphinan, W., R. Teye-Mensah, E. A. Evans, R. D. Ramsier, D. H. Reneker, and D. J. Smith (2003). "High-temperature fiber matrices: electrospinning and rare-earth modification." *Journal of Vacuum Science & Technology A: Vacuum, Surfaces, and Films* **21**(4):1574–1578.

Katta, P., M. Alessandro, R. D. Ramsier, and G. G. Chase (2004). "Continuous electrospinning of aligned polymer nanofibers onto a wire drum collector." *Nano Letters* **4**(11):2215–2218.

Katti, D. S., K. W. Robinson, F. K. Ko, and C. T. Laurencin (2004). "Bioresorbable nanofiber-based systems for wound healing and drug delivery: optimization of fabrication parameters." *Journal of Biomedical Materials Research Part B: Applied Biomaterials* **70B**(2):286–296.

Kedem, S., J. Schmidt, Y. Paz, and Y. Cohen (2005). "Composite polymer nanofibers with carbon nanotubes and titanium dioxide particles." *Langmuir* **21**(12):5600–5604.

Kenawy, E. R. and Y. R. Abdel-Fattah (2002). "Antimicrobial properties of modified and electrospun poly(vinyl phenol)." *Macromolecular Bioscience* **2**(6):261–266.

Kenawy, E. R., G. L. Bowlin, K. Mansfield, J. Layman, D. G. Simpson, E. H. Sanders, and G. E. Wnek (2002). "Release of tetracycline hydrochloride from electrospun poly(ethylene-*co*-vinylacetate), poly(lactic acid), and a blend." *Journal of Controlled Release* **81**(1–2):57–64.

Kenawy, E. R., J. M. Layman, J. R. Watkins, G. L. Bowlin, J. A. Matthews, D. G. Simpson, and G. E. Wnek (2003). "Electrospinning of poly(ethylene-*co*-vinyl alcohol) fibers." *Biomaterials* **24**(6):907–913.

Kessick, R., J. Fenn, and G. Tepper (2004). "The use of AC potentials in electrospraying and electrospinning processes." *Polymer* **45**(9):2981–2984.

Kessick, R. and G. Tepper (2006). "Electrospun polymer composite fiber arrays for the detection and identification of volatile organic compounds." *Sensors and Actuators B: Chemical* **117**(1):205–210.

Khil, M. S., D. I. Cha, H. Y. Kim, I. S. Kim, and N. Bhattarai (2003). "Electrospun nanofibrous polyurethane membrane as wound dressing." *Journal of Biomedical Materials Research Part B: Applied Biomaterials* **67B**(2):675–679.

Khil, M. S., H. Y. Kim, M. S. Kim, S. Y. Park, and D.-R. Lee (2004). "Nanofibrous mats of poly(trimethylene terephthalate) via electrospinning." *Polymer* **45**(1):295–301.

Khil, M.-S., S. R. Bhattarai, H.-Y. Kim, S.-Z. Kim, and K.-H. Lee (2005). "Novel fabricated matrix via electrospinning for tissue engineering." *Journal of Biomedical Materials Research Part B: Applied Biomaterials* **72B**:117–124.

Ki, C. S., D. H. Baek, K. D. Gang, K. H. Lee, I. C. Um, and Y. H. Park (2005). "Characterization of gelatin nanofiber prepared from gelatin–formic acid solution." *Polymer* **46**(14):5094–5102.

Kidoaki, S., I. K. Kwon, and T. Matsuda (2005). "Mesoscopic spatial designs of nano- and microfiber meshes for tissue-engineering matrix and scaffold based on newly devised multilayering and mixing electrospinning techniques." *Biomaterials* **26**(1):37–46.

Kidoaki, S., K. Kwon, and T. Matsuda (2006). "Structural features and mechanical properties of in situ-bonded meshes of segmented polyurethane electrospun from mixed solvents." *Journal of Biomedical Materials Research Part B: Applied Biomaterials* **76B**(1):219–229.

Kim, B., H. Park, S. H. Lee, and W. M. Sigmund (2005). "Poly(acrylic acid) nanofibers by electrospinning." *Materials Letters* **59**(7):829–832.

Kim, B. C., S. Nair, J. B. Kim, J. H. Kwak, J. W. Grate, S. H. Kim, and M. B. Gu (2005). "Preparation of biocatalytic nanofibres with high activity and stability via enzyme aggregate coating on polymer nanofibres." *Nanotechnology* **16**(7):S382–S388.

Kim, C. and K. S. Yang (2003). "Electrochemical properties of carbon nanofiber web as an electrode for supercapacitor prepared by electrospinning." *Applied Physics Letters* **83**(6):1216–1218.

Kim, C., Y.-O. Choi, W.-J. Lee, and K.-S. Yang (2004a). "Supercapacitor performances of activated carbon fiber webs prepared by electrospinning of PMDA-ODA poly(amic acid) solutions." *Electrochimica Acta* **50**(2–3):883–887.

Kim, C., Y.-J. Kim, and Y.-A. Kim (2004b). "Fabrication and structural characterization of electro-spun polybenzimidazol-derived carbon nanofiber by graphitization." *Solid State Communications* **132**(8):567–571.

Kim, C., J.-S. Kim, S.-J. Kim, W.-J. Lee, and K.-S. Yang (2004c). "Supercapacitors prepared from carbon nanofibers electrospun from polybenzimidazol." *Journal of the Electrochemical Society* **151**(5):A769–A773.

Kim, C., S.-H. Park, J.-I. Cho, D.-Y. Lee, T.-J. Park, W.-J. Lee, and K.-S. Yang (2004d). "Raman spectroscopic evaluation of polyacrylonitrile-based carbon nanofibers prepared by electrospinning." *Journal of Raman Spectroscopy* **35**(11):928–933.

Kim, C., S.-H. Park, W.-J. Lee, and K.-S. Yang (2004e). "Characteristics of super capacitor electrodes of PBI-based carbon nanofiber web prepared by electro spinning." *Electrochimica Acta* **50**(2–3):877–881.

Kim, C., K.-S. Yang, and W.-J. Lee (2004f). "The use of carbon nanofiber electrodes prepared by electrospinning for electrochemical supercapacitors." *Electrochemical and Solid State Letters* **7**(11):A397–A399.

Kim, C. (2005). "Electrochemical characterization of electrospun activated carbon nanofibres as an electrode in supercapacitors." *Journal of Power Sources* **142**(1–2):382–388.

Kim, C. H., Y. H. Jung, H. Y. Kim, D. R. Lee, N. Dharmaraj, and K. E. Choi (2006). "Effect of collector temperature on the porous structure of electrospun fibers." *Macromolecular Research* **14**(1):59–65.

Kim, C. W., M. W. Frey, M. Marquez, and Y. L. Joo (2005). "Preparation of submicron-scale, electrospun cellulose fibers via direct dissolution." *Journal of Polymer Science Part B: Polymer Physics* **43**(13):1673–1683.

Kim, G.-H. (2006). "Electrospinning process using field-controllable electrodes." *Journal of Polymer Science Part B: Polymer Physics* **44**(10):1426–1433.

Kim, G.-H. and W.-D. Kim (2006). "Nanofiber spraying method using a supplementary electrode." *Applied Physics Letters* **89**(1):013111.

Kim, G.-H., Y.-S. Cho, and W.-D. Kim (2006). "Stability analysis for multi-jets electrospinning process modified with a cylindrical electrode." *European Polymer Journal* **42**(9):2031–2038.

Kim, G.-M., R. Lach, G. H. Michler, and Y.-W. Chang (2005a). "The mechanical deformation process of electrospun polymer nanocomposite fibers." *Macromolecular Rapid Communications* **26**(9):728–733.

Kim, G.-M., G. H. Michler, and P. Pötschke (2005b). "Deformation processes of ultrahigh porous multiwalled carbon nanotubes/polycarbonate composite fibers prepared by electrospinning." *Polymer* **46**(18):7346–7351.

Kim, H. S., K. S. Kim, H. J. Jin, and I.-J. Chin (2005). "Morphological characterization of electrospun nano-fibrous membranes of biodegradable poly(L-lactide) and poly(lactide-*co*-glycolide)." *Macromolecular Symposia* **224**(1):145–154.

Kim, H. S., H.-J. Jin, S. J. Myung, M. S. Kang, and I.-J. Chin (2006). "Carbon nanotube-adsorbed electrospun nanofibrous membranes of nylon 6." *Macromolecular Rapid Communications* **27**(2):146–151.

Kim, H.-W., J.-H. Song, and H.-E. Kim (2005). "Nanofiber generation of gelatin-hydroxyapatite biomimetics for guided tissue regeneration." *Advanced Functional Materials* **15**(12):1988–1994.

Kim, H. Y., M. S. Khil, H. J. Kim, Y. H. Jung, and D. R. Lee (2003). "Preparation of porous filament via electrospinning." *IEEE Nanotechnology* **2**:801–803.

Kim, I.-D., A. Rothschild, B. H. Lee, D. Y. Kim, S. M. Jo, and H. L. Tuller (2006). "Ultrasensitive chemiresistors based on electrospun TiO_2 nanofibers." *Nano Letters* **6**(9):2009–2013.

Kim, J. R., S. W. Choi, S. M. Jo, W. S. Lee, and B. C. Kim (2004). "Electrospun PVdF-based fibrous polymer electrolytes for lithium ion polymer batteries." *Electrochimica Acta* **50**(1):69–75.

Kim, J. R., S. W. Choi, S. M. Jo, W. S. Lee, and B. C. Kim (2005). "Characterization and properties of P(VdF-HFP)-based fibrous polymer electrolyte membrane prepared by electrospinning." *Journal of the Electrochemical Society* **152**(2): A295–A300.

Kim, J.-S. and D. H. Reneker (1999a). "Mechanical properties of composites using ultrafine electrospun fibers." *Polymer Composites* **20**(1):124–131.

Kim, J.-S. and D. H. Reneker (1999b). "Polybenzimidazole nanofiber produced by electrospinning." *Polymer Engineering and Science* **39**(5):849–854.

Kim, J.-S. and D. S. Lee (2000). "Thermal properties of electrospun polyesters." *Polymer Journal* **32**(7):616–618.

Kim, K.-H., L. Jeong, H.-N. Park, S.-Y. Shin, W.-H. Park, S.-C. Lee, T.-I. Kim, Y.-J. Park, Y.-J. Seol, Y.-M. Lee, Y. Ku, I.-C. Rhyu, S.-B. Han, and C.-P. Chung (2005). "Biological efficacy of silk fibroin nanofiber membranes for guided bone regeneration." *Journal of Biotechnology* **120**(3):327–339.

Kim, K. S., M. K. Yu, X. H. Zong, J. Chiu, D. F. Fang, Y.-S. Seo, B. S. Hsiao, B. Chu, and M. Hadjiargyrou (2003). "Control of degradation rate and hydrophilicity in electrospun non-woven poly(D,L-lactide) nanofiber scaffolds for biomedical applications." *Biomaterials* **24**(27):4977–4985.

Kim, K. S., Y. K. Luu, C. Chang, D. F. Fang, B. S. Hsiao, B. Chu, and M. Hadjiargyrou (2004). "Incorporation and controlled release of a hydrophilic antibiotic using poly(lactide-*co*-glycolide)-based electrospun nanofibrous scaffolds." *Journal of Controlled Release* **98**(1):47–56.

Kim, K. W., K. H. Lee, M. S. Khil, Y. S. Ho, and H. Y. Kim (2004). "The effect of molecular weight and the linear velocity of drum surface on the properties of electrospun poly(ethylene terephthalate) nonwovens." *Fibers and Polymers* **5**(2):122–127.

Kim, K. W., K. H. Lee, B. S. Lee, Y. S. Ho, S. J. Oh, and H. Y Kim (2005). "Effects of blend ratio and heat treatment on the properties of the electrospun poly(ethylene terephthlate) nonwovens." *Fibers and Polymers* **6**(2):121–126.

Kim, S. H., Y. S. Nam, T. S. Lee, and W. H. Park (2003). "Silk fibroin nanofiber. Electrospinning, properties, and structure." *Polymer Journal* **35**(2): 185–190.

Kim, S. H., S.-H. Kim, S. Nair, and E. Moore (2005). "Reactive electrospinning of cross-linked poly(2-hydroxyethyl methacrylate) nanofibers and elastic properties of individual hydrogel nanofibers in aqueous solutions." *Macromolecules* **38**(9):3719–3723.

Kim, S. J., C. K. Lee, and S. I. Kim (2005). "Effect of ionic salts on the processing of poly(2-acrylamido-2-methyl-1-propane sulfonic acid) nanofibers." *Journal of Applied Polymer Science* **96**(4):1388–1393.

Kinley, C. E. and A. E. Marble (1980). "Compliance: a continuing problem with vascular grafts." *Journal of Cardiovascular Surgery* **21**:163–170.

Kissel, T., M. A. Rummelt, and H. P. Bier (1993). "Wirkstoffreisetzung aus bio abbaubaren mikropartikeln." *Deutsche-Apotheker-Ztg* **133**:29–32.

Kitazawa, M., R. Ohta, J. Tanaka, and M. Tanemura (2007). "Electrical properties of single carbon nanofibers grown on tips of scanning probe microscope cantilevers by ion irradiation." *Japanese Journal of Applied Physics* **46**(8B): 5607–5610.

Kitazono, E., H. Kaneko, T. Miyoshi, and K. Miyamoto (2004). "Tissue engineering using nanofiber." *Journal of Synthetic Organic Chemistry (Japan)* **62**(5): 514–519.

Klata, E., K. Babel, and I. Krucinska (2005). "Preliminary investigation into carbon nanofibres for electrochemical capacitors." *Fibres & Textiles in Eastern Europe* **13**(1):32–34.

Ko, F., Y. Gogotsi, A. Ali, N. Naguib, H. Ye, G. L. Yang, C. Li, and P. Willis (2003). "Electrospinning of continuous carbon nanotube-filled nanofiber yarns." *Advanced Materials* **15**(14):1161–1165.

Ko, F., M. Gandhi, and C. Karatzas (2004). "Carbon nanotube reinforced spider silk." 19th Annual Technical Conference of the American Society for Composites, American Chemical Society, Atlanta, GA.

Ko, F. K., M. D. Borden, and C. T. Laurencin (2001). "The role of fiber architecture in biocomposites: the tissue engineering approach." Proceedings of ICCM'13, Beijing, China, June 25–28, 2001.

Ko, F. K., H. Lam, N. Titchenal, H. Ye, and Y. Gogotsi (2006). "Coelectrospinning of carbon nanotube reinforced nanocomposite fibrils." In: *Polymeric Nanofibers*. ACS Symposium Series 918. Edited by D. H. Reneker and H. Fong. Oxford University Press (USA), pp. 231–245.

Koombhongse, S., W.-X. Liu, and D. H. Reneker (2001). "Flat polymer ribbons and other shapes by electrospinning." *Journal of Polymer Science, Part B: Polymer Physics* **39**(21):2598–2606.

Koski, A., K. Yim, and S. Shivkumar (2004). "Effect of molecular weight on fibrous PVA produced by electrospinning." *Materials Letters* **58**(3–4):493–497.

Krause, W. E., J. S. Tan, and R. H. Colby (1999). "Semidilute solution rheology of polyelectrolytes with no added salt." *Journal of Polymer Science Part B: Polymer Physics* **37**(24):3429–3437.

Krishnappa, R. V. N., K. Desai, and C. Sung (2003). "Morphological study of electrospun polycarbonates as a function of the solvent and processing voltage." *Journal of Materials Science* **38**(11):2357–2365.

Kuboki, Y., Q. Jin, and H. Takita (2001). "Geometry of carriers controlling phenotypic expression in BMP-induced osteogenesis and chondrogenesis." *Journal of Bone and Joint Surgery (American)* **83**(Suppl. 1):S105–S115.

Kulicke, W.-M., and C. Clasen (2004). *Viscosimetry of Polymers and Polyelectrolytes*. Springer-Verlag, Berlin, Germany.

Kulpinski, P. (2005). "Cellulose nanofibers prepared by the *N*-methylmorpholine-*N*-oxide method." *Journal of Applied Polymer Science* **98**(4):1855–1859.

Kumar, S., T. D. Dang, F. E. Arnold, A. R. Bhattacharyya, B. G. Min, X. Zhang, R. A. Vaia, C. Park, W. W. Adams, R. H. Hauge, R. E. Smalley, S. Ramesh, and P. A. Willis (2002). "Synthesis, structure, and properties of PBO/SWNT composites." *Macromolecules* **35**(24):9039–9043.

Kumar, S., T. Rath, R. N. Mahaling, C. S. Reddy, C. K. Das, K. N. Pandey, R. B. Srivastava, and S. B. Yadaw (2007). "Study on mechanical, morphological and electrical properties of carbon nanofiber/polyetherimide composites." *Materials Science and Engineering: B* **141**(1–2):61–70.

Kwon, I. K. and T. Matsuda (2005). "Co-electrospun nanofiber fabrics of poly(L-lactide-*co*-ε-caprolactone) with type I collagen or heparin." *Biomacromolecules* **6**(4):2096–2105.

Kwon, I. K., S. Kidoaki, and T. Matsuda (2005). "Electrospun nano- to microfiber fabrics made of biodegradable copolyesters: structural characteristics, mechanical properties and cell adhesion potential." *Biomaterials* **26**(18):3929–3939.

Kwoun, S. J., R. M. Lee, B. Han, and F. K. Ko (2001). "A novel polymer nanofiber interface for chemical and biochemical sensor applications." Nanotech 2000 Volume 1—Technical Proceedings of the 2001 International Conference on Modeling and Simulation of Microsystems, pp. 338–341.

Larrondo, L. and R. St. John Manley (1981a). "Electrostatic fiber spinning from polymer melts. I. Experimental observations of fiber formation and properties." *Journal of Polymer Science: Polymer Physics* **19**(6):909–920.

Larrondo, L. and R. St. John Manley (1981b). "Electrostatic fiber spinning from polymer melts. II. Examination of the flow field in an electrically driven jet." *Journal of Polymer Science: Polymer Physics* **19**(6):921–932.

Larrondo, L. and R. St. John Manley (1981c). "Electrostatic fiber spinning from polymer melts. III. Electrostatic deformation of a pendant drop of polymer melt." *Journal of Polymer Science: Polymer Physics* **19**(6):933–940.

Larsen, G., R. Velarde-Ortiz, K. Minchow, A. Barrero, and I. G. Loscertales (2003). "A method for making inorganic and hybrid (organic/inorganic) fibers and vesicles with diameters in the submicrometer and micrometer range via sol-gel chemistry and electrically forced liquid jets." *Journal of the American Chemical Society* **125**(5):1154–1155.

Larsen, G., S. Noriega, R. Spretz, and R. Velarde-Ortiz (2004a). "Electrohydrodynamics and hierarchical structure control: submicron-thick silica ribbons with an ordered hexagonal mesoporous structure." *Journal of Materials Chemistry* **14**(15):2372–2373.

Larsen, G., R. Spretz, and R. Velarde-Ortiz (2004b). "Use of coaxial gas jackets to stabilize Taylor cones of volatile solutions and to induce particle-to-fiber transitions." *Advanced Materials* **16**(2):166–169.

Laurencin, C. T., A. M. A. Ambrosio, M. D. Borden, and J. A. Cooper, Jr. (1999). "Tissue engineering: orthopedic applications." *Annual Review of Biomedical Engineering* **1**:19–46.

Lee, C. H., H. J. Shin, I. H. Cho, Y.-M. Kang, I. A. Kim, K.-D. Park, and J.-W. Shin (2005). "Nanofiber alignment and direction of mechanical strain affect the ECM production of human ACL fibroblast." *Biomaterials* **26**(11):1261–1270.

Lee, C. K., S. I. Kim, and S. J. Kim (2005). "The influence of added ionic salt on nanofiber uniformity for electrospinning of electrolyte polymer." *Synthetic Metals* **154**(1–3):209–212.

Lee, C. K., S. J. Kim, S. I. Kim, B.-J. Yi, and S. Y. Han (2006). "Preparation of chitosan microfibres using electro-wet-spinning and their electroactuation properties." *Smart Materials and Structures* **15**(2):607–611.

Lee, E. R. (2003). *Microdrop Generation*. CRC Press, Boca Raton, FL.

Lee, H. K., E. H. Jeong, C. K. Baek, and J. H. Youk (2005). "One-step preparation of ultrafine poly(acrylonitrile) fibers containing silver nanoparticles." *Materials Letters* **59**(23):2977–2980.

Lee, I. S., O. H. Kwon, W. Meng, I.-K. Kang, and Y. Ito (2004). "Nanofabrication of microbial polyester by electrospinning promotes cell attachment." *Macromolecular Research* **12**(4):374–378.

Lee, J. S., K. H. Choi, H. D. Ghim, S. S. Kim, D. H. Chun, H. Y. Kim, and W. S. Lyoo (2004). "Role of molecular weight of atactic poly(vinyl alcohol) (PVA) in the structure and properties of PVA nanofabric prepared by electrospinning." *Journal of Applied Polymer Science* **93**(4):1638–1646.

Lee, K. H., H. Y. Kim, Y. M. La, D. R. Lee, and N. H. Sung (2002). "Influence of a mixing solvent with tetrahydrofuran and *N,N*-dimethylformamide on electrospun poly(vinyl chloride) nonwoven mats." *Journal of Polymer Science Part B: Polymer Physics* **40**(19):2259–2268.

Lee, K. H., H. Y. Kim, H. J. Bang, Y. H. Jung, and S. G. Lee (2003a). "The change of bead morphology formed on electrospun polystyrene fibers." *Polymer* **44**(14):4029–4034.

Lee, K. H., H. Y. Kim, M. S. Khil, Y. M. Ra, and D. R. Lee (2003b). "Characterization of nano-structured poly(ε-caprolactone) nonwoven mats via electrospinning." *Polymer* **44**(4):1287–1294.

Lee, K. H., H. Y. Kim, Y. J. Ryu, K. W. Kim, and S. W. Choi (2003c). "Mechanical behavior of electrospun fiber mats of poly(vinyl chloride)/polyurethane polyblends." *Journal of Polymer Science Part B: Polymer Physics* **41**(11):1256–1262.

Lee, K. H., C. S. Ki, D. H. Baek, G. D. Kang, D.-W. Ihm, and Y. H. Park (2005). "Application of electrospun silk fibroin nanofibers as an immobilization support of enzyme." *Fibers and Polymers* **6**(3):181–185.

Lee, S. B., Y. H. Kim, M. S. Chong, S. W. Hong, and Y. M. Lee (2005). "Study of gelatin-containing artificial skin V: fabrication of gelatin scaffolds using a salt-leaching method." *Biomaterials* **26**(14):1961–1968.

Lee, S. C., H. Y. Kim, D. R. Lee, D. Bin, and S. J. Park (2002). "Morphological characteristics of electrospun poly(vinyl alcohol) nonwoven." *Journal of the Korean Fiber Society* **39**(3):316.

Lee, S.-G., S.-S. Choi, and C. W. Joo (2002). "Nanofiber formation of poly(ether imide) under various electrospinning conditions." *Journal of the Korean Fiber Society* **39**(1):1–13.

Lee, S.-H., C. Tekmen, and W. M. Sigmund (2005). "Three-point bending of electrospun TiO_2 nanofibers." *Materials Science and Engineering: A—Structural Materials: Properties, Microstructure and Processing* **398**(1–2):77–81.

Lee, S.-H. and W. M. Sigmund (2006). "Synthesis of anatase–silver nanocomposite fibers via electrospinning." *Journal of Nanoscience and Nanotechnology* **6**(2):554–557.

Lee, S.-W. and A. M. Belcher (2004). "Virus-based fabrication of micro- and nanofibers using electrospinning." *Nano Letters* **4**(3):387–390.

Lee, Y. H., J. H. Lee, I.-G. An, C. Kim, D. S. Lee, Y. K. Lee, and J.-D. Nam (2005). "Electrospun dual-porosity structure and biodegradation morphology of Montmorillonite reinforced PLLA nanocomposite scaffolds." *Biomaterials* **26**(16):3165–3172.

Leisen, J., H. W. Beckham, and P. Farber (2007). "Void structure in textiles by nuclear magnetic resonance, Part I. Imaging of imbibed fluids and image analysis by calculation of fluid density autocorrelation functions." National Textile Center Research Briefs–Fabrication Competency, June 2007. NTC Project: F04-GT05, (www.ntcresearch.org/pdf-rpts/Bref0607/F04-GT05-07.pdf).

Levit, N. and G. Tepper (2004). "Supercritical CO_2-assisted electrospinning." *Journal of Supercritical Fluids* **31**(3):329–333.

Lewis, L. N. (1993). "Chemical catalysis by colloids and clusters." *Chemical Reviews* **93**(8):2693–2730.

Li, C. M., H.-J. Jin, G. D. Botsaris, and D. L. Kaplan (2005). "Silk apatite composites from electrospun fibers." *Journal of Materials Research* **20**(12):3374–3384.

Li, C. M., C. Vepari, H.-J. Jin, H. J. Kim, and D. L. Kaplan (2006). "Electrospun silk-BMP-2 scaffolds for bone tissue engineering." *Biomaterials* **27**(16): 3115–3124.

Li, D., T. Herricks, and Y. N. Xia (2003a). "Magnetic nanofibers of nickel ferrite prepared by electrospinning." *Applied Physics Letters* **83**(22):4586–4588.

Li, D., Y. L. Wang, and Y. N. Xia (2003b). "Electrospinning of polymeric and ceramic nanofibers as uniaxially aligned arrays." *Nano Letters* **3**(8):1167–1171.

Li, D. and Y. N. Xia (2004). "Direct fabrication of composite and ceramic hollow nanofibers by electrospinning." *Nano Letters* **4**(5):933–938.

Li, D., A. Babel, S. A. Jenekhe, and Y. Xia (2004a). "Nanofibers of conjugated polymers prepared by electrospinning with a two-capillary spinneret." *Advanced Materials* **16**(22):2062–2066.

Li, D., J. T. McCann, M. Gratt, and Y. N. Xia (2004b). "Photocatalytic deposition of gold nanoparticles on electrospun nanofibers of titania." *Chemical Physics Letters* **394**(4–6):387–391.

Li, D., Y. Wang, and Y. Xia (2004c). "Electrospinning nanofibers as uniaxially aligned arrays and layer-by-layer stacked films." *Advanced Materials* **16**(4):361–366.

Li, D., J. T. McCann, and Y. Xia (2005a). "Use of electrospinning to directly fabricate hollow nanofibers with functionalized inner and outer surfaces." *Small* **1**(1):83–86.

Li, D., G. Ouyang, J. T. McCann, and Y. Xia (2005b). "Collecting electrospun nanofibers with patterned electrodes." *Nano Letters* **5**(5):913–916.

Li, J. L. (2005). "Formation and stabilization of an EHD jet from a nozzle with an inserted non-conductive fibre." *Aerosol Science* **36**(3):373–386.

Li, J. X., A. H. He, C. C. Han, D. F. Fang, B. S. Hsiao, and B. Chu (2006). "Electrospinning of hyaluronic acid (HA) and HA/gelatin blends." *Macromolecular Rapid Communications* **27**(2):114–120.

Li, L. and Y.-L. Hsieh (2005a). "Ultra-fine polyelectrolyte fibers from electrospinning of poly(acrylic acid)." *Polymer* **46**(14):5133–5139.

Li, L. and Y.-L. Hsieh (2005b). "Ultra-fine polyelectrolyte hydrogel fibres from poly(acrylic acid)/poly(vinyl alcohol)." *Nanotechnology* **16**(12):2852–2860.

Li, L. and Y.-L. Hsieh (2006). "Chitosan bicomponent nanofibers and nanoporous fibers." *Carbohydrate Research* **341**(3):374–381.

Li, M. Y., M. J. Mondrinos, M. R. Gandhi, F. K. Ko, A. S. Weiss, and P. I. Lelkes (2005). "Electrospun protein fibers as matrices for tissue engineering." *Biomaterials* **26**(30):5999–6008.

Li, M. Y., Y. Guo, Y. Wei, A. G. MacDiarmid, and P. I. Lelkes (2006). "Electrospinning polyaniline-contained gelatin nanofibers for tissue engineering applications." *Biomaterials* **27**(13):2705–2715.

Li, W.-J., C. T. Laurencin, E. J. Caterson, R. S. Tuan, and F. K. Ko (2002). "Electrospun nanofibrous structure: a novel scaffold for tissue engineering." *Journal of Biomedical Materials Research* **60**(4):613–621.

Li, W.-J., K. G. Danielson, P. G. Alexander, and R. S. Tuan (2003). "Biological response of chondrocytes cultured in three-dimensional nanofibrous poly(ε-caprolactone) scaffolds." *Journal of Biomedical Materials Research Part A* **67A**(4):1105–1114.

Li, W.-J., R. L. Mauck, and R. S. Tuan (2005a). "Electrospun nanofibrous scaffolds: production, characterization, and applications for tissue engineering and drug delivery." *Journal of Biomedical Nanotechnology* **1**(3):259–275.

Li, W.-J., R. Tuli, X. X. Huang, P. Laquerriere, and R. S. Tuan (2005b). "Multilineage differentiation of human mesenchymal stem cells in a three-dimensional nanofibrous scaffold." *Biomaterials* **26**(25):5158–5166.

Li, W.-J., R. Tuli, C. Okafor, A. Derfoul, K. G. Danielson, D. J. Hall, and R. S. Tuan (2005c). "A three-dimensional nanofibrous scaffold for cartilage tissue engineering using human mesenchymal stem cells." *Biomaterials* **26**(6):599–609.

Li, W.-J., J. A. Cooper, Jr., R. L. Mauck, and R. S. Tuan (2006). "Fabrication and characterization of six electrospun poly(α-hydroxy ester)-based fibrous scaffolds for tissue engineering applications." *Acta Biomaterialia* **2**(4):377–385.

Li, W.-J., R. L. Mauck, J. A. Cooper, X. Yuan, and R. S. Tuan (2007). "Engineering controllable anisotropy in electrospun biodegradable nanofibrous scaffolds for musculoskeletal tissue engineering." *Journal of Biomechanics* **40**(8): 1686–1693.

Li, X. S. and G. Y. Nie (2004). "Chemical catalysis by colloids and clusters." *Chinese Science Bulletin* **49**(22):2368–2371.

Li, Z. Y., H. M. Huang, T. C. Shang, F. Yang, W. Zheng, C. Wang, and S. K. Manohar (2006a). "Facile synthesis of single-crystal and controllable sized silver nanoparticles on the surfaces of polyacrylonitrile nanofibres." *Nanotechnology* **17**(3):917–920.

Li, Z. Y., H. M. Huang, and C. Wang (2006b). "Electrostatic forces induce poly(vinyl alcohol)-protected copper nanoparticles to form copper/poly(vinyl alcohol) nanocables via electrospinning." *Macromolecular Rapid Communications* **27**(2):152–155.

Liang, D. H., Y. K. Luu, K. S. Kim, B. S. Hsiao, M. Hadjiargyrou, and B. Chu (2005). "*In vitro* non-viral gene delivery with nanofibrous scaffolds." *Nucleic Acids Research* **33**(19):e170.

Liang, D. H., B. S. Hsiao, and B. Chu (2007). "Functional electrospun nanofibrous scaffolds for biomedical applications." *Advanced Drug Delivery Reviews* **59**(14):1392–1412.

Liao, I. C., S. Y. Chew, and K. W. Leong (2006). "Aligned core-shell nanofibers delivering bioactive proteins." *Nanomedicine: Nanotechnology, Biology and Medicine* **1**(4):465–471.

Lim, T. C., M. Kotaki, T. K. J. Yong, F. Yang, K. Fujihara, and S. Ramakrishna (2004). "Recent advances in tissue engineering applications of electrospun nanofibers." *Materials Technology* **19**(1):20–27.

Lin, D. Y. and D. C. Martin (2006). "Orientation development in electrospun liquid-crystalline polymer nanofibers." In: *Polymeric Nanofibers*. ACS Symposium Series 918. Edited by D. H. Reneker and H. Fong. Oxford University Press (USA).

Lin, H. F. (1997). "The TAO of stem cells in the germline." *Annual Review of Genetics* **31**:455–491.

Lin, T., H. X. Wang, H. M. Wang, and X. G. Wang (2004). "The charge effect of cationic surfactants on the elimination of fibre beads in the electrospinning of polystyrene." *Nanotechnology* **15**(9):1375–1381.

Lin, T., H. X. Wang, H. M. Wang, and X. G. Wang (2005a). "Effects of polymer concentration and cationic surfactant on the morphology of electrospun polyacrylonitrile nanofibres." *Journal of Materials Science & Technology* **21**:9–12.

Lin, T., H. X. Wang, and X. G. Wang (2005b). "Self-crimping bicomponent nano fibers electrospun from polyacrylonitrile and elastomeric polyurethane." *Advanced Materials* **17**(22):2699–2703.

Liu, B. Y. H., and K. L. Rubow (1986). "Air filtration by fibrous media." In: *Fluid Filtration: Gas, Volume I*. Edited by R. R. Raber. American Society for Testing and Materials, Philadelphia, PA.

Liu, H. Q. and Y. L. Hsieh (2002). "Ultrafine fibrous cellulose membranes from electrospinning of cellulose acetate." *Journal of Polymer Science Part B: Polymer Physics* **40**(18):2119–2129.

Liu, H. Q., J. Kameoka, D. A. Czaplewski, and H. G. Craighead (2004). "Polymeric nanowire chemical sensor." *Nano Letters* **4**(4):671–675.

Liu, H. Q., C. H. Reccius, and H. G. Craighead (2005). "Single electrospun regioregular poly(3-hexylthiophene) nanofiber field-effect transistor." *Applied Physics Letters* **87**(25):253106.

Liu, J. and S. Kumar (2005). "Microscopic polymer cups by electrospinning." *Polymer* **46**(10):3211–3214.

Liu, J., T. Wang, T. Uchida, and S. Kumar (2005). "Carbon nanotube core-polymer shell nanofibers." *Journal of Applied Polymer Science* **96**(5):1992–1995.

Liu, L.-Q., D. Tasis, M. Prato, and H. D. Wagner (2007). "Tensile mechanics of electrospun multiwalled nanotube/poly(methyl methacrylate) nanofibers." *Advanced Materials* **19**(9):1228–1233.

Liu, T. Q. (2004). "Preparation of novel micro/nanotubes via electrospun fiber as a template." *Journal of Materials Science and Technology* **20**(5):613–616.

Liu, W., Z. Wu, and D. H. Reneker (2000). "Structure and morphology of poly(metaphenylene isophthalamide) nanofibers produced by electrospinning." *Polymer Preprints* **41**(2):1193–1194.

Liu, W. X., M. Graham, E. A. Evans, and D. H. Reneker (2002). "Poly(meta-phenylene isophthalamide) nanofibers: coating and post processing." *Journal of Materials Research* **17**(12):3206–3212.

Liu, Y., L. Cui, F. X. Guan, Y. Gao, N. E. Hedin, L. Zhu, and H. Fong (2007). "Crystalline morphology and polymorphic phase transitions in electrospun nylon-6 nanofibers." *Macromolecules* **40**(17):6283–6290.

Liu, Y., J. He, L. Xu, and J. Yu (2007). "Effect of voltage." *Int. Journal Electrospun Nanofibers and Applications* **1**(1):7–15.

Loeb, L. B., A. F. Kip, G. G. Hudson, and W. H. Bennett (1941). "Pulses in negative point-to-plane corona." *Physical Review* **60**(10):714–722.

Loscertales, I. G., A. Barrero, I. Guerrero, T. Cortijo, M. Márquez, and A. M. Gañán-Calvo (2002). "Micro/nano encapsulation via electrified coaxial liquid jets." *Science* **295**(5560):1695–1698.

Loscertales, I. G., A. Barrero, M. Márquez, R. Spretz, R. Velarde-Ortiz, and G. Larsen (2004). "Electrically forced coaxial nanojets for one-step hollow nanofiber design." *Journal of the American Chemical Society* **126**(17):5376–5377.

Lu, C., P. Chen, J. F. Li, and Y. J. Zhang (2006). "Computer simulation of electrospinning. Part I. Effect of solvent in electrospinning." *Polymer* **47**(3):915–921.

Lu, X., Y. Zhao, and C. Wang (2005). "Fabrication of PbS nanoparticles in polymer-fiber matrices by electrospinning." *Advanced Materials* **17**(20):2485–2488.

Lu, X. B., J. H. Zhou, W. Lu, Q. Liu, and J. H. Li (2008). "Carbon nanofiber-based composites for the construction of mediator-free biosensors." *Biosensors and Bioelectronics* **23**(8):1236–1243.

Lu, X. F., L. L. Li, W. J. Zhang, and C. Wang (2005a). "Preparation and characterization of Ag_2S nanoparticles embedded in polymer fibre matrices by electro spinning." *Nanotechnology* **16**(10):2233–2237.

Lu, X. F., Y. Y. Zhao, C. Wang, and Y. Wei (2005b). "Fabrication of CdS nanorods in PVP fiber matrices by electrospinning." *Macromolecular Rapid Communications* **26**(16):1325–1329.

Lu, X. F., D. L. Zhang, Q. D. Zhao, C. Wang, W. J. Zhang, and Y. Wei (2006). "Large-scale synthesis of necklace-like single-crystalline $PbTiO_3$ nanowires." *Macromolecular Rapid Communications* **27**(1):76–80.

Lu, Y. and S. C. Chen (2004). "Micro and nano-fabrication of biodegradable polymers for drug delivery." *Advanced Drug Delivery Reviews* **56**(11):1621–1633.

Luong-Van, E., L. Grøndahl, K. N. Chua, K. W. Leong, V. Nurcombe, and S. M. Cool (2006). "Controlled release of heparin from poly(ε-caprolactone) electrospun fibers." *Biomaterials* **27**(9):2042–2050.

Luu, Y. K., K. Kim, B. S. Hsiao, B. Chu, and M. Hadjiargyrou (2003). "Development of a nanostructured DNA delivery scaffold via electrospinning of PLGA and PLA PEG block copolymers." *Journal of Controlled Release* **89**(2):341–353.

Lyons, J., C. Li, and F. Ko (2004). "Melt-electrospinning part I: processing parameters and geometric properties." *Polymer* **45**(22):7597–7603.

Lyoo, W. S., J. H. Youk, S. W. Lee, and W. H. Park (2005). "Preparation of porous ultra-fine poly(vinyl cinnamate) fibers." *Materials Letters* **59**(28):3558–3562.

Ma, M. L., R. M. Hill, J. L. Lowery, S. V. Fridrikh, and G. C. Rutledge (2005a). "Electrospun poly(styrene-*block*-dimethylsiloxane) block copolymer fibers exhibiting superhydrophobicity." *Langmuir* **21**(12):5549–5554.

Ma, M. L., Y. Mao, M. Gupta, K. K Gleason, and G. C. Rutledge (2005b). "Superhydrophobic fabrics produced by electrospinning and chemical vapor deposition." *Macromolecules* **38**(23):9742–9748.

Ma, P. X. (2004). "Scaffolds for tissue fabrication." *Materials Today* **7**(5):30–40.

Ma, Z. W., W. He, T. Yong, and S. Ramakrishna (2005a). "Grafting of gelatin on electrospun poly(caprolactone) nanofibers to improve endothelial cell spreading and proliferation and to control cell orientation." *Tissue Engineering* **11**(7–8):1149–1158.

Ma, Z. W., M. Kotaki, R. Inai, and S. Ramakrishna (2005b). "Potential of nanofiber matrix as tissue-engineering scaffolds." *Tissue Engineering* **11**(1–2):101–109.

Ma, Z. W., M. Kotaki, and S. Ramakrishna (2005c). "Electrospun cellulose nanofiber as affinity membrane." *Journal of Membrane Science* **265**(1–2):115–123.

Ma, Z. W., M. Kotaki, T. Yong, W. He, and S. Ramakrishna (2005d). "Surface engineering of electrospun polyethylene terephthalate (PET) nanofibers towards development of a new material for blood vessel engineering." *Biomaterials* **26**(15):2527–2536.

Ma, Z. W., M. Kotaki, and S. Ramakrishna (2006). "Surface modified nonwoven polysulphone (PSU) fiber mesh by electrospinning: a novel affinity membrane." *Journal of Membrane Science* **272**(1–2):179–187.

MacDiarmid, A. G., W. E. Jones, Jr., I. D. Norris, J. Gao, A. T. Johnson, Jr., N. J. Pinto, J. Hone, B. Han, F. K. Ko, H. Okuzaki, and M. Llaguno (2001). "Electrostatically-generated nanofibers of electronic polymers." *Synthetic Metals* **119**(1–3):27–30.

Macías, M., A. Chacko, J. P. Ferraris, and K. J. Balkus, Jr. (2005). "Electrospun mesoporous metal oxide fibers." *Microporous and Mesoporous Materials* **86**(1–3):1–13.

Mack, J. J., L. M. Viculis, A. Ali, R. Luoh, G. L. Yang, H. T. Hahn, F. K. Ko, and R. B. Kaner (2005). "Graphite nanoplatelet reinforcement of electrospun polyacrylonitrile nanofibers." *Advanced Materials* **17**(1):77–80.

Macky, W. A. (1931). "Some investigations on the deformation and breaking of water drops in strong electric fields." *Proceedings of the Royal Society of London. Series A, Containing Papers of a Mathematical and Physical Character (1905–1934)* **133**(822):565–587.

Madhugiri, S., A. Dalton, J. Gutierrez, J. P. Ferraris, and K. J. Balkus, Jr. (2003). "Electrospun MEH-PPV/SBA-15 composite nanofibers using a dual syringe method." *Journal of the American Chemical Society* **125**(47): 14531–14538.

Madhugiri, S., B. Sun, P. G. Smirniotis, J. P. Ferraris, and K. J. Balkus, Jr. (2004). "Electrospun mesoporous titanium dioxide fibers." *Microporous and Mesoporous Materials* **69**(1–2):77–83.

Mamedov, A. F., N. A. Kotov, M. Prato, D. M. Guldi, J. P. Wicksted, and A. Hirsch (2002). "Molecular design of strong single-wall carbon nanotube/polyelectrolyte multilayer composites." *Nature Materials* **1**(3):190–194.

Marcos, M., P. Cano, P. Fantazzini, C. Garavaglia, S. Gomez, and L. Garrido (2006). "NMR relaxometry and imaging of water absorbed in biodegradable polymer scaffolds." *Magnetic Resonance Imaging* **24**(1):89–95.

Martin, A. F. (1951). "Toward a referee viscosity method for cellulose." *TAPPI* **34**:363.

Mathur, R. B., O. P. Bahl, and J. Mittal (1992). "A new approach to thermal stabilisation of PAN fibres." *Carbon* **30**(4):657–663.

Matsuda, K., S. Suzuki, N. Isshiki, and Y. Ikada (1993). "Re-freeze dried bilayer artificial skin." *Biomaterials* **14**(13):1030–1035.

Matteson, M. J. and C. Orr (Editors) (1987). *Filtration: Principles and Practices*. Marcel-Dekker, Inc., New York.

Matthew, G., J. P. Hong, J. M. Rhee, H. S. Lee, and C. Nah (2005). "Preparation and characterization of properties of electrospun poly(butylene terephthalate) nano fibers filled with carbon nanotubes." *Polymer Testing* **24**(6):712–717.

Matthews, J. A., G. E. Wnek, D. G. Simpson, and G. L. Bowlin (2002). "Electrospinning of collagen nanofibers." *Biomacromolecules* **3**(2):232–238.

Matthews, J. A., E. D. Boland, G. E. Wnek, D. G. Simpson, and G. L. Bowlin (2003). "Electrospinning of collagen type II: a feasibility study." *Journal of Bioactive and Compatible Polymers* **18**(2):125–134.

McCann, J. T., D. Li, and Y. N. Xia (2005). "Electrospinning of nanofibers with core-sheath, hollow, or porous structures." *Journal of Materials Chemistry* **15**(7):735–738.

McCann, J. T., M. Marquez, and Y. N. Xia (2006). "Highly porous fibers by electrospinning into a cryogenic liquid." *Journal of the American Chemical Society* **128**(5):1436–1437.

McKee, M. G., C. L. Elkins, and T. E. Long (2004a). "Influence of self-complementary hydrogen bonding on solution rheology/electrospinning relationships." *Polymer* **45**(26):8705–8715.

McKee, M. G., G. L. Wilkes, R. H. Colby, and T. E. Long (2004b). "Correlations of solution rheology with electrospun fiber formation of linear and branched polyesters." *Macromolecules* **37**(5):1760–1767.

McKee, M. G., T. Park, S. Unal, I. Yilgor, and T. E. Long (2005). "Electrospinning of linear and highly branched segmented poly(urethane urea)s." *Polymer* **46**(7):2011–2015.

McKee, M. G., M. T. Hunley, J. M. Layman, and T. E. Long (2006a). "Solution rheological behavior and electrospinning of cationic polyelectrolytes." *Macromolecules* **39**(2):575.

McKee, M. G., J. M. Layman, M. P. Cashion, and T. E. Long (2006b). "Phospholipid nonwoven electrospun membranes." *Science* **311**(5759):353–355.

McManus, M. C., E. D. Boland, H. P. Koo, C. P. Barnes, K. J. Pawlowski, G. E. Wnek, D. G. Simpson, and G. L. Bowlin (2006). "Mechanical properties of electrospun fibrinogen structures." *Acta Biomaterialia* **2**(1):19–28.

Megelski, S., J. S. Stephens, D. B. Chase, and J. F. Rabolt (2002). "Micro- and nanostructured surface morphology on electrospun polymer fibers." *Macromolecules* **35**(22):8456–8466.

Melcher, J. R. (1972). "Electrohydrodynamics." In: *Proceedings of the 13th International Congress of Theoretical and Applied Mechanics*, Moscow, USSR, August 21–26, 1972. Edited by E. Becker and G. K. Mikhailov. Springer, 1973, pp. 240–263.

Meyer, H. K., P. Lorenz, B. Böhl-Kuhn, and P. Klobes (1994). "Porous solids and their characterization methods of investigation and application." *Crystal Research and Technology* **29**(7):903–930.

Miller, D. C., T. J. Webster, and K. M. Haberstroh (2004). "Technological advances in nanoscale biomaterials: the future of synthetic vascular graft design." *Expert Review of Medical Devices* **1**(2):259–268.

Min, B.-M., L. Jeong, Y. S. Nam, J.-M. Kim, J. Y. Kim, and W. H. Park (2004a). "Formation of silk fibroin matrices with different texture and its cellular response to normal human keratinocytes." *International Journal of Biological Macromolecules* **34**(5):223–230.

Min, B.-M., G. Lee, S. H. Kim, Y. S. Nam, T. S. Lee, and W. H. Park (2004b). "Electrospinning of silk fibroin nanofibers and its effect on the adhesion and spreading of normal human keratinocytes and fibroblasts in vitro." *Biomaterials* **25**(7–8):1289–1297.

Min, B.-M., S. W. Lee, J. N. Lim, Y. You, T. S. Lee, P. H. Kang, and W. H. Park (2004c). "Chitin and chitosan nanofibers: electrospinning of chitin and deacetylation of chitin nanofibers." *Polymer* **45**(21):7137–7142.

Min, B.-M., Y. You, J. M. Kim, S. J. Lee, and W. H. Park (2004d). "Formation of nanostructured poly(lactic-co-glycolic acid)/chitin matrix and its cellular response to normal human keratinocytes and fibroblasts." *Carbohydrate Polymers* **57**(3):285.

Mincheva, R., N. Manolova, D. Paneva, and I. Rashkov (2005). "Preparation of polyelectrolyte-containing nanofibers by electrospinning in the presence of a non-ionogenic water-soluble polymer." *Journal of Bioactive and Compatible Polymers* **20**(5):419–435.

Mitchell, S. B. and J. E. Sanders (2006). "A unique device for controlled electrospinning." *Journal of Biomedical Materials Research Part A* **78A**(1):110–120.

Mit-uppatham, C., M. Nithitanakul, and P. Supaphol (2004a). "Effects of solution concentration, emitting electrode polarity, solvent type, and salt addition on electrospun polyamide-6 fibers: a preliminary report." *Macromolecular Symposia* **216**(1):293–300.

Mit-uppatham, C., M. Nithitanakul, and P. Supaphol (2004b). "Ultrafine electrospun polyamide-6 fibers: effect of solution conditions on morphology and average fiber diameter." *Macromolecular Chemistry and Physics* **205**(17):2327–2338.

Mo, X. M. and H.-J. Weber (2004). "Electrospinning P(LLA-CL) nanofiber: a tubular scaffold fabrication with circumferential alignment." *Macromolecular Symposia* **217**(1):413–416.

Mo, X. M., C. Y. Xu, M. Kotaki, and S. Ramakrishna (2004). "Electrospun P(LLA-CL) nanofiber: a biomimetic extracellular matrix for smooth muscle cell and endothelial cell proliferation." *Biomaterials* **25**(10):1883–1890.

Moore, E. M., D. L. Ortiz, V. T. Marla, R. L. Shambaugh, and B. P. Grady (2004). "Enhancing the strength of polypropylene fibers with carbon nanotubes." *Journal of Applied Polymer Science* **93**(6):2926–2933.

Morota, K., H. Matsumoto, T. Mizukoshi, Y. Konosu, M. Minagawa, A. Tanioka, Y. Yamagata, and K. Inoue (2004). "Poly(ethylene oxide) thin films produced by electrospray deposition: morphology control and additive effects of alcohols on nanostructure." *Journal of Colloid and Interface Science* **279**(2):484–492.

Morozov, V. N., T. Y. Morozova, and N. R. Kallenbach (1998). "Atomic force microscopy of structures produced by electrospraying polymer solutions." *International Journal of Mass Spectrometry* **178**(3):143–159.

Motoo, F., Z. Xing, X. Huaqing, A. Hiroki, T. Koji, I. Tatsuya, A. Hidekazu, and S. Tetsuo (2005). "Measurements of thermal conductivity of individual carbon nanotubes." *Thermophysical Properties* **26**:26–28.

Murugan, R. and S. Ramakrishna (2006). "Nano-featured scaffolds for tissue engineering: a review of spinning methodologies." *Tissue Engineering* **12**(3):435–447.

Naebe, M., T. Lin, W. Tian, L. M. Dai, and X. G. Wang (2007). "Effects of MWNT nanofillers on structures and properties of PVA electrospun nanofibres." *Nanotechnology* **18**(22):225605.

Nagapudi, K., W. T. Brinkman, J. E. Leisen, L. Huang, R. A. McMillan, R. P. Apkarian, V. P. Conticello, and E. L. Chaikof (2002). "Photomediated solid-state cross-linking of an elastin-mimetic recombinant protein polymer." *Macromolecules* **35**(5):1730–1737.

Nair, L. S., S. Bhattacharyya, J. D. Bender, Y. E. Greish, P. W. Brown, H. R. Allcock, and C. T. Laurencin (2004). "Fabrication and optimization of methylphenoxy substituted polyphosphazene nanofibers for biomedical applications." *Biomacromolecules* **5**(6):2212–2220.

Nair, S., S. Natarajan, and S. H. Kim (2005). "Fabrication of electrically conducting polypyrrole-poly(ethylene oxide) composite nanofibers." *Macromolecular Rapid Communications* **26**(20):1599–1603.

Nakane, K., T. Ogihara, N. Ogata, and S. Yamaguchi (2005). *Sen-I Gakkaishi* **61**(12):313–316.

Nikolovski, J. and D. J. Mooney (2000). "Smooth muscle cell adhesion to tissue engineering scaffolds." *Biomaterials* **21**(20):2025–2032.

Noh, H. K., S. W. Lee, J.-M. Kim, J.-E. Oh, K.-H. Kim, C.-P. Chung, S.-C. Choi, W. H. Park, and B.-M. Min (2006). "Electrospinning of chitin nanofibers: degradation behavior and cellular response to normal human keratinocytes and fibroblasts." *Biomaterials* **27**(21):3934–3944.

Norman, J. J. and T. A. Desai (2006). "Methods for fabrication of nanoscale topography for tissue engineering scaffolds." *Annals of Biomedical Engineering* **34**(1):89–101.

Norris, I. D., M. M. Shaker, F. K. Ko, and A. G. MacDiarmid (2000). "Electrostatic fabrication of ultrafine conducting fibers: polyaniline/polyethylene oxide blends." *Synthetic Metals* **114**(2):109–114.

Ochanda, F. and W. E. Jones, Jr. (2005). "Sub-micrometer-sized metal tubes from electrospun fiber templates." *Langmuir* **21**(23):10791–10796.

Ohgo, K., C. H. Zhao, M. Kobayashi, and T. Asakura (2003). "Preparation of non-woven nanofibers of *Bombyx mori* silk, *Samia Cynthia ricini* silk and recombinant hybrid silk with electrospinning method." *Polymer* **44**(3):841–846.

Ohkawa, K., D. I. Cha, H. Kim, A. Nishida, and H. Yamamoto (2004a). "Electrospinning of chitosan." *Macromolecular Rapid Communications* **25**(18):1600–1605.

Ohkawa, K., H. Kim, and K. H. Lee (2004b). "Biodegradation of electrospun poly(ε-caprolactone) non-woven fabrics by pure-cultured soil filamentous fungi." *Journal of Polymers and the Environment* **12**(4):211–218.

Ohkawa, K., K.-I. Minato, G. Kumagai, S. Hayashi, and H. Yamamoto (2006). "Chitosan nanofiber." *Biomacromolecules* **7**(11):3291–3294.

Okano, T. and T. Matsuda (1998). "Muscular tissue engineering: capillary-incorporated hybrid muscular tissues in vivo tissue culture." *Cell Transplantation* **7**(5):435–442.

Onozuka, K., B. Ding, Y. Tsuge, T. Naka, M. Yamazaki, S. Sugi, S. Ohno, M. Yoshikawa, and S. Shiratori (2006). "Electrospinning processed nanofibrous TiO_2 membranes for photovoltaic applications." *Nanotechnology* **17**(4):1026–1031.

Orwall, R. A. and Arnold, P. A. (2007). "Polymer–solvent interaction parameter." In: *Physical Properties of Polymers Handbook*. Second Edition. Edited by J. E. Mark. Springer, p. 324.

Pan, H. and H. L. Jiang (2006). "Interaction of dermal fibroblasts with electrospun composite polymer scaffolds prepared from dextran and poly lactide-co-glycolide." *Biomaterials* **27**(17):3209–3220.

Park, C., Z. Ounaies, K. A. Watson, R. E. Crooks, J. Smith, Jr., S. E. Lowther, J. W. Connell, E. J. Siochi, J. S. Harrison, and T. L. St. Clair (2002). "Polymer single wall carbon nanotube composites for spacecraft applications." *Chemical Physics Letters* **364**(3–4):303–308.

Park, H.-S. and Y. O. Park (2005). "Filtration properties of electrospun ultrafine fiber webs." *Korean Journal of Chemical Engineering* **22**(1):165–172.

Park, K. and R. J. Mrsny (Eds.) (2000). "Controlled drug delivery: designing technologies for the future." In: ACS Symposium Series, Volume 752, American Chemical Society, Washington, DC.

Park, K. E., H. K. Kang, S. J. Lee, B.-M. Min, and W. H. Park (2006). "Biomimetic nanofibrous scaffolds: preparation and characterization of PGA/chitin blend nanofibers." *Biomacromolecules* **7**(2):635–643.

Park, S. H., C. Kim, Y. O. Choi, and K. S. Yang (2003). "Preparations of pitch-based CF/ACF webs by electrospinning." *Carbon* **41**(13):2655–2657.

Park, S. H., C. Kim, Y. I. Jeong, D. Y. Lim, Y. E. Lee, and K. S. Yang (2004a). "Activation behaviors of isotropic pitch-based carbon fibers from electrospinning and meltspinning." *Synthetic Metals* **146**(2):207–212.

Park, S. H., C. Kim, and K. S. Yang (2004b). "Preparation of carbonized fiber web from electrospinning of isotropic pitch." *Synthetic Metals* **143**(2):175–179.

Park, S. H., S. M. Jo, D. Y. Kim, W. S. Lee, and B. C. Kim (2005). "Effects of iron catalyst on the formation of crystalline domain during carbonization of electrospun acrylic nanofiber." *Synthetic Metals* **150**(3):265–270.

Park, W. H., L. Jeong, D. I. Yoo, and S. Hudson (2004). "Effect of chitosan on morphology and conformation of electrospun silk fibroin nanofibers." *Polymer* **45**(21):7151–7157.

Patel, A. C., S. X. Li, J.-M. Yuan, and Y. Wei (2006). "In situ encapsulation of horseradish peroxidase in electrospun porous silica fibers for potential biosensor applications." *Nano Letters* **6**(5):1042–1046.

Pavlov, M. P., J. F. Mano, N. M. Neves, and R. L. Reis (2004). "Fibers and 3D mesh scaffolds from biodegradable starch-based blends: production and characterization." *Macromolecular Bioscience* **4**(8):776–784.

Pedicini, A. and R. J. Farris (2003). "Mechanical behavior of electrospun polyurethane." *Polymer* **44**(22):6857–6862.

Pedicini, A. and R. J. Farris (2004). "Thermally induced color change in electrospun fiber mats." *Journal of Polymer Science Part B: Polymer Physics* **42**(5):752–757.

Pham, Q. P., U. Sharma, and A. G. Mikos (2006). "Electrospinning of polymeric nanofibers for tissue engineering applications: a review." *Tissue Engineering* **12**(5):1197–1211.

Pich, J. (1987). "Gas filtration theory." In: *Filtration. Principles and Practice*. Edited by M. J. Matteson and C. Orr. Marcel-Dekker, New York.

Picha, A. and N. Schiemenza (2006). "Thermoreversible gelation of biodegradable polyester (PHBV) in toluene." *Polymer (Korea)* **47**(2):553–560.

Pinto, N. J., A. T. Johnson, Jr., A. G. MacDiarmid, C. H. Mueller, N. Theofylaktos, D. C. Robinson, and F. A. Miranda (2003). "Electrospun polyaniline/polyethylene oxide nanofiber field-effect transistor." *Applied Physics Letters* **83**(20):4244–4246.

Pinto, N. J., P. Carrión, and J. X. Quiñones (2004). "Electroless deposition of nickel on electrospun fibers of 2-acrylamido-2-methyl-1-propanesulfonic acid doped polyaniline." *Materials Science and Engineering A* **366**(1):1–5.

Pinto, N. J., P. L. Carrión, A. M. Ayala, and M. Ortiz-Marciales (2005). "Temperature dependence of the resistance of self-assembled polyaniline nanotubes doped with 2-acrylamido-2-methyl-1-propanesulfonic acid." *Synthetic Metals* **148**(3):271–274.

Podgórski, A., A. Bałazy, and L. Gradoń (2006). "Application of nanofibers to improve the filtration efficiency of the most penetrating aerosol particles in fibrous filters." *Chemical Engineering Science* **61**(20):6804–6815.

Pornsopone, V., P. Supaphol, R. Rangkupan, and S. Tantayanon (2005). "Electrospinning of methacrylate-based copolymers: effects of solution concentration and applied electrical potential on morphological appearance of as-spun fibers." *Polymer Engineering and Science* **45**(8):1073–1080.

Pukánszky, B. (2005). "Interfaces and interphases in multicomponent materials: past, present, future." *European Polymer Journal* **41**(4):645–662.

Qian, Y., K. Willeke, S. A. Grinshpun, J. Donnelly, and C. C. Coffey (1998). "Performance of N95 respirators: filtration efficiency for airborne microbial and inert particles." *American Industrial Hygiene Association Journal* **59**(2):128–132.

Qin, X.-H., Y.-Q. Wan, J.-H. He, J. Zhang, J.-Y. Yu, and S.-Y. Wang (2004). "Effect of LiCl on electrospinning of PAN polymer solution: theoretical analysis and experimental verification." *Polymer* **45**(18):6409–6413.

Qin, X.-H., S.-Y. Wang, T. Sandra, and D. Lukas (2005). "Effect of LiCl on the stability length of electrospinning jet by PAN polymer solution." *Materials Letters* **59**(24–25):3102–3105.

Qin, X.-H. and S.-Y. Wang (2006). "Filtration properties of electrospinning nanofibers." *Journal of Applied Polymer Science* **102**(2):1285–1290.

Ra, E. J., K. H. An, K. K. Kim, S. Y. Jeong, and Y. H. Lee (2005). "Anisotropic electrical conductivity of MWCNT/PAN nanofiber paper." *Chemical Physics Letters* **413**(1–3):188–193.

Ramakrishna, S., K. Fujihara, W.-E. Teo, T.-C. Lim, and Z. W. Ma (2005). *An Introduction to Electrospinning and Nanofibers.* World Scientific Publishing, Singapore.

Ramakrishna, S., T. C. Lim, R. Inai, and K. Fujihara (2006). "Modified Halpin–Tsai equation for clay-reinforced polymer nanofiber." *Mechanics of Advanced Materials and Structures* **13**(1):77–81.

Ramaseshan, R., S. Sundarrajan, Y. J. Liu, R. S. Barhate, N. L. Lala, and S. Ramakrishna (2006). "Functionalized polymer nanofiber membranes for protection from chemical warfare stimulants." *Nanotechnology* **17**(2):2947–2953.

Ran, S. F., C. Burger, I. Sics, K. W. Yoon, D. F. Fang, K. Kim, C. Avila-Orta, J. K. Keum, B. Chu, B. S. Hsiao, D. Cookson, D. Shultz, M. Lee, J. Viccaro, and Y. Ohta (2004). "In situ synchrotron SAXS/WAXD studies during melt spinning of modified carbon nanofiber and isotactic polypropylene nanocomposite." *Colloid and Polymer Science* **282**(8):802–809.

Rangkupan, R. and D. H. Reneker (2003). "Electrospinning process of molten polypropylene in vacuum." *Journal of Metals, Materials and Minerals* **12**(2):81–87.

Rathbone, M. J., J. Hadgraft, and M. S. Roberts (Editors) (2003). *Modified-Release Drug Delivery Technology.* Drugs and the Pharmaceutical Sciences Series, Volume 126. Marcel-Dekker, Inc., New York.

Rayleigh, Lord J. W. S. (1882). "On the equilibrium of liquid conducting mass charged with electricity." *Philosophical Magazine* **14**(5th Series):184–186.

Reneker, D. H. and I. Chun (1996). "Nanometre diameter fibres of polymer, produced by electrospinning." *Nanotechnology* **7**(3):216–223.

Reneker, D. H., A. L. Yarin, H. Fong, and S. Koombhongse (2000). "Bending instability of electrically charged liquid jets of polymer solutions in electrospinning." *Journal of Applied Physics* **87**(9):4531–4547.

Reneker, D. H., W. Kataphinan, A. Theron, E. Zussman, and A. L. Yarin (2002). "Nanofiber garlands of polycaprolactone by electrospinning." *Polymer* **43**(25):6785–6794.

Reneker, D. H. and H. Fong (2006). "Polymeric nanofibers: Introduction." In: *Polymeric Nanofibers.* ACS Symposium Series 918. Edited by D. H. Reneker and H. Fong. Oxford University Press (USA). p. 430.

Rho, K. S., L. Jeong, G. Lee, B.-M. Seo, Y. J. Park, S.-D. Hong, S. Roh, J. J. Cho, W. H. Park, and B.-M. Min (2006). "Electrospinning of collagen nanofibers: effects on the behavior of normal human keratinocytes and early-stage wound healing." *Biomaterials* **27**(8):1452–1461.

Riboldi, S. A., M. Sampaolesi, P. Neuenschwander, G. Cossu, and S. Mantero (2005). "Electrospun degradable polyesterurethane membranes: potential scaffolds for skeletal muscle tissue engineering." *Biomaterials* **26**(22):4606–4615.

Rigbi, Z. (1978). "Prediction of swelling of polymers in 2 and 3 component solvent mixtures." *Polymer* **19**(10):1229–1232.

Ristolainen, N., P. Heikkilä, A. Harlin, and J. Seppälä (2006). "Poly(vinyl alcohol) and polyamide-66 nanocomposites prepared by electrospinning." *Macromolecular Materials and Engineering* **291**(2):114–122.

Ritger, P. L. and N. A. Peppas (1987). "A simple equation for description of solute release. I. Fickian and non-fickian release from non-swellable devices in the form of slabs, spheres, cylinders or discs." *Journal of Controlled Release* **5**(1):23–36.

Roseman, T. J. (1972). "Release of steroids from a silicone polymer." *Journal of Pharmaceutical Science* **61**(1):46–50.

Russ, J. C. (2002). *The Image Processing Handbook.* 4th Edition. CRC Press, Boca Raton, FL.

Rutledge, G. and C. S. B. Warner (2003). Electrostatic spinning and properties of ultrafine fibers (NTC Project M01-MD22). National Textile Center Research Briefs — Materials Competency: June 2003:1.

Ryu, Y. J., H. Y. Kim, K. H. Lee, H. C. Park, and D. R. Lee (2003). "Transport properties of electrospun nylon 6 nonwoven mats." *European Polymer Journal* **39**(9):1883–1889.

Salalha, W., Y. Dror, R. L. Khalfin, Y. Cohen, A. L. Yarin, and E. Zussman (2004). "Single-walled carbon nanotubes embedded in oriented polymeric nanofibers by electrospinning." *Langmuir* **20**(22):9852–9855.

Samatham, R. and K. J. Kim (2006). "Electric current as a control variable in the electrospinning process." *Polymer Engineering and Science* **46**(7): 954–959.

Saw, S. H., K. Wang, T. Yong, and S. Ramakrishna (2006). "Polymeric nanofiber for tissue engineering." In: *Nanotechnologies for the Life Sciences, Volume 9: Tissue, Cell and Organ Engineering.* Edited by Challa S. S. R. Kumar. Wiley-VCH Verlag GmbH.

Sawicka, K. M., A. K. Prasad, and P. I. Gouma (2005). "Metal oxide nanowires for use in chemical sensing applications." *Sensor Letters* **3**(1):31–35.

Schindler, M., I. Ahmed, J. Kamal, A. Nur-E-Kamal, T. H. Grafe, H. Y. Chung, and S. Meiners (2005). "A synthetic nanofibrillar matrix promotes in vivo-like organization and morphogenesis for cells in culture." *Biomaterials* **26**(28):5624–5631.

Schreuder-Gibson, H., P. Gibson, K. Senecal, M. Sennett, J. Walker, W. Yeomans, and D. Ziegler (2002). "Protective textile materials based on electrospun nano fibers." *Journal of Advanced Materials* **34**(3):44.

Schreuder-Gibson, H. and P. Gibson (2006). "Applications of electrospun nanofibers in current and future materials." In: *Polymer Nanofibers*. ACS Symposium Series 918. Edited by D. H. Reneker and H. Fong. Oxford University Press (USA), p. 430.

Seidel, A., O. Liivak, and L. W. Jelinski (1998). "Artificial spinning of spider silk." *Macromolecules* **31**(19):6733–6736.

Seki, T. and Y. Okahata (1984). "pH-sensitive permeation of ionic fluorescent probes from nylon capsule membranes." *Macromolecules* **17**(9):1880–1882.

Seoul, C., Y.-T. Kim, and C.-K. Baek (2003). "Electrospinning of poly(vinylidene fluoride)/dimethylformamide solutions with carbon nanotubes." *Journal of Polymer Science Part B: Polymer Physics* **41**(13):1572–1577.

Shao, C. L., H.-Y. Kim, J. Gong, B. Ding, D.-R. Lee, and S.-J. Park (2003). "Fiber mats of poly(vinyl alcohol)/silica composite via electrospinning." *Materials Letters* **57**(9–10):1579–1584.

Shao, C. L., H. Y. Guan, Y. C. Liu, J. Gong, N. Yu, and X. H. Yang (2004a). "A novel method for making ZrO_2 nanofibres via an electrospinning technique." *Journal of Crystal Growth* **267**(1–2):380–384.

Shao, C. L., H. Y. Guan, Y. C. Liu, X. L. Li, and X. H. Yang (2004b). "Preparation of Mn_2O_3 and Mn_3O_4 nanofibers via an electrospinning technique." *Journal of Solid State Chemistry* **177**(7):2628–2631.

Shao, C. L., H. Y. Guan, S. B. Wen, B. Chen, D. X. Han, J. Gong, X. H. Yang, and Y. C. Liu (2004c). "Preparation of alpha-Fe_2O_3 nanofiber via electrospinning process." *Chemical Journal of Chinese Universities (Chinese)* **25**(6):1013–1015.

Shao, C. L., H. Y. Guan, S. B. Wen, B. Chen, X. H. Yang, and J. Gong (2004d). "A novel method for making NiO nanofibres via an electrospinning technique." *Chinese Chemical Letters* **15**(3):365–367.

Shao, C. L., H. Y. Guan, S. B. Wen, B. Chen, X. H. Yang, and J. Gong (2004e). "Preparation of Mn_3O_4 nanofibres via an electrospinning technique." *Chinese Chemical Letters* **15**(4):471–474.

Shao, C. L., H. Y. Guan, S. B. Wen, B. Chen, X. H. Yang, J. Gong, and Y. C. Liu (2004f). "Preparation of Co_3O_4 nanofibers via an electrospinning technique." *Chinese Chemical Letters* **15**(4):492–494.

Shao, C. L., X. H. Yang, H. Y. Guan, Y. C. Liu, and J. Gong (2004g). "Electrospun nanofibers of NiO/ZnO composite." *Inorganic Chemistry Communications* **7**(5):625–627.

Shawon, J. and C. M. Sung (2004). "Electrospinning of polycarbonate nanofibers with solvent mixtures THF and DMF." *Journal of Materials Science* **39**(14):4605–4613.

Shen, J. F., W. S. Huang, L. P. Wu, Y. Z. Hu, and M. X. Ye (2007). "The reinforcement role of different amino-functionalized multi-walled carbon nanotubes in epoxy nanocomposites." *Composites Science and Technology* **67**(15–16):3041–3050.

Shenoy, S. L., W. D. Bates, and G. Wnek (2005a). "Correlations between electrospinnability and physical gelation." *Polymer* **46**(21):8990–9004.

Shenoy, S. L., W. D. Bates, H. L. Frisch, and G. E. Wnek (2005b). "Role of chain entanglements on fiber formation during electrospinning of polymer solutions: good solvent, non-specific polymer–polymer interaction limit." *Polymer* **46**(10):3372–3384.

Shin, C. and G. G. Chase (2005). "Nanofibers from recycle waste expanded polystyrene using natural solvent." *Polymer Bulletin* **55**(3):209–215.

Shin, C., G. G. Chase, and D. H. Reneker (2005). "Recycled expanded polystyrene nanofibers applied in filter media." *Colloids and Surfaces A: Physicochemical and Engineering Aspects* **262**(1–3):211–215.

Shin, C. and G. G. Chase (2006). "Separation of water-in-oil emulsions using glass fiber media augmented with polymer nanofibers." *Journal of Dispersion Science and Technology* **27**(4):517–522.

Shin, H. J., C. H. Lee, I. H. Chu, Y.-J. Kim, Y.-J. Lee, I. A. Kim, K.-D. Park, N. Yui, and J.-W. Shin (2006). "Electrospun PLGA nanofiber scaffolds for articular cartilage reconstruction: mechanical stability, degradation and cellular responses under mechanical stimulation *in vitro*." *Journal of Biomaterials Science — Polymer Edition* **17**(1–2):103–119.

Shin, M., O. Ishii, T. Sueda, and J. P. Vacanti (2004a). "Contractile cardiac grafts using a novel nanofibrous mesh." *Biomaterials* **25**(17):3717–3723.

Shin, M., H. Yoshimoto, and J. P. Vacanti (2004b). "*In vivo* bone tissue engineering using mesenchymal stem cells on a novel electrospun nanofibrous scaffold." *Tissue Engineering* **10**(1–2):33–41.

Shin, M. K., S. I. Kim, S. J. Kim, S.-K. Kim, and H. Lee (2006). "Reinforcement of polymeric nanofibers by ferritin nanoparticles." *Applied Physics Letters* **88**(19):193901.

Shin, Y. M., M. M. Hohman, M. P. Brenner, and G. C. Rutledge (2001a). "Electrospinning: a whipping fluid jet generates submicron polymer fibers." *Applied Physics Letters* **78**(8):1149–1151.

Shin, Y. M., M. M. Hohman, M. P. Brenner, and G. C. Rutledge (2001b). "Experimental characterization of electrospinning: the electrically forced jet and instabilities." *Polymer* **42**(25): 9955–9967.

Shukla, S., E. Brinley, H. J. Cho, and S. Seal (2005). "Electrospinning of hydroxypropyl cellulose fibers and their application in synthesis of nano and submicron tin oxide fibers." *Polymer* **46**(26):12130–12145.

Simha, R. and J. L. Zakin (1962). "Solution viscosities of linear flexible high polymers." *Colloid Science* **17**(3):270–287.

Simpson, D. G. (2006). "Dermal templates and the wound-healing paradigm: the promise of tissue regeneration." *Expert Review of Medical Devices* **3**(4):471–484.

Singh, A., L. Steely, and H. R. Allcock (2005). "Poly[bis(2,2,2-trifluoroethoxy)phosphazene]superhydrophobic nanofibers." *Langmuir* **21**(25):11604–11607.

Siochi, E. J., D. C. Working, C. Park, P. T. Lillehei, J. H. Rouse, C. C. Topping, A. R. Bhattacharyya, and S. Kumar (2004). "Melt processing of SWCNT-polyimide nanocomposite fibers." *Composites Part B: Engineering* **35**(5):439–446.

Sittinger, M., D. Reitzel, M. Dauner, H. Hierlemann, C. Hammer, E. Kastenbauer, H. Planck, G. R. Burmester, and J. Bujia (1996). "Resorbable polyesters in cartilage engineering: affinity and biocompatibility of polymer fiber structures to chondrocytes." *Journal of Biomedical Materials Research Part B: Applied Biomaterials* **33**(2):57–63.

Smit, E., U. Buttner, and R. D. Sanderson (2005). "Continuous yarns from electrospun fibers." *Polymer* **46**(8):2419–2423.

Smith, L. A. and P. X. Ma (2004). "Nano-fibrous scaffolds for tissue engineering." *Colloids and Surfaces B: Biointerfaces* **39**(3):125–131.

Sombatmankhong, K., Sanchavanakit, P. Pavasant, and P. Supaphol (2007). "Bone scaffolds from electrospun fiber mats of poly(3-hydroxybutyrate), poly(3-hydroxybutyrate-*co*-3-hydroxyvalerate) and their blend." *Polymer* **48**(5): 1419–1427.

Son, W. K., J. H. Youk, T. S. Lee, and W. H. Park (2004a). "Electrospinning of ultrafine cellulose acetate fibers: studies of a new solvent system and deacetylation of ultrafine cellulose acetate fibers." *Journal of Polymer Science Part B: Polymer Physics* **42**(1):5–11.

Son, W. K., J. H. Youk, T. S. Lee, and W. H. Park (2004b). "Preparation of antimicrobial ultrafine cellulose acetate fibers with silver nanoparticles." *Macromolecular Rapid Communications* **25**(18):1632–1637.

Son, W. K., J. H. Youk, and W. H. Park (2004c). "Preparation of ultrafine oxidized cellulose mats via electrospinning." *Biomacromolecules* **5**(1):197–201.

Son, W. K., J. H. Youk, T. S. Lee, and W. H. Park (2004d). "The effects of solution properties and polyelectrolyte on electrospinning of ultrafine poly(ethylene oxide) fibers." *Polymer* **45**(9):2959–2966.

Son, W. K., J. H. Youk, T. S. Lee, and W. H. Park (2005). "Effect of pH on electrospinning of poly(vinyl alcohol)." *Materials Letters* **59**(12):1571–1575.

Song, M. Y., D. K. Kim, K. J. Ihn, S. M. Jo, and D. Y. Kim (2005). "New application of electrospun TiO_2 electrode to solid-state dye-sensitized solar cells." *Synthetic Metals* **153**(1–3):77–80.

Song, T., Y. Z. Zhang, T. J. Zhou, C. T. Lim, S. Ramakrishna, and B. Liu (2005). "Encapsulation of self-assembled FePt magnetic nanoparticles in PCL nanofibers by coaxial electrospinning." *Chemical Physics Letters* **415**(4–6): 317–322.

Spasova, M., N. Manolova, D. Paneva, and I. Rashkov (2004). "Preparation of chitosan-containing nanofibres by electrospinning of chitosan/poly(ethylene oxide) blend solutions." *e-polymers*, no. 056, www.e-polymers.org.

Spasova, M., O. Stoilova, N. Manolova, I. Rashkov, and G. Altankov (2007). "Preparation of PLIA/PEG nanofibers by electrospinning and potential applications." *Journal of Bioactive and Compatible Polymers* **22**(1):62–76.

Spivak, A. F. and Y. A. Dzenis (1998). "Asymptotic decay of radius of a weakly conductive viscous jet in an external electric field." *Applied Physics Letters* **73**(21):3067–3069.

Spivak, A. F. and Y. A. Dzenis (1999). "A condition of the existence of a conductive liquid meniscus in an external electric field." *Journal of Applied Mechanics* **66**(4):1026–1028.

Spivak, A. F., Y. A. Dzenis, and D. H. Reneker (2000). "A model of steady state jet in the electrospinning process." *Mechanics Research Communications* **27**(1):37–42.

Srinivasan, G. and D. H. Reneker (1995). "Structure and morphology of small diameter electrospun aramid fibers." *Polymer International* **36**(2):195–201.

Srinivasarao, M., D. Collings, A. Philips, and S. Patel (2001). "Three-dimensionally ordered array of air bubbles in a polymer film." *Science* **292**(5514):79–83.

Stankus, J. J., J. J. Guan, K. Fujimoto, and W. R. Wagner (2006). "Microintegrating smooth muscle cells into a biodegradable, elastomeric fiber matrix." *Biomaterials* **27**(5):735–744.

Stephens, J. S., S. Frisk, S. Megelski, J. F. Rabolt, and D. B. Chase (2001). "'Real time' Raman studies of electrospun fibers." *Applied Spectroscopy* **55**(10):1287–1290.

Stephens, J. S., D. B. Chase, and J. F. Rabolt (2004). "Effect of the electrospinning process on polymer crystallization chain conformation in nylon-6 and nylon-12." *Macromolecules* **37**(3):877–881.

Stephens, J. S., S. R. Fahnestock, R. S. Farmer, K. L. Kiick, D. B. Chase, and J. F. Rabolt (2005). "Effects of electrospinning and solution casting protocols on the secondary structure of a genetically engineered dragline spider silk analogue investigated via Fourier transform Raman spectroscopy." *Biomacromolecules* **6**(3):1405–1413.

Stitzel, J., J. Liu, S. J. Lee, M. Komura, J. Berry, S. Soker, G. Lim, M. van Dyke, R. Czerw, J. J. Yoo, and A. Atala (2006). "Controlled fabrication of a biological vascular substitute." *Biomaterials* **27**(7):1088–1094.

Subbiah, T., G. S. Bhat, R. W. Tock, S. Parameswaran, and S. S. Ramkumar (2005). "Electrospinning of nanofibers." *Journal of Applied Polymer Science* **96**(2):557–569.

Subramanian, A., D. Vu, G. F. Larsen, and H. Y. Lin (2005). "Preparation and evaluation of the electrospun chitosan/PEO fibers for potential applications in cartilage tissue engineering." *Journal of Biomaterials Science: Polymer Edition* **16**(7):861–873.

Sukigara, S., M. Gandhi, J. Ayutsede, M. Micklus, and F. Ko (2003). "Regeneration of *Bombyx mori* silk by electrospinning. Part 1. Processing parameters and geometric properties." *Polymer* **44**(19):5721–5727.

Sukigara, S., M. Gandhi, J. Ayutsede, M. Micklus, and F. Ko (2004). "Regeneration of *Bombyx mori* silk by electrospinning. Part 2. Process optimization and empirical modeling using response surface methodology." *Polymer* **45**(11):3701–3708.

Sun, T., S. M. Mai, D. Norton, J. W. Haycock, A. J. Ryan, and S. MacNeil (2005). "Self-organization of skin cells in three-dimensional electrospun polystyrene scaffolds." *Tissue Engineering* **11**(7–8):1023–1033.

Sun, Z. C., E. Zussman, A. L. Yarin, J. H. Wendorff, and A. Greiner (2003). "Compound core-shell polymer nanofibers by co-electrospinning." *Advanced Materials* **15**(22):1929–1932.

Sundaray, B., V. Subramanian, T. S. Natarajan, R.-Z. Xiang, C.-C. Chang, and W.-S. Fann (2004). "Electrospinning of continuous aligned polymer fibers." *Applied Physics Letters* **84**(7):1222–1224.

Sundaray, B. (2006). "Preparation and electrical characterization of electrospun fibers of carbon nanotube-polymer nanocomposite." PhD Thesis, Department of Physics, Indian Institute of Technology, Chennai, India.

Sundaray, B., V. Subramanian, T. S. Natarajan, and K. Krishnamurthy (2006). "Electrical conductivity of a single electrospun fiber of poly(methyl methacrylate) and multiwalled carbon nanotube nanocomposite." *Applied Physics Letters* **88**(14):143114.

Sung, J. H., H. S. Kim, H.-J. Jin, H. J. Choi, and I.-J. Chin (2004). "Nanofibrous membranes prepared by multiwalled carbon nanotube/poly(methyl methacrylate) composites." *Macromolecules* **37**(26):9899–9902.

Supaphol, P., C. Mit-uppatham, and M. Nithitanakul (2005a). "Ultrafine electrospun polyamide-6 fibers: effect of emitting electrode polarity on morphology and average fiber diameter." *Journal of Polymer Science Part B: Polymer Physics* **43**(24):3699–3712.

Supaphol, P., C. Mit-uppatham, and M. Nithitanakul (2005b). "Ultrafine electrospun polyamide-6 fibers: effects of solvent system and emitting electrode polarity on morphology and average fiber diameter." *Macromolecular Materials and Engineering* **290**(9):933–942.

Takahashi, T., M. Taniguchi, and T. Kawai (2005). "Fabrication of DNA nanofibers on a planar surface by electrospinning." *Japanese Journal of Applied Physics Part 2: Letters & Express Letters* **44**(24–27):L860.

Tan, E. P. S. and C. T. Lim (2005). "Nanoindentation study of nanofibers." *Applied Physics Letters* **87**(12):123106.

Tan, E. P. S., C. N. Goh, C. H. Sow, and C. T. Lim (2005a). "Tensile test of a single nanofiber using an atomic force microscope tip." *Applied Physics Letters* **86**(7):073115.

Tan, E. P. S., S. Y. Ng, and C. T. Lim (2005b). "Tensile testing of a single ultrafine polymeric fiber." *Biomaterials* **26**(13):1453–1456.

Tan, E. P. S. and C. T. Lim (2006). "Mechanical characterization of nanofibers – a review." *Composites Science and Technology* **66**(9):1102–1111.

Tan, S.-H., R. Inai, M. Kotaki, and S. Ramakrishna (2005). "Systematic parameter study for ultra-fine fiber fabrication via electrospinning process." *Polymer* **46**(16):6128–6134.

Tan, S. T., J. H. Wendorff, C. Pietzonka, Z. H. Jia, and G. Q. Wang (2005). "Biocompatible and biodegradable polymer nanofibers displaying superparamagnetic properties." *ChemPhysChem* **6**(8):1461–1465.

Taylor, G. (1964). "Disintegration of water drops in an electric field." *Proceedings of the Royal Society of London. Series A, Mathematical and Physical Sciences* **280**(1382):383–397.

Taylor, G. (1969). "Electrically driven jets." *Proceedings of the Royal Society of London. Series A, Mathematical and Physical Sciences* **313**(1515):453–475.

Teas, J. P. (1968). "Graphic analysis of resin solubilities." *Journal of Paint Technology* **40**(516):19–25.

Telemeco, T. A., C. Ayres, G. L. Bowlin, G. E. Wnek, E. D. Boland, N. Cohen, C. M. Baumgarten, J. Mathews, and D. G. Simpson (2005). "Regulation of cellular infiltration into tissue engineering scaffolds composed of submicron diameter fibrils produced by electrospinning." *Acta Biomaterialia* **1**(4):377–385.

Teo, W. E. and S. Ramakrishna (2006). "A review on electrospinning design and nanofibre assemblies." *Nanotechnology* **17**(14):R89–R104.

Teo, W. E. and S. Ramakrishna (2005). "Electrospun fibre bundle made of aligned nanofibres over two fixed points." *Nanotechnology* **16**(9):1878–1884.

Teo, W. E., M. Kotaki, X. M. Mo, and S. Ramakrishna (2005). "Porous tubular structures with controlled fibre orientation using a modified electrospinning method." *Nanotechnology* **16**(6):918–924.

Theron, A., E. Zussman, and A. L. Yarin (2001). "Electrostatic field-assisted alignment of electrospun nanofibres." *Nanotechnology* **12**(3):384–390.

Theron, S. A., E. Zussman, and A. L. Yarin (2004). "Experimental investigation of the governing parameters in the electrospinning of polymer solutions." *Polymer* **45**(6):2017–2030.

Theron, S. A., A. L. Yarin, E. Zussman, and E. Kroll (2005). "Multiple jets in electrospinning: experiment and modeling." *Polymer* **46**(9):2889–2899.

Thomas, V., S. Jagani, K. Johnson, M. V. Jose, D. R. Dean, Y. K. Vohra, and E. Nyairo (2006). "Electrospun bioactive nanocomposite scaffolds of polycaprolactone and nanohydroxyapatite for bone tissue engineering." *Journal of Nanoscience and Nanotechnology* **6**(2):487–493.

Tomaszewski, W. and M. Szadkowski (2005). "Investigation of electrospinning with the use of a multi-jet electrospinning head." *Fibres & Textiles in Eastern Europe* **13**(4):22–26.

Tomczak, N., N. F. van Hulst, and G. J. Vancso (2005). "Beaded electrospun fibers for photonic applications." *Macromolecules* **38**(18):7863–7866.

Tomer, V., R. Teye-Mensah, J. C. Tokash, N. Stojilovic, W. Kataphinan, E. A. Evans, G. G. Chase, R. D. Ramsier, D. J. Smith, and D. H. Reneker (2005). "Selective emitters for thermophotovoltaics: erbia-modified electrospun titania nanofibers." *Solar Energy Materials and Solar Cells* **85**(4):477–488.

Tomlins, P., P. Grant, P. Vadgama, S. L. James, and S. V. Mikhalovsky (2004). "Structural characterisation of polymer based tissue scaffolds." National Physical Laboratory, Report Number DEPC-MN-002, Teddington, Middlesex, United Kingdom.

Tsai, P. P., H. Schreuder-Gibson, and P. Gibson (2002). "Different electro static methods for making electret filters." *Journal of Electrostatics* **54**(3–4): 333–341.

Tuan, R. S., G. Boland, and R. Tuli (2003). "Adult mesenchymal stem cells and cell-based tissue engineering." *Arthritis Research and Therapy* **5**(1):32–45.

Tuzlakoglu, K., N. Bolgen, A. J. Salgado, M. E. Gomes, E. Piskin, and R. L. Reis (2005). "Nano- and micro-fiber combined scaffolds: a new architecture for bone tissue engineering." *Journal of Materials Science: Materials in Medicine* **16**(12):1099–1104.

Um, I. C., D. F. Fang, B. S. Hsiao, A. Okamoto, and B. Chu (2004). "Electro-spinning and electro-blowing of hyaluronic acid." *Biomacromolecules* **5**(4):1428–1436.

Vaz, C. M., S. van Tuijl, C. V. C. Bouten, and F. P. T. Baaijens (2005). "Design of scaffolds for blood vessel tissue engineering using a multi-layering electro-spinning technique." *Acta Biomaterialia* **1**(5):575–582.

Veluru, J. B., K. K. Satheesh, D. C. Trivedi, M. V. Ramakrishna, and N. T. Srinivasan (2007). "Electrical properties of electrospun fibers of PANI-PMMA composites." *Journal of Engineered Fibers and Fabrics* **2**(2):25–31.

Venugopal, J. and S. Ramakrishna (2005). "Applications of polymer nanofibers in biomedicine and biotechnology." *Applied Biochemistry and Biotechnology* **125**(3):147–157.

Venugopal, J., L. L. Ma, T. Yong, and S. Ramakrishna (2005a). "In vitro study of smooth muscle cells on polycaprolactone and collagen nanofibrous matrices." *Cell Biology International* **29**(10):861–867.

Venugopal, J., Y. Z. Zhang, and S. Ramakrishna (2005b). "Fabrication of modified and functionalized polycaprolactone nanofibre scaffolds for vascular tissue engineering." *Nanotechnology* **16**(10):2138–2142.

Venugopal, J. R., Y. Z. Zhang, and S. Ramakrishna (2006). "In vitro culture of human dermal fibroblasts on electrospun polycaprolactone collagen nanofibrous membrane." *Artificial Organs* **30**(6):440–446.

Verreck, G., I. Chun, J. Peeters, J. Rosenblatt, and M. E. Brewster (2003a). "Preparation and characterization of nanofibers containing amorphous drug dispersions generated by electrostatic spinning." *Pharmaceutical Research* **20**(5):810–817.

Verreck, G., I. Chun, J. Rosenblatt, J. Peeters, A. Van Dijck, J. Mensch, M. Noppe, and M. E. Brewster (2003b). "Incorporation of drugs in an amorphous state into electrospun nanofibers composed of a water-insoluble, nonbiodegradable polymer." *Journal of Controlled Release* **92**(3):349–360.

Viriyabanthorn, N., R. G. Stacer, C. M. Sung, and J. L. Mead (2006). "Effect of carbon black loading on electrospun butyl rubber nonwoven mats." In: *Polymeric Nanofibers*. ACS Symposium Series 918. Edited by D. H. Reneker and H. Fong. Oxford University Press (USA).

Virji, S., J. X. Huang, R. B. Kaner, and B. H. Weiller (2004). "Polyaniline nanofiber gas sensors: examination of response mechanisms." *Nano Letters* **4**(3):491–496.

Viswanathamurthi, P., N. Bhattarai, H. Y. Kim, and D. R. Lee (2003a). "Vanadium pentoxide nanofibers by electrospinning." *Scripta Materialia* **49**(6):577–581.

Viswanathamurthi, P., N. Bhattarai, H. Y. Kim, D. R. Lee, S. R. Kim, and M. A. Morris (2003b). "Preparation and morphology of niobium oxide fibres by electrospinning." *Chemical Physics Letters* **374**(1–2):79–84.

Viswanathamurthi, P., N. Bhattarai, H. Y. Kim, D. I. Cha, and D. R. Lee (2004a). "Preparation and morphology of palladium oxide fibers via electrospinning." *Materials Letters* **58**(26):3368–3372.

Viswanathamurthi, P., N. Bhattarai, H. Y. Kim, and D. R. Lee (2004b). "The photoluminescence properties of zinc oxide nanofibres prepared by electrospinning." *Nanotechnology* **15**(3):320–323.

Von Recum, A. F., C. E. Shannon, C. E. Cannon, K. J. Long, T. G. Van Kooten, and J. Meyle (1996). "Surface roughness, porosity, and texture as modifiers of cellular adhesion." *Tissue Engineering* **2**(4):241–253.

Wan, L.-S., J. Wu, and Z.-K. Xu (2006). "Porphyrinated nanofibers via copolymerization and electrospinning." *Macromolecular Rapid Communications* **27**(18):1533–1538.

Wan, Y.-Q., J.-H. He, Y. Wu, and J.-Y. Yu (2006). "Vibrorheological effect on electrospun polyacrylonitrile (PAN) nanofibers." *Materials Letters* **60**(27): 3296–3300.

Wan, Y.-Q., J.-H. He, Y. Wu, and J.-Y. Yu (2007). "Vibration-electrospinning for high-concentration poly(butylene succinate)/chloroform solution." *International Journal of Electrospun Nanofibers and Applications* **1**(1):17–28.

Wang, C., C. H. Hsu, and J. H. Lin (2006). "Scaling laws in electrospinning of polystyrene solutions." *Macromolecules* **39**(22):7662–7672.

Wang, H., Y. P. Zhang, H. L. Shao, and X. C. Hu (2005). "Electrospun ultra-fine silk fibroin fibers from aqueous solutions." *Journal of Materials Science* **40**(20):5359–5363.

Wang, M., H.-J. Jin, D. L. Kaplan, and G. C. Rutledge (2004a). "Mechanical properties of electrospun silk fibers." *Macromolecules* **37**(18):6856–6864.

Wang, M., H. Singh, T. A. Hatton, and G. C. Rutledge (2004b). "Field-responsive superparamagnetic composite nanofibers by electrospinning." *Polymer* **45**(16):5505–5514.

Wang, M., A. J. Hsieh, and G. C. Rutledge (2005). "Electrospinning of poly(MMA-*co*-MAA) copolymers and their layered silicate nanocomposites for improved thermal properties." *Polymer* **46**(10):3407–3418.

Wang, M., J. H. Yu, D. L. Kaplan, and G. C. Rutledge (2006). "Production of submicron diameter silk fibers under benign processing conditions by two-fluid electrospinning." *Macromolecules* **39**(3):1102–1107.

Wang, X., I. C. Um, D. Fang, A. Okamoto, B. S. Hsiao, and B. Chu (2005). "Formation of water-resistant hyaluronic acid nanofibers by blowing-assisted electro-spinning and non-toxic post treatments." *Polymer* **46**(13):4853–4867.

Wang, X. F., X. M. Chen, K. H. Yoon, D. F. Fang, B. S. Hsiao, and B. Chu (2005). "High flux filtration medium based on nanofibrous substrate with hydrophilic nanocomposite coating." *Environmental Science & Technology* **39**(19): 7684–7691.

Wang, X. Y., C. Drew, S.-H. Lee, K. J. Senecal, J. Kumar, and L. A. Samuelson (2002a). "Electrospinning technology: a novel approach to sensor application." *Journal of Macromolecular Science Part A: Pure and Applied Chemistry* **A39**(10):1251–1258.

Wang, X. Y., C. Drew, S.-H. Lee, K. J. Senecal, J. Kumar, and L. A. Samuelson (2002b). "Electrospun nanofibrous membranes for highly sensitive optical sensors." *Nano Letters* **2**(11):1273–1275.

Wang, X. Y., Y.-G. Kim, C. Drew, B.-C. Ku, J. Kumar, and L. A. Samuelson (2004). "Electrostatic assembly of conjugated polymer thin layers on electrospun nano fibrous membranes for biosensors." *Nano Letters* **4**(2):331–334.

Wang, Y., S. Serrano, and J. J. Santiago-Avilés (2003). "Raman characterization of carbon nanofibers prepared using electrospinning." *Synthetic Metals* **138**(3):423–427.

Wang, Y., M. Aponte, N. Leon, I. Ramos, R. Furlan, S. Evoy, and J. J. Santiago-Avilés (2004). "Synthesis and characterization of tin oxide microfibres electrospun from a simple precursor solution." *Semiconductor Science and Technology* **19**(8):1057–1060.

Wang, Y., M. Aponte, N. Leon, I. Ramos, R. Furlan, N. Pinto, S. Evoy, and J. J. Santiago-Avilés (2005). "Synthesis and characterization of ultra-fine tin oxide fibers using electrospinning." *Journal of the American Ceramic Society* **88**(8):2059–2063.

Wang, Y., I. Ramos, and J. J. Santiago-Avilés (2007). "Synthesis of ultra-fine porous tin oxide fibres and its process characterization." *Nanotechnology* **18**(29): 295601.

Wang, Y. H. and Y. L. Hsieh (2004). "Enzyme immobilization to ultra-fine cellulose fibers via amphiphilic polyethylene glycol spacers." *Journal of Polymer Science Part A: Polymer Chemistry* **42**(17):4289–4299.

Wang, Y. Z., Q. B. Yang, G. Y. Shan, C. Wang, J. S. Du, S. G. Wang, Y. X. Li, X. S. Chen, X. B. Jing, and Y. Wei (2005). "Preparation of silver nanoparticles dispersed in polyacrylonitrile nanofiber film spun by electrospinning." *Materials Letters* **59**(24–25):3046–3049.

Wang, Z.-G., Z.-K. Xu, L.-S. Wan, J. Wu, C. Innocent, and P. Seta (2006). "Nanofibrous membranes containing carbon nanotubes: electrospun for redox enzyme immobilization." *Macromolecular Rapid Communications* **27**(7):516–521.

Wannatong, L., A. Sirivat, and P. Supaphol (2004). "Effects of solvents on electrospun polymeric fibers: preliminary study on polystyrene." *Polymer International* **53**(11):1851–1859.

Warner, S. B., A. Buer, M. Grimler, S. C. Ugbolue, G. C. Rutledge, and M. Y. Shin (1998). "A fundamental investigation of the formation and properties of electrospun fibers." Annual Report, M98-D01, National Textile Center.

Watthanaarun, J., V. Pavarajarn, and P. Supaphol (2005). "Titanium (IV) oxide nanofibers by combined sol–gel and electrospinning techniques: preliminary report on effects of preparation conditions and secondary metal dopant." *Science and Technology of Advanced Materials* **6**(3–4):240–245.

Wayne, J. S., C. L. McDowell, K. J. Shields, and R. S. Tuan (2005). "*In vivo* response of polylactic acid–alginate scaffolds and bone marrow-derived cells for cartilage tissue engineering." *Tissue Engineering* **11**(5–6):953–963.

Wei, M., J. Lee, B. Kang, and J. Mead (2005). "Preparation of core-sheath nanofibers from conducting polymer blends." *Macromolecular Rapid Communications* **26**(14):1127–1132.

Wei, M., B. W. Kang, C. M. Sung, and J. Mead (2006a). "Core-sheath structure in electrospun nanofibers from polymer blends." *Macromolecular Materials and Engineering* **291**(11):1307–1314.

Wei, M., B. W. Kang, C. M. Sung, and J. Mead (2006b). "Preparation of nanofibers with controlled phase morphology from electrospinning of polybutadiene-polycarbonate blends." In: *Polymeric Nanofibers*. ACS Symposium Series 918. Edited by D. H. Reneker and H. Fong. Oxford University Press (USA), p. 149.

Wei, Q. F., H. Ye, D. Y. Hou, H. B. Wang, and W. D. Gao (2006). "Surface functionalization of polymer nanofibers by silver sputter coating." *Journal of Applied Polymer Science* **99**(5):2384–2388.

White, L. A. and C. Delhom (2004). "Cellulose-based nanocomposites: fiber production and characterization. In: *Polymeric Materials: Science and Engineering Preprints*, 227th American Chemical Society Meeting, Anaheim, CA, Volume 90, Number 2, pp. 45–50.

Williams, D. B. and C. B. Carter (2004). *Transmission Electron Microscopy: A Textbook for Materials Science* (4-Vol Set). Springer-Verlag, Berlin, Germany. p. 708.

Williamson, M. R., R. Black, and C. Kielty (2006). "PCL-PU composite vascular scaffold production for vascular tissue engineering: attachment, proliferation and bioactivity of human vascular endothelial cells." *Biomaterials* **27**(19):3608–3616.

Wilson, K. G. (1971a). "Renormalization group and critical phenomena. I. Renormalization group and the Kadanoff scaling picture." *Physical Review B* **4**(9):3174–3183.

Wilson, K. G. (1971b). "Renormalization group and critical phenomena. II. Phase-space cell analysis of critical behavior." *Physical Review B* **4**(9):3184–3205.

Wiltzius, P., H. R. Haller, D. S. Cannell, and D. W. Schaefer (1983). "Universality for static properties of polystyrenes in good and marginal solvents." *Physical Review Letters* **51**:1183–1186.

Wnek, G. E., M. E. Carr, D. G. Simpson, and G. L. Bowlin (2003). "Electrospinning of nanofiber fibrogen structures." *Nano Letters* **3**(2):213–216.

Woerdeman, D. L., P. Ye, S. Shenoy, R. S. Parnas, G. E. Wnek, and O. Trofimova (2005). "Electrospun fibers from wheat protein: investigation of the interplay between molecular structure and the fluid dynamics of the electrospinning process." *Biomacromolecules* **6**(2):707–712.

Wolf, B. A. and R. J. Molinari (1973). "True cosolvency. Acetone/diethylether/polystyrene." *Die Makromolekulare Chemie* **173**(1):241–245.

Wolf, B. A. and M. M. Willms (1978). "Measured and calculated solubility of polymers in mixed solvents: co-nonsolvency." *Die Makromolekulare Chemie* **179**(9):2265–2277.

Woo, K. M., V. J. Chen, and P. X. Ma (2003). "Nano-fibrous scaffolding architecture selectively enhances protein adsorption contributing to cell attachment." *Journal of Biomedical Materials Research Part A* **67A**(2):531–537.

Wu, H. and W. Pan (2006). "Preparation of zinc oxide nanofibers by electrospinning." *Journal of the American Ceramic Society* **89**(2):699–701.

Wu, L. L., X. Y. Yuan, and J. Sheng (2005). "Immobilization of cellulose in nanofibrous PVA membranes by electrospinning." *Journal of Membrane Science* **250**(1–2):167–173.

Wu, X. H., L. Wang, H. Yu, and Y. Huang (2005). "Effect of solvent on morphology of electrospinning ethyl cellulose fibers." *Journal of Applied Polymer Science* **97**(3):1292–1297.

Wu, X. H., L. G. Wang, and Y. Huang (2006). "Application of electrospun ethyl cellulose fibers in drug release systems." *Acta Polymerica Sinica* **2**:264–268.

Wu, Y. Q., L. L. Hench, J. Du, K.-L. Choy, and J. K. Guo (2004). "Preparation of hydroxyapatite fibers by electrospinning technique." *Journal of the American Ceramic Society* **87**(10):1988–1991.

Xie, J. B. and Y. L. Hsieh (2003). "Ultra-high surface fibrous membranes from electrospinning of natural proteins: casein and lipase enzyme." *Journal of Materials Science* **38**(10):2125–2133.

Xie, X.-L., Y.-W. Mai, and X.-P. Zhou (2005). "Dispersion and alignment of carbon nanotubes in polymer matrix: a review." *Materials Science and Engineering: R: Reports* **49**(4):89–112.

Xin, Y., Z. H. Huang, E. Y. Yan, W. Zhang, and Q. Zhao (2006). "Controlling poly(*p*-phenylene vinylene)/poly(vinyl pyrrolidone) composite nanofibers in different morphologies by electrospinning." *Applied Physics Letters* **89**(5):053101.

Xin, X. J., M. Hussain, and J. J. Mao (2007). "Continuing differentiation of human mesen-chymal stem cells and induced chondrogenic and osteogenic lineages in electrospun PLGA nanofiber scaffold." *Biomaterials* **28**(2):316–325.

Xu, C. Y., R. Inai, M. Kotaki, and S. Ramakrishna (2004a). "Aligned biodegradable nanofibrous structure: a potential scaffold for blood vessel engineering." *Biomaterials* **25**(5):877–886.

Xu, C. Y., R. Inai, M. Kotaki, and S. Ramakrishna (2004b). "Electrospun nanofiber fabrication as synthetic extracellular matrix and its potential for vascular tissue engineering." *Tissue Engineering* **10**(7–8):1160–1168.

Xu, H. and D. H. Reneker (2006). "Characterization of electrospinning jets using interference color technique." In: *Polymeric Nanofibers*. ACS Symposium Series 918. Edited by D. H. Reneker and H. Fong. Oxford University Press (USA), pp. 21–33.

Xu, X. L., L. X. Yang, X. Y. Xu, X. Wang, X. S. Chen, Q. Z. Liang, J. Zeng, and X. B. Jing (2005). "Ultrafine medicated fibers electrospun from W/O emulsions." *Journal of Controlled Release* **108**(1):33–42.

Yang, F., C. Y. Xu, M. Kotaki, S. Wang, and S. Ramakrishna (2004). "Characterization of neural stem cells on electrospun poly(L-lactic acid) nanofibrous scaffold." *Journal of Biomaterials Science-Polymer Edition* **15**(12): 1483–1497.

Yang, F., R. Murugan, S. Wang, and S. Ramakrishna (2005). "Electrospinning of nano/micro scale poly(L-lactic acid) aligned fibers and their potential in neural tissue engineering." *Biomaterials* **26**(15):2603–2610.

Yang, G. C., Y. Pan, J. Gong, C. L. Shao, S. B. Wen, C. Shao, and L. Y. Qu (2004). "Beaded fiber mats of PVA containing unsaturated heteropoly salt." *Chinese Chemical Letters* **15**(10):1212–1214.

Yang, G. C., Y. Pan, F. M. Gao, J. Gong, X. J. Cui, C. G. Shao, Y. H. Guo, and L. Y. Qu (2005). "A novel photochromic PVA fiber aggregates contained $H_4SiW_{12}O_{40}$." *Materials Letters* **59**(4):450–455.

Yang, Q. B., D. M. Li, Y. L. Hong, Z. Y. Li, C. Wang, S. L. Qui, and Y. Wei (2003). "Preparation and characterization of a PAN nanofibre containing Ag nanoparticles via electrospinning." *Synthetic Metals* **137**(1–3):973–974.

Yang, Q. B., Z. Y. Li, Y. L. Hong, Y. Y. Zhao, S. L. Qiu, C. Wang, and Y. Wei (2004). "Influence of solvents on the formation of ultrathin uniform poly(vinyl pyrrolidone) nanofibers with electrospinning." *Journal of Polymer Science Part B: Polymer Physics* **42**(20):3721–3726.

Yang, X. H., C. L. Shao, H. Y. Guan, X. L. Li, and J. Gong (2004). "Preparation and characterization of ZnO nanofibers by using electrospun PVA/zinc acetate composite fiber as precursor." *Inorganic Chemistry Communications* **7**(2): 176–178.

Yang, X. H., C. L. Shao, Y. C. Liu, R. X. Mu, and H. Y. Guan (2005). “Nanofibers of CeO_2 via an electrospinning technique.” *Thin Solid Films* **478**(1–2):228–231.

Yao, L., T. W. Haas, A. Guiseppi-Elie, G. L. Bowlin, D. G. Simpson, and G. E. Wnek (2003). “Electrospinning and stabilization of fully hydrolyzed poly(vinyl alcohol) fibers.” *Chemistry of Materials* **15**(9):1860–1864.

Yao, Y. Y., P. X. Zhu, H. Ye, A. J. Niu, X. S. Gao, and D. C. Wu (2005). “Polysulfone nanofibers prepared by electrospinning and gas-jet/electrospinning.” *Acta Polymerica Sinica* (5):687–692.

Yarin, A. L., S. Koombhongse, and D. H. Reneker (2001a). “Bending instability in electrospinning of nanofibers.” *Journal of Applied Physics* **89**(5): 3018–3026.

Yarin, A. L., S. Koombhongse, and D. H. Reneker (2001b). “Taylor cone and jetting from liquid droplets in electrospinning of nanofibers.” *Journal of Applied Physics* **90**(9):4836–4846.

Yarin, A. L. and E. Zussman (2004). “Upward needleless electrospinning of multiple nanofibers.” *Polymer* **45**(9):2977–2980.

Yarin, A. L., W. Kataphinan, and D. H. Reneker (2005). “Branching in electrospinning of nanofibres.” *Journal of Applied Physics* **98**(6):064501.

Ye, H. H., H. Lam, N. Titchenal, Y. Gogotsi, and F. Ko (2004). “Reinforcement and rupture behavior of carbon nanotubes–polymer nanofibers.” *Applied Physics Letters* **85**(10):1775–1777.

Ye, P., Z.-K. Xu, J. Wu, C. Innocent, and P. Seta (2006). “Nanofibrous membranes containing reactive groups: electrospinning from poly(acrylonitrile-*co*-maleic acid) for lipase immobilization.” *Macromolecules* **39**(3):1041–1045.

Yeo, S. Y., H. J. Lee, and S. H. Jeong (2003). “Preparation of nanocomposite fibers for permanent antibacterial effect.” *Journal of Materials Science* **38**(10): 2143–2147.

Yi, F., Z.-X. Guo, P. Hu, Z.-X. Fang, J. Yu, and Q. Li (2004). “Mimetics of eggshell membrane protein fibers by electrospinning.” *Macromolecular Rapid Communications* **25**(10):1038–1043.

Yoo, H. S., E. A. Lee, J. J. Yoon, and T. G. Park (2005). “Hyaluronic acid modified biodegradable scaffolds for cartilage tissue engineering.” *Biomaterials* **26**(14):1925–1933.

Yoon, K.-H., M. B. Polk, B. G. Min, and D. A. Schiraldi (2004). “Structure and property study of nylon-6/clay nanocomposite fiber.” *Polymer International* **53**(12):2072–2078.

Yoon, K. H., K. S. Kim, X. F. Wang, D. F. Fang, B. S. Hsiao, and B. Chu (2006). “High flux ultrafiltration membranes based on electrospun nanofibrous PAN scaffolds and chitosan coating.” *Polymer* **47**(7):2434–2441.

Yoshimoto, H., Y. M. Shin, H. Terai, and J. P. Vacanti (2003). “A biodegradable nanofiber scaffold by electrospinning and its potential for bone tissue engineering.” *Biomaterials* **24**(12):2077–2082.

You, Y., S. W. Lee, J. H. Youk, B.-M. Min, S. J. Lee, and W. Ho (2005a). "In vitro degradation behavior of non-porous ultra-fine poly(glycolic acid)/poly(L-lactic acid) fibres and porous ultra-fine poly(glycolic acid) fibres." *Polymer Degradation and Stability* **90**(3):441–448.

You, Y., B.-M. Min, S. J. Lee, T. S. Lee, and W. H. Park (2005b). "*In vitro* degradation behavior of electrospun polyglycolide, polylactide, and poly(lactide-*co*-glycolide)." *Journal of Applied Polymer Science* **95**(2):193–200.

You, Y., S. J. Lee, B.-M. Min, and W. H. Park (2006a). "Effect of solution properties on nanofibrous structure of electrospun poly(lactic-co-glycolic acid)." *Journal of Applied Polymer Science* **99**(3):1214–1221.

You, Y., J. H. Youk, S. W. Lee, B.-M. Min, S. J. Lee, and W. H. Park (2006b). "Preparation of porous ultrafine PGA fibers via selective dissolution of electrospun PGA/PLA blend fibers." *Materials Letters* **60**(6):757–760.

Yu, J. H., S. V. Fridrikh, and G. C. Rutledge (2004). "Production of submicrometer diameter fibers by two-fluid electrospinning." *Advanced Materials* **16**(17): 1562–1566.

Yu, J. H., S. V. Fridrikh, and G. C. Rutledge (2006). "The role of elasticity in the formation of electrospun fibers." *Polymer* **47**(13):4789–4797.

Yuan, X. Y., Y. Y. Zhang, C. H. Dong, and J. Sheng (2004). "Morphology of ultra-fine polysulfone fibers prepared by electrospinning." *Polymer International* **53**(11):1704–1710.

Yuh, J. H., J. C. Nino, and W. M. Sigmund (2005). "Synthesis of barium titanate ($BaTiO_3$) nanofibers via electrospinning." *Materials Letters* **59**(28): 3645–3647.

Yuwono, V. M. and J. D. Hartgerink (2007). "Peptide amphiphile nanofibers template and catalyze silica nanotube formation." *Langmuir* **23**(9):5033–5038.

Zarkoob, S., R. K. Eby, D. H. Reneker, S. D. Hudson, D. Ertley, and W. W. Adams (2004). "Structure and morphology of electrospun silk nanofibers." *Polymer* **45**(11):3973–3977.

Zeleny, J. (1935). "The role of surface instability in electrical discharges from drops of alcohol and water in air at atmospheric pressure." *Journal of the Franklyn Institute* **219**:659–675.

Zeng, J., X. S. Chen, X. Y. Xu, Q. Z. Liang, X. C. Bian, L. X. Yang, and X. B. Jing (2003a). "Ultrafine fibers electrospun from biodegradable polymers." *Journal of Applied Polymer Science* **89**(4):1085–1092.

Zeng, J., X. Y. Xu, X. S. Chen, Q. Z. Liang, X. C. Bian, L. X. Yang, and X. B. Jing (2003b). "Biodegradable electrospun fibers for drug delivery." *Journal of Controlled Release* **92**(3):227–231.

Zeng, J., X. S. Chen, Q. Z. Liang, X. L. Xu, and X. B. Jing (2004a). "Enzymatic degradation of poly(L-lactide) and poly(ε-caprolactone) electrospun fibers." *Macromolecular Bioscience* **4**(12):1118–1125.

Zeng, J., H. Q. Hou, J. H. Wendorff, and A. Greiner (2004b). "Electrospun poly(vinyl alcohol)/poly(acrylic acid) fibres with excellent water-stability." *e-Polymers*, 078.

Zeng, J., A. Aigner, F. Czubayko, T. Kissel, J. H. Wendorff, and A. Greiner (2005a). "Poly(vinyl alcohol) nanofibers by electrospinning as a protein delivery system and the retardation of enzyme release by additional polymer coatings." *Biomacromolecules* **6**(3):1484–1488.

Zeng, J., H. Q. Hou, J. H. Wendorff, and A. Greiner (2005b). "Photo-induced solid-state crosslinking of electrospun poly(vinyl alcohol) fibers." *Macromolecular Rapid Communications* **26**(19):1557–1562.

Zeng, J., S. Kumar, S. Iyer, D. A. Schiraldi, and R. I. Gonzalez (2005c). "Reinforcement of poly(ethylene terephthalate) fibers with polyhedral oligomeric silsesquioxanes (POSS)." *High Performance Polymers* **17**(3):403–424.

Zeng, J., L. X. Yang, Q. Z. Liang, X. F. Zhang, H. L. Guan, X. L. Xu, X. S. Chen, and X. B. Jing (2005d). "Influence of the drug compatibility with polymer solution on the release kinetics of electrospun fiber formation." *Journal of Controlled Release* **105**(1–2):43–51.

Zeng, W., Y. Du, Y. Xue, and H. L. Frisch (2007). "Solubility parameters." In: *Physical Properties of Polymers Handbook.* Second Edition. Edited by J. E. Mark. Springer-Verlag, New York p. 447.

Zhang, C. X., X. Y. Yuan, L. L. Wu, and J. Sheng (2005a). "Drug-loaded ultrafine poly(vinyl alcohol) fibre mats prepared by electrospinning." *e-Polymers* 072.

Zhang, C. X., X. Y. Yuan, L. L. Wu, and J. Sheng (2005b). "Study on morphology of electrospun poly(vinyl alcohol) mats." *European Polymer Journal* **41**(3): 423–432.

Zhang, G., W. Kataphinan, R. Teye-Mensah, P. Katta, L. Khatri, E. A. Evans, G. G. Chase, R. D. Ramsier, and D. H. Reneker (2005). "Electrospun nanofibers for potential space-based applications." *Materials Science and Engineering: B* **116**(3):353–358.

Zhang, Y. Z., Z. M. Huang, X. J. Xu, C. T. Lim, and S. Ramakrishna (2004). "Preparation of core-shell structured PCL-r-gelatin bi-component nanofibers by coaxial electrospinning." *Chemistry of Materials* **16**(18):3406–3409.

Zhang, Y. Z., C. T. Lim, S. Ramakrishna, and Z.-M. Huang (2005a). "Recent development of polymer nanofibers for biomedical and biotechnological applications." *Journal of Materials Science: Materials in Medicine* **16**(10):933–946.

Zhang, Y. Z., H. W. Ouyang, C. T. Lim, S. Ramakrishna, and Z.-M. Huang (2005b). "Electrospinning of gelatine fibers and gelatin/PCL composite fibrous scaffolds." *Journal of Biomedical Materials Research* **72B**(1):156–165.

Zhang, Y. Z., J. Venugopal, Z.-M. Huang, C. T. Lim, and S. Ramakrishna (2005c). "Characterization of the surface biocompatibility of the electrospun PCL-collagen nanofibers using fibroblasts." *Biomacromolecules* **6**(5):2583–2589.

Zhang, Y. Z., Y. Feng, Z.-M. Huang, S. Ramakrishna, and C. T. Lim (2006a). "Fabrication of porous electrospun nanofibres." *Nanotechnology* **17**(3): 901–908.

Zhang, Y. Z., J. Venugopal, Z.-M. Huang, C. T. Lim, and S. Ramakrishna (2006b). "Crosslinking of the electrospun gelatin nanofibers." *Polymer* **47**(8): 2911–2917.

Zhang, Y. Z., X. Wang, Y. Feng, J. Li, C. T. Lim, and S. Ramakrishna (2006c). "Coaxial electrospinning of (fluorescein isothiocyanate-conjugated bovine serum albumin)-encapsulated poly(ε-caprolactone) nanofibers for sustained release." *Biomacromolecules* **7**(4):1049–1057.

Zhao, S. L., X. H. Wu, L. Wang, and Y. Huang (2003). "Electrostatically generated fibers of ethyl-cyanoethyl cellulose." *Cellulose* **10**(4):405–409.

Zhao, S. L., X. H. Wu, L. Wang, and Y. Huang (2004). "Electrospinning of ethyl-cyanoethyl cellulose/tetrahydrofuran solutions." *Journal of Applied Polymer Science* **91**(1):242–246.

Zhao, Y. Y., Q. B. Yang, X. F. Lu, C. Wang, and Y. Wei (2005). "Study on correlation of morphology of electrospun products of polyacrylamide with ultrahigh molecular weight." *Journal of Polymer Science Part B: Polymer Physics* **43**(16):2190–2195.

Zhao, Z. Z., J. Q. Li, X. Y. Yuan, X. Li, Y. Y. Zhang, and J. Sheng (2005). "Preparation and properties of electrospun poly(vinylidene fluoride) membranes." *Journal of Applied Polymer Science* **97**(2):466–474.

Zhong, S. P., W. E. Teo, X. Zhu, R. Beuerman, S. Ramakrishna, and L. Y. L. Yung (2005). "Formation of collagen-glycosaminoglycan blended nanofibrous scaffolds and their biological properties." *Biomacromolecules* **6**(6):2998–3004.

Zhong, S. P., W. E. Teo, X. Zhu, R. W. Beuerman, S. Ramakrishna, and L. Y. L. Yung (2006). "An aligned nanofibrous collagen scaffold by electrospinning and its effects on *in vitro* fibroblast culture." *Journal of Biomedical Materials Research Part A* **79A**:(3):456–463.

Zhou, H., K. Kim, E. P. Giannelis, and Y. L. Joo (2006). "Nanofibers from polylactic acid nanocomposites: effect of nanoclays on molecular structure." In: *Polymeric Nanofibers*. ACS Symposium Series 918. Edited by D. H. Reneker and H. Fong. Oxford University Press (USA).

Zhou, W. P., Y. L. Wu, F. Wei, G. H. Luo, and W. Z. Qian (2005). "Elastic deformation of multiwalled carbon nanotubes in electrospun MWCNTs-PEO and MWCNTs-PVA nanofibers." *Polymer* **46**(26):12689–12695.

Zhou, Y. X., M. Freitag, J. Hone, C. Staii, A. T. Johnson, Jr., N. J. Pinto, and A. G. MacDiarmid (2003). "Fabrication and electrical characterization of polyaniline-based nanofibers with diameter below 30 nm." *Applied Physics Letters* **83**(18):3800–3802.

Zhu, Y., J. C. Zhang, Y. M. Zheng, J. Zhai, and L. Jiang (2006a). "Conducting PANI/PAN coaxial nanofibers with tuned wettability." *Chemical Journal of Chinese Universities (Chinese)* **27**(1):196–198.

Zhu, Y., J. C. Zhang, J. Zhai, Y. M. Zheng, L. Feng, and L. Jiang (2006b). "Multifunctional carbon nanofibers with conductive, magnetic and superhydrophobic properties." *ChemPhysChem* **7**(2):336–341.

Zhu, Y., J. Zhang, Y. Zheng, Z. Huang, L. Feng, and L. Jiang (2006c). "Stable, superhydrophobic, and conductive polyaniline/polystyrene films for corrosive environments." *Advanced Functional Materials* **16**(4):568–574.

Zhu, Y., M. F. Leong, W. F. Ong, M. B. Chan-Park, and K. S. Chian (2007). "Esophageal epithelium regeneration on fibronectin grafted poly(L-lactide-co-caprolactone) (PLLC) nanofiber scaffold." *Biomaterials* **28**(5):861–868.

Ziabari, M., V. Mottaghitalab, S. T. McGovern, and A. K. Haghi (2007). "A new image analysis based method for measuring electrospun nanofiber diameter." *Nanoscale Research Letters* **2**(12):597–600.

Zong, D. F., K. Kim, S. Ran, B. S. Hsiao, B. Chu, C. Brathwaite, S. Li, and E. Chen (2002). "Nonwoven nanofiber membranes of poly(lactide) and poly(glycolide-co-lactide) via electrospinning and application for anti-adhesions." *Polymer Preprints (American Chemical Society)* **43**:659–660.

Zong, X. H., K. Kim, D. Fang, S. F. Ran, B. S. Hsiao, and B. Chu (2002). "Structure and process relationship of electrospun bioabsorbable nanofiber membranes." *Polymer* **43**(16):4403–4412.

Zong, X. H., S. F. Ran, D. Fang, B. S. Hsiao, and B. Chu (2003a). "Control of structure, morphology and property in electrospun poly(glycolide-co-lactide) non-woven membranes via post-draw treatments." *Polymer* **44**(17):4959–4967.

Zong, X. H., S. F. Ran, K.-S. Kim, D. F. Fang, B. S. Hsiao, and B. Chu (2003b). "Structure and morphology changes during in vitro degradation of electrospun poly (glycolide-*co*-lactide) nanofiber membrane." *Biomacromolecules* **4**(2): 416–423.

Zong, X. H., S. Li, E. Chen, B. Garlick, K.-S. Kim, D. F. Fang, J. Chiu, T. Zimmerman, C. Brathwaite, B. S. Hsiao, and B. Chu (2004). "Prevention of postsurgery-induced abdominal adhesions by electrospun bioabsorbable nano fibrous poly(lactide-co-glycolide)-based membranes." *Annals of Surgery* **240**(5):910–915.

Zong, X. H., H. Bien, C.-Y. Chung, L. H. Yin, D. F. Fang, B. S. Hsiao, B. Chu, and E. Entcheva (2005). "Electrospun fine-textured scaffolds for heart tissue constructs." *Biomaterials* **26**(26):5330–5338.

Zuo, W. W., M. F. Zhu, W. Yang, H. Yu, Y. M. Chen, and Y. Zhang (2005). "Experimental study on relationship between jet instability and formation of beaded fibers during electrospinning." *Polymer Engineering and Science* **45**(5):704–709.

Zussman, E., D. Rittel, and A. L. Yarin (2003). "Failure modes of electrospun nanofibers." *Applied Physics Letters* **82**(22):3958–3960.

Zussman, E., X. Chen, W. Ding, L. Calabri, D. A. Dikin, J. P. Quintana, and R. S. Ruoff (2005). "Mechanical and structural characterization of electrospun PAN-derived carbon nanofibers." *Carbon* **43**(10):2175–2185.

Zussman, E., A. L. Yarin, A. V. Bazilevsky, R. Avrahami, and M. Feldman (2006). "Electrospun polyaniline/poly(methyl methacrylate)-derived turbostratic carbon micro-/nanotubes." *Advanced Materials* **18**(3):348–353.

INDEX

Science and Technology of Polymer Nanofibers. By Anthony L. Andrady